SAT MATH

Authors: Spyros Kalomitsines & Fotios I. Travlopanos

Title: SAT Math: Applications of two basic techniques to all chapters

Cover designer: Vassilis Ioannou

Publisher: Spyros Kalomitsines & Fotios I. Travlopanos

Production supervisor: Platon Malliagkas
www.mediterrabooks.com

E-mail: spyroskalap@hotmail.com & ftravlo@gmail.com

ISBN 978-618-00-0022-1

Spyros Kalomitsines & Fotios I. Travlopanos

SAT Math

Applications of two basic techniques to all chapters

An easier way to learn and practice

Two Basic Techniques for Solving Problems

Examples for mathematical competitions

Contents

To the Reader

We believe that the present work should help considerably the students to their maths examinations, either for the mathematical competitions as well as in SAT math Level 2 exams.
In case you are going to take the SAT math Level 2 examinations, then:

- 1. Read some pages from the first part of this book. You will learn techniques on *How you can help your mind generate more ideas in problem solving.*
  Then go to the other part.

- 2. Here, there is a selection of problems for SAT Mathematics examinations, from all chapters.
  The solutions are presented in a very easy way, using simple and effective techniques.

- 3. We apply, in a very simple and instructive way, knowledge from *Cognitive Psychology* (see: *How the mind works,* 1980) which is the easiest and fastest way to learn and remember.
  In few days, or in two weeks, you can go through this point, and then you will gain much self - confidence in solving problems.

- 4. If you already have any textbook, use it parallel to this part. You will very soon be able to achieve a high score in the examinations. The more you practise, the greater the speed of the brain increases. Mathematics will be very easy for you, it will be a pleasure ...

Preface

Why are **techniques** useful?
Techniques are like machines which render manual work unnecessary. Without techniques every single time, someone is trying to solve a problem, should have to reproduce the whole thought process, as if the problem under consideration was a new one, although in the majority of cases it is not.

Techniques are not a substitute for the mathematical talent. On the contrary, the talent, the inspiration, the passion and the personal efforts of the mathematician give a new boost to the techniques. The talent can be characterized as the *oil* needed for the engine.

Techniques do not prevent the mathematicians from using *trial and error*. Instead, they help them to generate more ideas and ways of attacking the problems. It is actually at this point where the intuition of the mathematician starts to get involved. The techniques assist both, intuition and talent, to faster and efficiently find and choose the more promising ways and paths which lead to the solution.

Alexandra Kalomitsines

Acknowledgements

A number of people have assisted and supported me throughout the course of writing this book. I express my graditude to all of them.

Professor W. Wickelgreen encouraged me to plung into the field of *problem solving* after he read my ideas. His book *How to solve problems* provided me with new hints and convinced me that my work was on the right track.

My former high school student, now Professor at the Technical University of Crete, Yiannis Phillis, encouraged me to come to contact with American researchers on problem solving, so that my work reached its present level of maturity. Without his encouragement my ideas would probably have filled a few scraps of paper.

Professor J. Greeno at LRDC of Pittsburgh University, Stanford later, gave me several suggestions thus assisting me in formulating rigorously my ideas. I am grateful to him.

At Carnegie Mellon University I attended several lectures of the late Herbert Simon (Nobel Price 1978) on Cognitive Psychology. These lectures provided me with a vis-á-vis perspective of my problem solving directions. I also recall gratefully the discussions I held with the wise old man about my subject.

My wife Matina who insisted that I had to accept the invitation to go to USA. As we already had two young children, I did not want to leave my family for a certain time and go far away from my country. She assured me that she could take care of them. I feel deeply grateful to her.

Spyros Kalomitsines.

What is it about?

This book consists of two parts:
In the first part we present answers to the two questions:

- Is there any technique that can help someone who is thoroughly stuck with a problem and no idea seems to come in mind?
- Is there any technique that can help someone who is trapped into a loop and makes circles around the problem, since the solution lies elsewhere?

In this part of this book we present easy and efficient techniques which improve our ability to overcome the above difficulties. The techniques have been tested many times at the Experimental School of the University of Athens, where the first of the authors was the director. They helped students to become better problem solvers (Read comments at the end of this part) There are also many examples where we apply the techniques so that they will be completely clarified.

This part is written by Dr. Spyros Kalomitsines.

The other part is written by Dr. Spyros Kalomitsines and Dr. Fotios I. Travlopanos.

Part I

An introduction
Theory & Examples

by Spyros Kalomitsines

Chapter 1

Introduction

Spyros Kalomitsines **A short true story:**
Three friends - two doctors and a mathematician were sitting at a coffee shop. At a moment one of the doctors, say doctor A, said: *In medicine we have made great progress in the diagnosis of diseases but less progress has been made in the therapy.* Then the other doctor, doctor B, answered: *Yes! However this difference between diagnosis and therapy, which shall always exist, is getting smaller and smaller as the years pass.* Finally the mathematician spoke: *Oh, I see, that what we are talking about, has an analogy to problem solving difficulties. Therapy is, for us, the continuous improvement of problem solving techniques or the invention of new ones.*
-This is actually what we are trying to do in this book

How to become a better problem solver

After a long time of teaching mathematics, and while trying to find an answer to the above question, I came to the conclusion that we have difficulties not only with the problems we deal with but also with our own mind and our way of thinking.

Unjustified phobias, indecision in generating ideas, mind blockage, loops in thought: all these obstacles intermingle and do not allow us to give our best effort. The presence of these difficulties is not only related to mathematical problems but is also so common to other fields where problems do appear in a daily basis of our life.

Here we shall present two techniques that help people to generate more ideas for the solution of problems.

In this book you can read some comments, concerning the initial form of the techniques, made by professors who have worked on problem solving. You can also read comments for the final form of the techniques, as they will be presented here.

Technique A

A dynamic use of short notes

What is Technique A? If someone is thoroughly stuck with a problem, then Technique A helps in generating ideas. Let us start presenting it using an example.

Example (1). *If in the statement of the problem there is the phrase ... triangle* $\triangle ABC$ *with* $\hat{A} = 90^o$ ... *and you have got stuck, then continue as follows:*

- ***Where is the subtle point?***
- ***Describe, using short notes, whatever you see, for instance:***
- $\hat{A} = 90^o$ - ***relationship to something?***
  Yes, Pythagorean Theorem

$$AB^2 + AC^2 = BC^2$$

- ***Also***

$$\hat{B} + \hat{C} = 90^o$$

  What else does a triangle remind me of? Recall any relationships of a triangle to something
  So you increase the probability to find the crucial idea.
  Set your mind free.

- ***Median, heights etc.*** *Let us* ***draw*** *median* AM

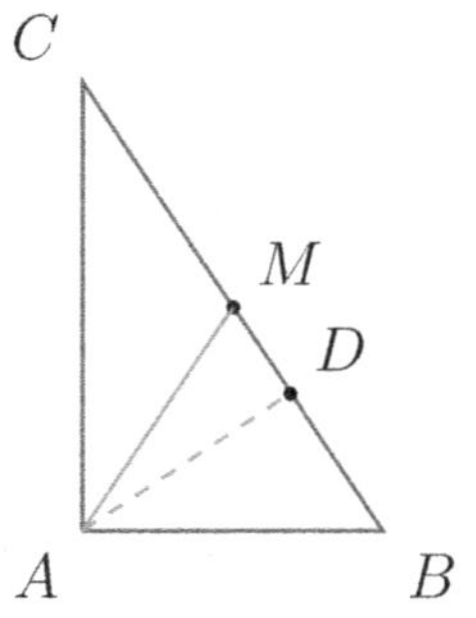

- ***Any quantification? Yes***

$$AM = MB = MC$$

- ***Any qualification?***
  Look at the figure, the ***triangles*** $\triangle AMB$, $\triangle AMC$ *are* ***isosceles,***
  Properties of isosceles triangles ?
- ***Two equal sides***
- ***Two equal angles***
- ***Two equal heights or medians or bisectors etc.***
- ***Now, stop and ask yourself? Is there something useful for the solution of the problem. If yes, try to apply it!***
- ***If you have not found the crucial idea up to now then, go on....***
- ***Any other relationship here ?***

- ***Let us draw the height*** AD

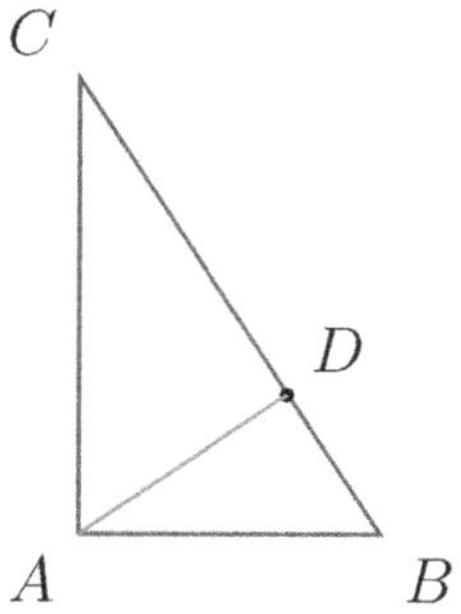

- ***Any formula? Any quantification?***
- *Try to remember, or have a look at your textbook*

$$AD^2 = CD \cdot DB,$$

$$AB^2 = BC \cdot BD,$$

$$AC^2 = BC \cdot CD \text{ etc}$$

Problem (1). *If a, b, c are real numbers such that $bc > 0$, $ab = c^2$ then*

- *[A]* $a^3 + 3b^5 + 7c^{11} \neq 0$
- *[B]* $a^3 + 3b^5 + 7c^{11} < 0$
- *[C]* $a^3 + 3b^5 + 7c^{11} > 0$
- *[D]* $a^3 + 3b^5 + 7c^{11} = 0$
- *[E] none of the above*

Solution:
Let use use short notes (write as more as you can):

- $bc > 0$
- ***Quality and Quantity:*** $\rightarrow$ *hence* $b > 0, \quad c > 0$ or $b < 0, \quad c < 0.$
- $ab = c^2$
- ***Quality and Quantity:***, $\rightarrow$ $c^2 > 0$
- *so* $ab > 0$
- $\rightarrow$ *hence either* $a > 0, b > 0$
- $\rightarrow$ *or* $a < 0, b < 0$
- ***Relationship between them:*** *Let us look now at everything we have written:*
- ***Finally:*** *either* $a > 0, b > 0, c > 0$
- *or* $a < 0, b < 0, c < 0$

- ***Quantity now:*** $a^3 + 3b^5 + 7c^{11} =???????$
- ***Now the answer is very easy:***
- *If* $a > 0,\ b > 0,\ c > 0$ *then*

$$a^3 + 3b^5 + 7c^{11} > 0$$

- *If* $a < 0,\ b < 0,\ c < 0$ *then*

$$a^3 + 3b^5 + 7c^{11} < 0$$

Hence, [A] is the correct answer.

> In the above procedure we repeat simple phrases as: **relationship to something, quantity, quality etc**. These phrases, or, words help our mind to generate more ideas giving certain directions. Other useful words are: **place, time, posture etc**. All these help us to achieve a more systematic and productive way of thinking. Also, the use of **short notes**, helps our mind to generate more ideas.
> Let me emphasize that short notes and graphs help the mind to work fastrer. I will repeat it many times here.
> These useful phrases, or, words, like **relationship to something, etc.** are included in **Aristotle's logic**, where he analyses the meaning of any term, and are called by the term of **Categories**. Later on we shall present them completely.
> **In this book, we try our best to adapt and apply, in a simple way, the Aristotle's categories to problem solving, using the technique of short notes**.

For this final form of the technique professor S. Krantz commented (May 2016) : ***quite interesting and different from everything that I have seen.*** Once in a while pause and look at everything you have written. If the time for the solution is limited, instead of writing, just think of most of them.
Now, ask yourself:
Is there anything useful for your problem?
If yes, use it.
If no, go on retrieving more information from other parts and details of the problem.
Retrieve more information from information that you have written down. Then ask yourself:
Is there, now, any similarity with some other problem that I already know how to solve?
If yes, try to use it.

Try to surpass yourself, so the crucial idea will emerge ...
Your mind will work faster if you write down
anything you recall in short notes,
as if you write down a childish poem, or song.
Set your mind free – do not hesitate.

If you can draw diagrams representing parts of the problem, then the mind works even faster.
With diagrams the mind works about fifteen times faster.
Do not forget that the most difficult thing in problem solving, is to see another aspect of the same detail.
The technique of short notes helps people find many aspects,
one of which may be the crucial idea for the solution of the problem.
It is easier for the mind to produce short notes and to elaborate them, as we have already said. The developement of human language started with short phrases because it raises the speed of the brain.
Your mind should be free, so forget momentarily the whole problem and focus on one part or a specific detail of it.
Do not forget: Diagrams, Simple symbols, Categories ......
Stop once more and have a look at everything that we have done - written.
Try to detect if something is useful for the solution.
If not, go on and try to find more aspects.
Let us start applying the technique to problems.
Be careful: If some problem seems difficult for you, skip it.
Go to the next one.
You will find other problems that are easier.

Note: In this book we shall repeat the same phrases many times as for example *use short notes* etc. We do that because in this way, students will have more practice in our techniques.

1.1 More applications of Technique A

Problem (2). *The least prime factor of an integer a is 3 and the least prime factor of an integer b is 7. Find the least prime factor of their sum $a+b$.*

Answer:

Try for a while to solve it before reading on.

Solution Procedure Let us describe everything using short notes.

Prime numbers? make it explicit:

Prime numbers are $2, 3, 5, 7, 11, 13, 17, 19, 23, \ldots$

Let us now describe a

Which category could help us?

Let us try:

Relationship to something? $a = 3x$

Quantity? $a > 2, \quad x > 2$

Quality? $x = 3$ or 5 or 3×5 etc

$\ldots$ so x is **odd**

Similarly we conclude that y is **odd**.

What about $a+b$?

It is : $a+b$ is odd + odd = even <u>**Crucial Idea**</u>

hence the **first prime factor** of $a+b$ is 2.

A more short proof could be:

a is odd, hence $a = 2k+1$ (use of simple symbols)
b is odd, hence $b = 2k'+1$ and thus

$$a+b = 2k+1+2k'+1 = 2(k+k')+2 = 2\cdot(k+k'+1)$$

i.e: $a+b$ is **even**, therefore 2 is its least prime factor.

Next problem is about powers' comparison.

Problem (3). *Given two numbers,* $a = 2^{3000}$ *and* $b = 3^{2000}$. *Find which is greater(compare them).*

Answer: *Try for a while to solve it, before reading on.*

Let us consider any relation between the basis 2, 3 as for example $2 < 3$
and between the exponents 3000 and 2000 ......
Any relationship between 3000 and 2000?

$$3000 = 3\times 1000, \ 2000 = 2\times 1000$$

so, there is a common factor 1000
Let us try to use it

$$2^{3000} = 2^{3\times 1000} = (2^3)^{1000} = 8^{1000}$$
$$3^{2000} = 3^{2\times 1000} = (3^2)^{1000} = 9^{1000}$$

hence

$$2^{3000} < 3^{2000}$$

Note: The application of the *category* relationship helped us in this problem.

Problem. *Prove that*

$$31^{11} < 17^{14},$$

without using a calculator.

Answer:

Try for a while to solve it before reading on.

The great mathematician D. Hilbert said that the most important part of mathematics concerns: ***inequalities!***

- Well, a very good special technique to prove an inequality of the type $A < B$ consists in:
- Try to spot some A_1 close to A, such as $A < A_1$
- Now try to spot another B_1 close to B, such that $B_1 < B$
- Then, try to prove that $A_1 \leq B_1$, hence

$$A < A_1 \leq B_1 < B \Rightarrow A < B$$

- Let us now try to apply this technique to our problem:

- 31 is close to 32, $31 < 32$ and 16 is close to 17, $16 < 17$, so in order to prove that

$$31^{11} < 17^{14}$$

we have to prove that

$$32^{11} < 16^{14}$$

Any relationship between 32 and 16? **Yes,** $32 = 2 \times 16$ or $32 = 2^5$, $16 = 2^4$
Let us use it

$$32^{11} < 16^{14} \Leftrightarrow (2^5)^{11} < (2^4)^{14} \Leftrightarrow 2^{55} < 2^{56}$$

which **is true** and the problem is solved.

Note-An advice to students:
Read carefully any special techniques for any chapter of mathematics. Also, read carefully the solution of difficult, original problems. Each solution should become, for us, a technique or, as we can say, an *exact method* for similar problems.

- Start always with the question: **Where is the subtle point which is hidden in the statement of the problem?**

- **Use diagrams and simple symbols, for instance:** a, b, c ... **are better symbols for our mind than** $a_1, a_2, a_3, \ldots\ldots$

- **Do not forget: By writing, ideas are generated!**

We shall repeat many times that we can help our mind by describing our thoughts using short notes. In a statement of a problem there often exist subtle points, *hidden* implicit information. The implicit character of the information given, in the statement of the problem under consideration, it is actually the main difficulty in the problem solving procedure: Implicit information should become explicit. **The solver is often obliged to behave like a police inspector who tries to detect a crime.**
The most difficult *thing* **in problem solving is to manage to see the other aspect or aspects of the same detail.**
If we use short notes, short phrases, our mind works faster, so we increase the probability to see more aspects of the problem we deal with. On the other hand, as we have already said, **language was initially developed with short phrases, that did the same, and this enabled people to generate more ideas.**
Complex phrases were developed later. Short notes, as it is the writing of a children's poem, raised the speed of the brain.
Anthropologists say that humans, after starting to use language, they invented a large number of tools of higher complexity.
This means that thought and language, toghether, help creativity.
So, think and use short notes - this is the best advice to our students. Also, use graphs, whenever this is possible.
There is also a formal computer model of this technique - see references. If a technique can become a computer program, using methods and concepts of Cognitive Psychology, then it is accepted, else we reject it, it is considered as *too general* to work. We have either to improve it or to abandon it. Our technique has been successfully tested by using computers.

This proved the efficacy of the technique. This model was designed and implemented at the LRDC of Pittsburgh University, under the supervision of professor J. Greeno, professor at Stanford University later.

A dynamical scheme of the Technique A

Solution Procedure

Data.	**Differences**	**Goal**
Think of the data and follow a forward procedure: • Set your mind free • Use the categories • Make several trials • Draw various conclusions from the data and write them down **using short notes.** Short notes make you work faster.	Locate Differences between the two columns and try to eliminate them	Place emphasis to the goal Try previous steps (backward procedure) or try equivalent steps to the goal. Any relationship to something?

Note: Use short notes not only in the first column (data), but also in the other two: **Locate differences** - and - **Goal.**

An application now:

Problem. *If a, b, c are the sides of a triangle, prove that*

$$(b^2 + c^2 - a^2)^2 < 4b^2c^2.$$

Try for a while to solve it before reading on.

Solution procedure

Data	**Differences**	**Goal**
Short notes:		**Emphasis here**
a, b, c sides, so		$(b^2 + c^2 - a^2)^2 < 4b^2c^2$
Any relationship ?	Differences between	An equivalent step?
Write it down	the two columns?	or, a previous step?
$a < b + c$		$(b^2 + c^2 - a^2)^2 - 4b^2c^2 < 0$
$b < a + c$	Eliminate	
$c < a + b$	differences	Short notes
or	How?	
$a - b - c < 0$	Factorize:	
$b - a - c < 0$		or
$c - a - b < 0$	Go on, trying to	$(b^2 + c^2 - a^2 - 2bc)(b^2 + c^2 - a^2 + 2bc) < 0$
or	eliminate differences:	$\{(b-c)^2 - a^2\}\{(b+c)^2 - a^2\} < 0$
$b + c - a > 0$	$b - c - a < 0$	or
$a + c - b > 0$	$b - c + a > 0$	
$a + b - c > 0$	$b + c - a > 0$	
	$b + c + a > 0$	
	hence it is true $\rightarrow$	
		$(b - c - a)(b - c + a)(b + c - a)(b + c + a) < 0$

Now you can write shortly the proof.

Comment: It seems like a battle. If we know only a few things about the enemy, it is unwise to go straight at him. We had better to go around the enemy first, try to gather any information we can and then try to discover his weak points. When dealing with problems we have to find the subtle point.

More Comments

Never lose courage. Overcome fears and disappointments.
If you reach a dead end but something seems
to have a relation to your problem, try it;
it may reduce differences between the data and the goal.
Be patient.
Reduction of differences often happens gradually, step by step.
Help your mind generate more ideas,
following our instructions.
If you want to write a program to solve a problem with a computer,
then provide the machine with all
the formulas, theorems and techniques
for solving similar problems.
For example, in our case,
provide the machine with the factorizing techniques.
That is, we should translate
all the experience of a good problem solver
into statements of the type *if, then.*
This way the computer performs something
equivalent to short notes.
Such a program was made by the author
using concepts and methods of cognitive psychology.
This proved the efficacy of the technique.

Note

Is problem solving a *combination of contradictions?*

We have observed many times that , trying
to express information in a computer language,
unexpected solutions are found.
We do not need to write a full program
which might be hard, but,
while trying to develop an algorithm
that requires concrete steps,
the crucial idea suddenly pops up.
All we have to do is to recognize it.
Finally, a combination of the pattern of short notes
(which means *freeing the mind*)
with a pattern of machine steps,

generate valuable ideas that
help us to solve the problem as we already saw.

Why does this happen?
Because the procedure of communicating with the computer
forces the mind to turn all implicit information
into explicit and to clarify the subtle points,
because the computer is only a machine
that requires mechanical information.
In such a way we increase the probability
to target on the crucial idea.
Let us call it *the computer way*
(see an example -
the checker board problem -
next paragraph) .

Chapter 2

Categories & a longer discussion on short notes

In Aristotle's logic we read:

<u>10 - CATEGORIES</u>

Every uncombined term indicates substance, or quantity, or quality, or relationship to something, or place, or time, or posture, or state, or doing something, or undergoing of something.

What does Aristotle mean?
Can we apply these categories to problem solving?
Yes we can as we have already seen. - Well, I stumbled upon Aristotle's categories few years ago. Then another idea on problem solving has emerged:

Why not adapt his categories to my technique? This might make it more efficient.

Let me emphasize here, that sometimes for chaotic problems a little *insight* or *intuition* is necessary. Our technique helps people find the solution, or come very close to it. Then a little *insight* or *intuition* is necessary.

Now an adaptation and extension of the categories minimizes the need for any *instinct*. We can generate more ideas, more short notes extracted from the details of the statement of the problem if for every detail we consider some of the following:

Aristotle's categories organized into groups:

1. Substance, quality, quantity
2. Relationship to something
3. Place, time, doing something, undergoing of a change
4. Posture, state

An extension of Categories: Further more,it is often useful to consider for every detail the following extension of the categories to problem solving which has a more dynamical character.

Dynamic Categories .

1. Find any change of quantity, find differences, changes of differences.
2. Look for invariant quantities. If we consider invariant a part of the data, how does the remaining part change?
3. Differences between the data and the goal: can you eliminate them?
4. Try to use the definitions of the various terms, formulas for parts of the statement of the problem. Emphasis on the goal: can you find previous steps? Equivalent steps? Can you generalize the problem?
5. Use special techniques for similar problems, find analogies, make intensive use of the data. Set subgoals. Try the computer way.
6. Forget momentarily the whole problem: Emphasize on what happens to the unit, or to the elementary Δx, Δy, or to one part or a variable of the problem. Then start working on the whole problem.

2.1 Applications - Problems

Solved problems follow that aim at improving the problem solving abilities of the students and practitioners.

Let me emphasize that the same techniques are efficient for solving problems at any level: puzzles, or High School, or University level, etc.
Problem (1). *Given a seven by seven checkerboard, as shown in the figure, draw a line through all the squares on the board, starting from the dot in the second square, first column, crossing each square only once, without exiting the check board and without entering a square at a vertex. Show how this can be done, or prove that this is impossible.*

Try for a while to solve it before reading on.

- After several trials you might feel that it is impossible.

- How can you prove that it impossible indeed?
- Perhaps there is a subtle point you are missing.
- Now follow the technique.
- For a more dynamic description of details, try some of the
- **Categories: time, quantity, quality ...**
- **Always place emphasis on the goal**
- **Do not forget : use simple symbols**

Time	I started from a white square
Quality	Let me symbolize it by W
	then a reached a black one,
	B and so on. Therefore,
	I developed a sequence $WBWB\ldots$
	Suppose you have finished,
Emphasis on	or you have reached a dead end.
the goal	How could this be?
	$WBWBWB\ldots\ldots WB$ or
	$WBWBWB\ldots\ldots WBW$
Category	How many squares are there in each case?
quantity now	Count:
	First case: $WBWB\ldots WB$, hence
	the same number of W and B
	But the checkerboard has
	$24W$ and $25B$ (impossible)
	Second case: $WBWB\ldots WBW$
	This means $25W$ and $24B$,
	that is more W than B, i,e: **Impossible**
	Comments:
	Without using the categories, we could get lost
	in a great number of trials ...
	So, the use of categories leads us
	to a smaller number of steps,
	and helps also a programmer to develop programs
	to solve such problems.
	We do not need *intuition* or *instinct*, or
	inspiration any more.
	All we need is practise and training.
	Let us imagine ourselves having to cross a wide river.
	Finding a few logs scattered on the water,
	can help you cross easily.
	In this example, the logs play the role of *categories*
	in problem solving.

Note: Is the use of *categories* an application of philosophy to science? Yes, it is.

Problem (2). ***The checkerboard problem:***

Consider the puzzle shown in Figure 1.

You are given a checkerboard and 32 *dominoes.*

Each domino covers exactly two adjacent squares on the board.

Thus, the 32 *dominoes can cover all* 64 *squares of the checkerboard.*

Now suppose one square is cut off from each of two diagonally opposite corners of the board.

***Is it possible** to place* 31 *dominoes on the board so that all of the remaining* 62 *squares are covered?.*

If so, show how it can be done; if not, prove that is impossible.

Many people find it hard to solve this problem. Let us use our technique:

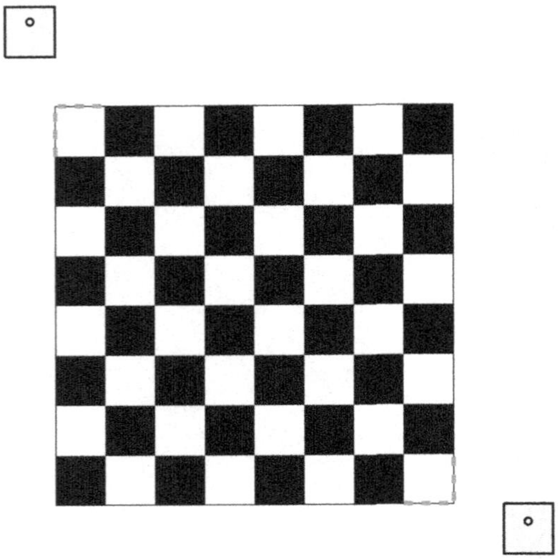

Try for a while to solve it before you go on reading solution

Data:

What do I see in this problem?
I have 62 squares
I have 31 dominoes
Each domino is rectangular
Properties of rectangles?
Parallel sides, equal angles, equal diagonals, etc.
What else do I see?
Let us focus on another detail
Each domino covers 2 squares

Categories now: quantity, quality . . .

The two squares have different colors.
So, one domino covers one black and one white square.
Here I see a simple correspondence.
1 domino covers 1 black square and 1 white square.
So, 31 dominoes cover 31 black squares and 31 white squares.

Differences:

Here is a contradiction between the two columns.
In the left column, 31 black and 31 white squares, but in the right column 32 back and 30 white squares . . .

Goal:

Assume that 31 dominoes can cover 62 squares.
Previous step?

(quantity, quality, time . . .)

- At first we use 1 domino.
We covered 1 black and 1 white square
. . . .
. . . end of time:

(quantity, quality . . .)

We have covered 30 white and 32 black squares.

So we have proved that it is impossible.

Note: Using this technique, solution is found very easily ...... Implicit information (...the color of the squares) becomes explicit. So, this procedure can be a preliminary step for programming a computer to solve our problem ...

An important observation:

How can someone solve the checkerboard problem, if all the squares are white?

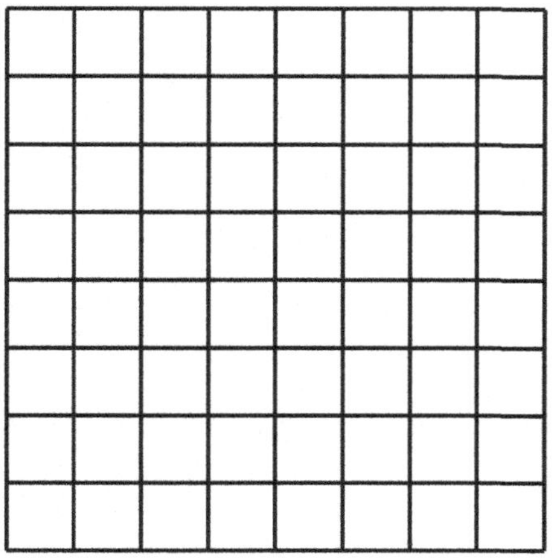

It seems to be very difficult. What can we do ?

Since we are completely stuck, let us try to use the computer way.

Up to now we have emphasized that freeing the mind of constraints helps to generate ideas (describe everything in short notes until you reach the crucial idea.)

But problem solving is often a kind of art.Instead of freeing the mind of constraints, you may constrain it to follow the steps of a computer. It naturally raises the question: What kind of instructions are given to a computer?

To solve a problem using a computer we start by first giving the input, the data. A computer is just a machine that lucks the human flexibility. The computer cannot see ... How can we give the data to a computer ? Let us try our best. A good way to input the data consists in using the matrix notation:

(1, 1)

(	(1,2)	(1,3)	(1,4)	(1,5)	(1,6)	(1,7)	(1,8)
(2,1)	(2,2)	(2,3)	(2,4)	(2,5)	(2,6)	(2,7)	(2,8)
(3,1)	(3,2)	(3,3)	(3,4)	(3,5)	(3,6)	(3,7)	(3,8)
(4,1)	(4,2)	(4,3)	(4,4)	(4,5)	(4,6)	(4,7)	(4,8)
(5,1)	(5,2)	(5,3)	(5,4)	(5,5)	(5,6)	(5,7)	(5,8)
(6,1)	(6,2)	(6,3)	(6,4)	(6,5)	(6,6)	(6,7)	(6,8)
(7,1)	(7,2)	(7,3)	(7,4)	(7,5)	(7,6)	(7,7)	(7,8)
(8,1)	(8,2)	(8,3)	(8,4)	(8,5)	(8,6)	(8,7)	)

(8, 8)

Now we may use the pattern of short notes:

Qualitative observation	Numbers are either odd $1, 3, 5, 7$ or even $2, 4, 6, 8$ So we can symbolize: $(1,1) = (o,o)$ both odd $(1,3) = (o,o)$ both odd etc. $(2,2) = (e,e)$ both even, $(2,4) = (e,e)$ both even, etc. $(1,2) = (o,e)$ $(2,3) = (e,o)$, etc.
Now quantify	(A) 16 squares are (o,o) 16 squares are (e,e) We cut $(1,1)$ and $(8,8)$, so there are 15 squares (o,o) and 15 squares (e,e) (B) 16 squares are (o,e) and 16 squares are (e,o)

Each domino covers one of the squares (A) and one of (B). So 31 dominoes must cover 31 of the squares of case (A) and 31 of that of case (B). So, we conclude that this is impossible, because there are 30 of case A and 32 of case B.
Next we present **a problem from physics:**

Problem (3). *A tank - bucket containing water balances a weight as shown in the figure below. Will the equilibrium be affected if you put your finger into the water without touching the tank?*

Try for a while to solve it before reading on.

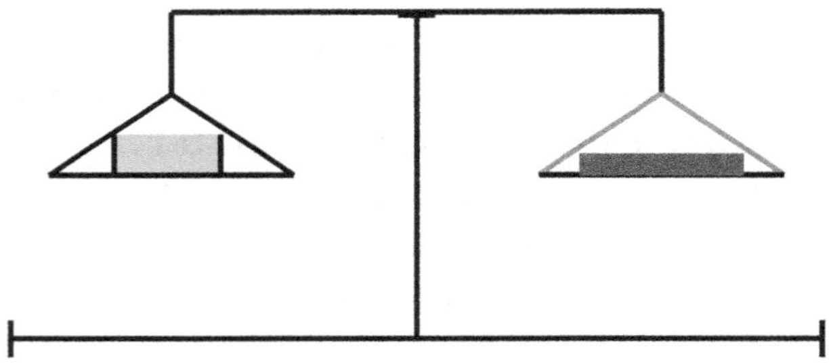

Categories:

Time	Before putting your finger into the water
Quantity	Weight of the water: Let it be B Weight of the finger: Let it be F Depth of the water: Let it be d Weight of the bucket: Let it be Bu
Time	After putting the finger into the water:
Quantity	Weight of the water, the same, B Weight of the finger, lighter $F_1 < F$ depth of the water, $d_1 > d$ weight of the bucket, pressure on the bottom increases so the force on the bottom increases This is due to an external force
Place	(you keep your finger into the water)
Quality	so the tank becomes **heavier** (Third Newton's Law).

Problem (4). *Prove that 5 is the only prime number that can be written under the form*

$$x^4 + 4y^4$$

where x, y are integers.

Try for a while to solve it before reading on.

Let us use the short notes technique:

Categories: $x^4 + 4y^4$ must be a prime.

Relationship to something What does *prime* mean?

If $x^4 + 4y^4 = a \times b$ then $a = 1$ or $b = 1$

Use of definition How can we factor $x^4 + 4y^4$?

Similar problem: An easy way is to transform in such a way that $x^4 + 4y^4 = (\ldots)^2 - (\ldots)^2$

Read your text book

Try to find

special techniques

So $x^4 + 4y^4 = x^4 + 4y^4 + 4x^2y^2 - 4x^2y^2$
$= (x^2 + 2y^2)^2 - (2xy)^2$
$= (x^2 + 2y^2 + 2xy)\,(x^2 + 2y^2 - 2xy)$

Hence, either $x^2 + 2y^2 + 2xy = 1$ (1)
or $x^2 + 2y^2 - 2xy = 1$ (2)

Short notes for (1)
$x^2 + 2y^2 + 2xy = 1$ it can be studied
as a second degree equation
$x^2 + 2y^2 + 2xy - 1 = 0$, where x, y are integers.

The discriminant is $D = 4(1 - y^2)$ therefore
$x = \frac{-2y \pm 2\sqrt{1-y^2}}{2} = -y \pm \sqrt{1 - y^2}$
so $y = 0$, or 1, or -1 hence $x = \pm 1$.
Finally $x^4 + 4y^4 = 1 + 4 = 5$.

If we work with (2) we obtain the same result.

Note:

Here are some examples that describe the use of the *short notes* technique.

1. **If you see the phrase *two consecutive numbers* write down x and $x + 1$.**
   If there are three consecutive numbers then write x, $x + 1$, $x + 2$ or $x - 1$, x, $x + 1$.

2. **If you see the phrase r *is a root of a polynomial* $P(x)$ then write down**

$$P(r) = 0$$

$$P(x) = (x - r)Q(x)$$

3. **If you have a problem on sequences, then:**
if it is an arithmetic sequence, write down the formulas:

$$u_n = u_1 + (n-1)\cdot d,$$
$$S_n = \frac{n}{2}\{2u_1 + (n-1)d\} = \frac{n}{2}(u_1 + u_n)$$

Look at these formulas now and use the data of your problem.
If it is a geometric sequence, write down the formulas:

$$u_n = u_1 r^n,$$
$$S_n = \frac{u_1(r^n - 1)}{r-1} = \frac{u_1(1-r^n)}{1-r}$$

If $|r| < 1$, then

$$S = \frac{u_1}{1-r}$$

(**i.e: the sum of an infinite series.**)
Attention: Do not only think of the formulas, but write them down! It takes only a few seconds, but looking at them makes your mind work faster. (by Alexandra Kalomitsines)

4. **If you see an equality** $xy = a^2$, notice that $a^2 \geq 0$, $xy \geq 0$, so either $x \geq 0$, $y \geq 0$ or $x \leq 0$, $y \leq 0$

5. **If you see the phrase** a, b, c **are the sides of a triangle, then write down:**

$$a < b + c,\ c < b + a,\ c < a + b,$$
$$\cos A = \frac{b^2 + c^2 - a^2}{2bc},\quad \frac{a}{\sin A} = \frac{b}{\sin B} = \frac{c}{\sin C},$$
$$\text{Area} = \frac{1}{2}ab\sin C$$

etc.

2.2 Word problems and the *reduction to one* technique

Symbolize by x the required quantity
Problem (1). *George is 48 years old. His age is twice the age that John was, when George was the age John is today. What is John's age?*

Try for a while to solve it before reading on.
Let us use the categories: time - quantity.

Today:	George 48 & John x
Then:	George x & John $\frac{48}{2} = 24$
Dynamic categories now :	
Differences now:	
Today:	George - John $\rightarrow 48 - x$
Then:	George - John $\rightarrow x - 24$

hence $48 - x = x - 24$
therefore $2x = 72$
so $x = 36$

Another Way:
Differences:
George:
John:

Today - Then: $\rightarrow 48 - x$
Today - Then: $\rightarrow x - 24$
hence $48 - x = x - 24$ etc.

Important!
Many word problems can be solved via a classical technique called *reduction to one*, or *reduction to the elementary.*

Example (1). *8 men can do a certain job in 12 days. How long will it take for 6 men to do the same job if they work at the same rate?*

Solution: Use short notes:

- 8 men 12 days
- **reduction to one**
- 1 man needs to work
$$12 \times 8 = 96 \text{ days}$$
- and thus 6 men need
$$\frac{96}{6} = 16 \text{ days}$$

Example (2). *The daily production of 40 oil wells is 16.000 barrels. This production decreases by 8 barrels for every new well. Find how many new wells must be opened so that we achieve maximum oil production.*

Solution: Use short notes:

Reduction to one and categories

- 40 wells produce 16.000 barrels, hence
- 1 well produces 400 barrels
- Let x be the number of new wells, then
- Every new well produces $400 - 8x$ barrels

- hence the total production shall be

$$(40 + x) \cdot (400 - 8x) = -8x^2 + 80x + 16.000$$

and from elementary algebra the maximum of production is obtained for

$$x = \frac{-b}{2a} = \frac{-80}{-16} = 5$$

new wells. [1]

[1] If $y = ax^2 + bx + c$ is the standard form for the second degree polynomial then its maximum-minimum is attained at the point $x = \frac{-b}{2a}$.

Chapter 3

Technique B

3.1 Technique B - How to get out of loops

Sometimes a problem solver gets trapped into loops because of his previous knowledge that introduces unnecessary constraints and makes circles around the problem.
However, the solution lies elsewhere.
Naturally, raises the question:
How can you get get out of loops?

You can work in the way it is described in what follows:

Description of attempts	**Negations**
Here you write in short notes all the attempts you have made.	Here you write all possible negations, all possible opposite statements and actions. Do your best to write as many negations as possible. Often, one of these negations includes the crucial idea for the solution.

Problem (1). *You are given six matches and you are asked to make four equilateral triangles with them.*

Description of attempts	**Negations**
Categories in short notes	
Time: First I make a triangle	I do not make a triangle but more
Then I make another with a common side	
Quantity: This leaves only one match	
Place, posture: I am trying to construct it on a table	Let me not try to construct it on a table Let me try to construct it not on a plane
	Let me try to construct it in space (Crucial idea)

The solution is to
construct a tetrahedron

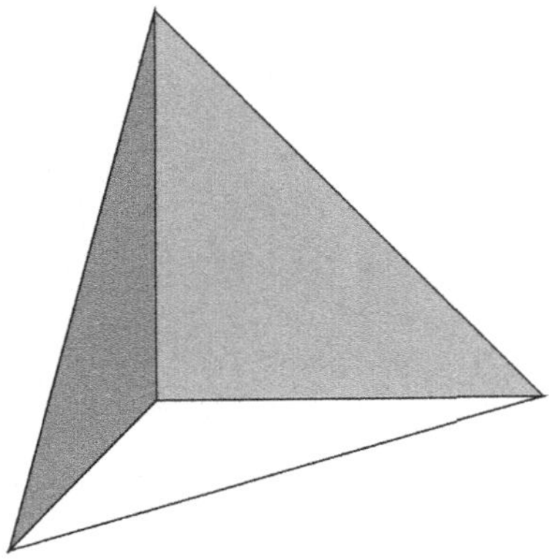

Another form of the method

- Take one sentence from the left side, i.e: *I am trying to construct it on a table.*
- Now repeat the sentence inserting the word *not* between the words. Some of the resulting sentences may have no meaning. However, others may reveal an important idea.

Let us do it.

- I am **not** trying to construct it on a table
- I am trying **not** to construct it on a table
- I am trying to **not** construct it on a table
- I am trying to construct it **not** on a table
- I am trying to construct it on **not** a table

Obviously, the fourth sentence contains the crucial idea. The chance to solve the problem is now very good.
Using this technique you may succeed in getting rid of a common hesitation to change procedures. Previously learned associations prevent your mind grasping the correct procedure to a solution due to unneccesary constraints that lead you to the wrong track. Perhaps, in problem 3.1, the word *triangle*, a planar figure, constrains the brain to the plane. Even great scientists sometimes get into loops because of their biases.
Problem (2). *If $x, y, z > 0$ and $x^2 + y^2 + z^2 = 3$ show that*

$$\frac{3}{2} < \frac{1+y^2}{x+2} + \frac{1+z^2}{y+2} + \frac{1+x^2}{z+2} < 3.$$

Category: Relationship to something

Let us try to use the classical inequalities which seem to be relevant here:

$$x^2 + y^2 + z^2 \geq xy + yz + zx,$$
$$1 + y^2 \geq 2y, \ \ x + 2 \geq 2\sqrt{2x},$$
$$1 + z^2 \geq 2z, \ \ y + 2 \geq 2\sqrt{2y},$$
$$1 + x^2 \geq 2x, \ \ z + 2 \geq 2\sqrt{2z},$$

...some trials, but ... dead end ...
It seems we get lost in these trials ...
We have gotten into a loop.

Now dynamic category Let us try to exit the loop.
- Do not use the classical inequalities.
Another aspect of $x^2 + y^2 + z^2 = 3$? Try your best ...
What about x? How does it vary? Find other aspects for x, y, z :
Yes! $x^2 + y^2 + z^2 = 3$, hence

$$0 < x < 2,\ 0 < y < z,\ 0 < z < 2$$

and let us use this fact:

$$\frac{1+y^2}{2+2} < \frac{1+y^2}{x+2} < \frac{1+y^2}{2},$$
$$\frac{1+z^2}{2+2} < \frac{1+z^2}{y+2} < \frac{1+z^2}{2},$$
$$\frac{1+x^2}{2+2} < \frac{1+x^2}{z+2} < \frac{1+x^2}{2}$$

Try to eliminate differences step by step.
Adding them to get:

$$\frac{3+y^2+z^2+x^2}{4} < \frac{1+y^2}{x+2} + \frac{1+z^2}{y+2} + \frac{1+x^2}{z+2} < \frac{3+y^2+z^2+x^2}{2}$$

or

$$\frac{3+3}{4} < \frac{1+y^2}{x+2} + \frac{1+z^2}{y+2} + \frac{1+x^2}{z+2} < \frac{3+3}{2}$$

so we have proved that

$$\frac{3}{2} < \frac{1+y^2}{x+2} + \frac{1+z^2}{y+2} + \frac{1+x^2}{z+2} < 3.$$

Example. ***An unusual daily example*** ... *This example is included in some comments from Mrs. Anna Vayaki: Anna Vayaki, researcher emeritus at the Greek Center Demokritos, having worked for* 35 *years in CERN experiments on elementary particles, said to Dr. Kalomitsines: Your techniques are very good for researchers. But I also use them in everyday activities. Yesterday my sister lost an accessory for her vacuum cleaner and I found it, following your technique of getting out of loops. She had been looking for an orange tube: negating it, not orange, not tube I found it. It was rolled up, black with a small orange button.*

Even great scientists sometimes get into a loop sometimes.

How to develop a computer program to solve problems

Here is the summary of the procedure

- **STEP 1.** Separate out the data of the problem. Separate out the goal.

- **STEP 2.** If the problem has a uniquely specified goal, then try the technique of working backward.
  This means that all the required knowledge will be given to the memory of the computer: definitions, theorems, special techniques.
  All these in the form of *if ... then.*
  Example: The definition of an arithmetic sequence *if* a, b, c *are consecutive terms of an arithmetic sequence then* $b - a = c - b$.
  Use the dynamic form of Technique A, in order to find previous steps, starting from the goal, until you reach the given information.

- **STEP 3.** Use the contradiction method if you must determine whether the goal is inconsistent with the given information.
  Apply the dynamic form of Technique A to the givens and the goal, until you find a contradiction.

- **STEP 4.** Use the most dynamic form of the technique (for difficult problems).
  Give to the memory of the computer everything: definitions, formulas, theorems, all the known techniques - all these in the form *if ... then.*

3.2 A few applications to problems for mathematical competitions

Problem (1). *If* x, y, z *are positive numbers,* $z < 1$ *such that*

$$xyz = 1 \text{ and } \mathrm{x} + \mathrm{y} + \mathrm{z} > \frac{1}{\mathrm{x}} + \frac{1}{\mathrm{y}} + \frac{1}{\mathrm{z}},$$

prove that

$$x > 1 \text{ or } y > 1.$$

Try for a while to solve it before you go on
Solution: We use short notes in a dynamic form. We write the data on the left and the goal on the right side.

Data

$x > 0$, $y > 0$, $z > 0$, $z < 1$

$xyz = 1$

$x + y + z > \frac{1}{x} + \frac{1}{y} + \frac{1}{z}$

$z = \frac{1}{xy}$

So

$x + y + \frac{1}{xy} > \frac{1}{x} + \frac{1}{y} + xy \Rightarrow$

$x^2y + xy^2 + 1$
$- y - x - x^2y^2 > 0$

We group terms so that $x - 1$ or $y - 1$ become factors.

Try

$x^2y + xy^2 + 1$
$- y - x - x^2y^2 > 0 \Leftrightarrow$
$y(x^2 - 1) + xy^2(1 - x)$
$+ (1 - x) =$
$(x - 1)(y(x + 1) - xy^2 - 1)$
$= (x - 1)[xy(1 - y) - (1 - y)]$
$= (x - 1)(1 - y)(xy - 1) > 0$

Hence either $y > 1$ or $x > 1$ is true.

Indeed, if $y > 1$ and $x > 1$ then $(x - 1)(1 - y)(xy - 1) < 0$ i.e: **a contradiction**.

Differences between the 2 columns?

There is no $x - 1$, $y - 1$ on the left side, and no z on the right side.
Eliminate z

How can we have $x - 1$, $y - 1$ in the left side?
How can we have something like $(x - 1)(y - 1)A > 0$ or < 0
Eliminate denominators transfer all terms to the left side and try to factor
We know that $x - 1$, $y - 1$ are factors;
The problem becomes straightforward: (look at the first column)

Goal

$x > 1$ or $y > 1$

A preceding step?
Prove $x - 1 > 0$, or $y - 1 > 0$

Problem (2). *Points A, B, C and M lie on a circle and MD, ME and MF are perpendicular lines to the sides of the triangle (see: figure). Prove that the points D, E, F are collinear.*

Proof:

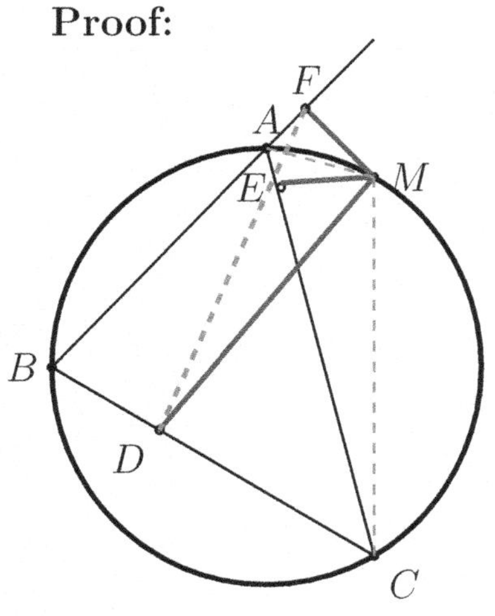

Solution (Applying Technique A):

Data

Forward procedure

Try to find more aspects for each detail

Describe everything using short notes.

$\widehat{MEC} = \widehat{MDC} = 90^o$

hence

M, E, D, C lie on the same circle.

Draw $MC \rightarrow$ (complete the figure $MEDC$).

Also M, E, A, F lie on the same circle.

Draw $AM \rightarrow$ (complete the figure $MEAF$)

Some more trials?

Relationship to something?

Look at the figure and draw any conclusions:

$\widehat{FEM} = \widehat{FAM}$

$\widehat{FAE} + \widehat{FME} = 180^o$

$\widehat{MED} + \widehat{MCD} = 180^o$

Try to eliminate differences between the two columns

-there is $\widehat{MED}$ (right)

Let us use it

$\widehat{MED} = 180^o - \widehat{MCD}$,

so if we prove that $\widehat{FAM} = \widehat{MCD}$, we have solved our problem.

Let us try:

Look at the figure:

B, A, M, C lie on the circle, so

$\widehat{FAM} = \widehat{MCB} = \widehat{MCD}$

Goal

Backward procedure

D, E, F collinear

previous step?

$\widehat{FEM} + \widehat{MED} = 180^o$ (1)

equivalent steps?

since $\widehat{FEM} = \widehat{FAM}$

we deduce that

$\widehat{FAM} + \widehat{MED} = 180^o$ (2)

equivalent step?

$\widehat{FAM} + 180^o - \widehat{MCD} = 180^o$ (3)

or

$\widehat{FAM} = \widehat{MCD}$ (4)

or

$\widehat{FAM} = \widehat{MCB}$

Now we can say that the procedure

$$(4) \Rightarrow (3) \Rightarrow (2) \Rightarrow (1)$$

is the proof.

Problem (3). *If x, y, z are* ***non negative*** *real numbers, prove that*

$$\frac{(x+y+z)^2}{3} \geq x\sqrt{yz} + y\sqrt{xz} + z\sqrt{xy}.$$

Solution:

Try to solve it before you go on reading.

Data

x, y, z real numbers

relevant formulas

$x + y \geq 2\sqrt{xy}$,

$x^2 + y^2 \geq 2xy$,

$x^2 + y^2 + z^2 \geq xy + yz + zx$

Differences!

Eliminate them

Let us use

$$x^2 + y^2 + z^2 \geq xy + yz + zx$$

so

$$\frac{x^2 + y^2 + z^2}{3} \geq \frac{xy + yz + zx}{3}$$

Goal

$$\frac{(x+y+z)^2}{3} \geq x\sqrt{yz} + y\sqrt{xz} + z\sqrt{xy}$$

Backward step?

$$\frac{x^2 + y^2 + z^2 + 2xy + 2yz + 2xz}{3} \geq x\sqrt{yz} + y\sqrt{xz} + z\sqrt{xy}$$

Another aspect?

$$\frac{x^2 + y^2 + z^2}{3} + \frac{2}{3}(xy + yz + zx) \geq x\sqrt{yz} + y\sqrt{xz} + z\sqrt{xy}$$

Another aspect?

Backward step

It is sufficient to prove that

$$\frac{xy + yz + zx}{3} + \frac{2}{3}(xy + yz + zx) \geq x\sqrt{yz} + y\sqrt{xz} + z\sqrt{xy}$$

or

$$xy + yz + zx \geq x\sqrt{yz} + y\sqrt{xz} + z\sqrt{xy}$$

Is it true?

Yes, because if we apply the inequality

$$a^2 + b^2 + c^2 \geq ab + bc + ca$$

and set

$$a = \sqrt{xy},\ b = \sqrt{yz},\ c = \sqrt{xz}$$

then we find that

$$xy + yz + zx \geq x\sqrt{yz} + y\sqrt{xz} + z\sqrt{xy}.$$

Problem (4). ***Olympiad Problem:*** *Given a square ABCD and nine straight lines, where each one of the lines divides the square into two quadrilaterals, one being triple in area in relation to the other, prove that at least three of the straight lines cross the same point.*

Solution:

Try for a while to solve it before you go on reading

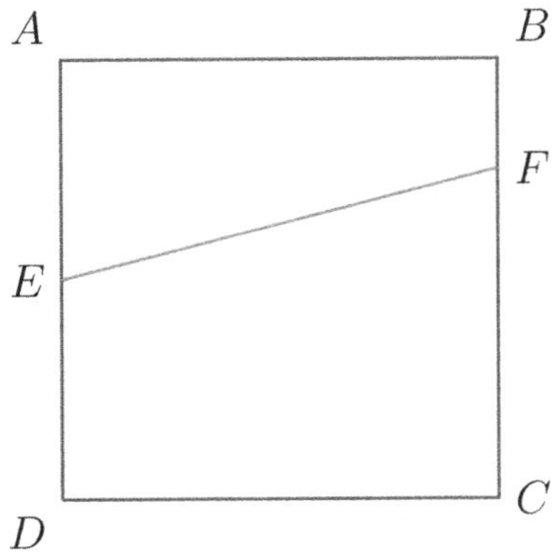

- **Reduction to one** (short notes) and categories:
- Let EF be such a straight line.
- Any backward procedure?
- We shall look for a constant point on EF.
- Describe in detail.
- The area of $EFCD$.
- How can it be found?
- It is a trapezoid.
- **Use of formulas:**
- Then area is

$$\mathcal{A}_{EFCD} = \frac{FC + DE}{2} \cdot DC$$

- What does $\frac{FC+DE}{2}$ remind us of?
- **Look at your text book!**
- From a text book we see that

$$\frac{FC + DE}{2} = \text{median of the trapezoid}$$

in other words

$$\frac{FC + DE}{2} = MN$$

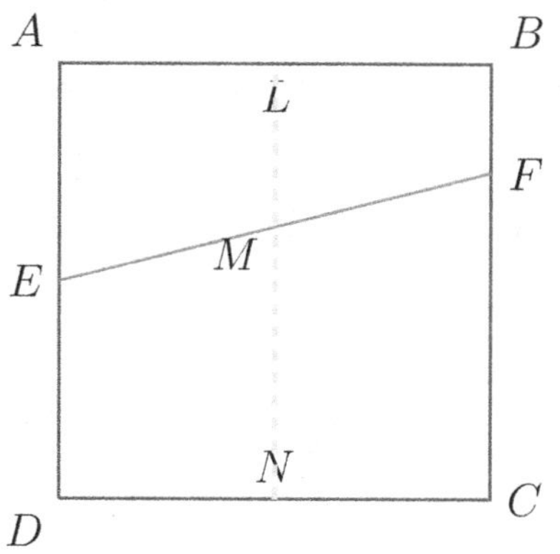

M middle of EF
N middle of DC
MN parallel to FC

- Continue describing
- L,M, N are collinear
- LN is parallel to AD
- **Use of formulas**
- Area of $EFCD = (MN) \times (DC)$
- Area of $ABFE = (LM) \times (AB)$ hence
- $(MN) \times (DC) = 3(LM) \times (AB)$
- where
$$AB = DC = BC = AD = LN$$
and thus
-
$$MN = 3LM$$
therefore M **is constant**.
- If the upper part is $\frac{1}{3}$ of the lower one then
- all EF cross the constant point M.

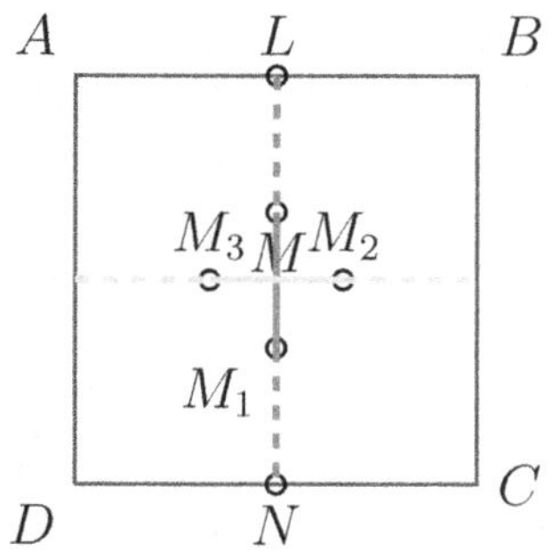

- How many of these mat exist? **Four**
- M_1, M_2, M_3, M_4 as in the figure.
- Consequently, 9 straight lines cross through these four points M_1, M_2, M_3, M_4.

- This means that at least 3 lines cross through the same point, otherwise we deduce a contradiction: *if only two lines cross each point then there will be no more that* 8 *such straight lines.*

- This is an application of the **Pigeonhole Principle**: How can 9 pigeons stay in 4 nests?

- At least 3 of them will be in the same nest.

Note:
Good knowledge of the Pigeonhole Principle led to a correct statement. The most important factor here consists in the use of short notes that led to the crucial ideas.

Problem (5). *Prove that the number*

$$5555^{2222}+2222^{5555}$$

is divisible by 7.

Solution (using short notes):

Any similarity to something known to us?
- Let us try to use known formulas as

$$a^m+b^m=(a+b)\left(a^{m-1}-a^{m-2}b+a^{m-3}b^2-\cdots\cdots-ab^{m-2}+b^{m-1}\right)$$

and try your best to eliminate differences.

- Write

$$\begin{aligned}5555^{2222}+2222^{5555}&=(5\times 1111)^{2\times 1111}+(2\times 1111)^{5\times 1111}\\&=(5^2\times 1111^2)^{1111}+(2^5\times 1111^5)^{1111}\end{aligned}$$

- so

$$\begin{aligned}5555^{2222}+2222^{5555}=&\left(5^2\times 1111^2+2^5\times 1111^5\right)\\&\times\left(5555^{2\times 1110}-5555^{2\times 1109}\times 2222+\cdots\cdots+2^{5\times 1110}\right)\end{aligned}$$

- where for the term $5^2\times 1111^2+2^5\times 1111^5$ we have:

$$5^2\times 1111^2+2^5\times 1111^5=5555^2+2222^5$$

- It is easy to prove that 5555^2+2222^5 is divisible by 7. An easy way is to start by writing

$$5555=5\times 11\times 101,\ \ 2222=2\times 11\times 101$$

to obtain

$$5555^2+2222^5=11^2\times 101^2\times\left[5^2+2^5\times 11^3\times 101^3\right]$$

- where $7 \nmid 11$ and $7 \nmid 101$ but it can be easily verified that

$$7|(5^2 + 2^5 \times 11^3 \times 101^3)$$

and the proof has been completed.

Another Approach - using a little a bit of Number Theory Relationship to something:

- Use the fact that if

$$x \equiv a \mod n$$

then

$$x^q \equiv a^q \mod n$$

and that if the

$$x \equiv a \mod n \text{ and } y \equiv b \mod n$$

occur, then

$$x + y \equiv a + b \mod n$$

Let us use that $n = 7$:

- Easily verified is that

$$2222 \equiv 3 \mod 7 \text{ and } 5555 \equiv 4 \equiv -3 \mod 7$$

hence

$$2222^3 \equiv 3^3 \equiv 6 \equiv (-1) \mod 7$$
$$5555^3 \equiv (-3)^3 \equiv 1 \mod 7$$

- Now, observe that since $5555 = 3 \times 1850 + 5$ and $2222 = 3 \times 740 + 2$ then:

$$2222^{5555} = 2222^{3\times 1850} \times 2222^5 \equiv (-1)^{1850} \times (-1)^5 \equiv (-1) \mod 7$$

and also

$$5555^{2222} = 5555^{3\times 740} \times 5555^2 \equiv 1^{740} \times (1)^5 \equiv 1 \mod 7$$

- Adding we finally deduce

$$2222^{5555} + 5555^{2222} \equiv (-1 + 1) \equiv 0 \mod 7 \Leftrightarrow 7|(2222^{5555} + 5555^{2222})$$

and the claim has been proved.

Problem (6). *Find the values of the integers x, y for which the*

$$x + y + x^2 + y^2 = p$$

holds true where p is a ***prime*** *number.*

Solution: Try for a while to solve it before reading on.

If you are stuck, go on as follows:

- **Where is the subtle point?**
- Describe completely using short notes all the parts of the problem statement.
- Use Short Notes and the Categories.
- Forget the whole problem and put emphasis on the details. Write as more as you can. In this way you inverse to hit on the subtle point.
- Prime numbers are $\{2, 3, 5, 7, \ldots\ldots\}$
- Where is the subtle point?
  Category : quantity, or, quality of x

Way 1:

- Let x be an odd number, say $x = 2k + 1$
  Relation to x^2? It is
  $$x^2 = 4k^2 + 4k + 1 = 2(2k^2 + 2k) + 1$$
  so x^2 is and odd number too.
- Hence
  $$x^2 + x$$
  is an even number
- Similarly if y is an odd number then
  $$y^2 + y$$
  is an even number.
- Let x be an even number, say $x = 2m$ then $x^2 = 4m^2$ and
  $$x^2 + x = 2(2m^2 + m)$$
  is an even number
- Similarly for y even,
  $$y^2 + y$$
  is an even number.
- Finally, for any choice of x, y we get that
  $$x + y + x^2 + y^2$$
  is an even number and since the only prime number is 2 it follows that
  $$x + y + x^2 + y^2 = 2$$
  This is the crucial idea - step

- We can easily conclude that

$$(x, y) \in \{(1,0),(1,-1),(-2,0),(-2,-1)\}$$

and also

$$(x, y) \in \{(0,1),(-1,1),(0,-2),(-1,-2)\}$$

Way 2:

- See $x+y+x^2+y^2$, another aspect?

$$x+x^2+y+y^2 = x(x+1)+y(y+1)$$

- **Quality** of $x(x+1)$ and $y(y+1)$?
Consecutive integers, hence

$$x(x+1),\ y(y+1)$$

are even numbers

- Therefore

$$x(x+1)+y(y+1)$$

is an even number

- Relation to p?
so p is an even number and a prime number too, so $p=2$ and we go back to the last part of Way 1.

The techniques in short notes
Read it if you like some rythm
A

- [1] I take the words one by one
- one by one each piece of the problem, small or large
I use simple and least number of symbols
- by writing, ideas are generated
I try to use the categories ...

- [2] I write down (or I retrieve in my mind)
-definitions-theorems-similar problems-techniques
-everything that seems to have a relation
small or large, to my problem
-by writing, ideas are generated ...

- [3] I am not afraid to think,
or to write everything down
- All I am looking for is a good idea
-All I am wrestling with is for a crucial idea
-That is why I set my imagination quite free
-I do my best to recognize the subtle point
-There is always a subtle clever point
How shall I spot it?
-By generating related or unrelated ideas
By describing what is given and the goal

- [4] I find more aspects
  -every detail may have more aspects

- [5] I let my imagination loose
  I stop caring about the problem
  I care only for the detail
  The detail itself -what does it remind me of?
  Have I ever seen it anywhere else?

- [6] I get stubbornly angry
  something is hidden from me
  something is fooling me
  something is fooling me, it seems trivial
  -However, this one might be all I am looking for
  Let me relate it
  Let me develop it
  Let me write down, let me remember
  more ideas, related or unrelated.

B

I go on
Not only thought, but thought and action
I continue with anger and persistence

- [7] I activate everything
  If there is some expression?
  I change it again and again
  I factorize
  I solve for some letter a or b ...
  I separate it into smaller expressions
  I symbolize it using better symbols
  I am searching for *hidden identities* ...
  or hidden inequalities
  or hidden relations

- [8] I search intensively for
  ideas related or unrelated
  I am not afraid to write
  I change again the expression
  I modify the expression

- [9] I concentrate on the goal
  Are there any previous steps?
  Any previous situations?

- [10] How does the goal differ from the given data?
  By what property or proposition
  can I eliminate the difference
  or a part of the difference?

- [11] Courage-Courage and imagination
  That is the secret for the solution

- [12] I lay siege the problem, cleverly and smoothly,
  around - around the given and the goal
  I work on the differences little by little until they are eliminated.

C

I go on
Not only thought, but thought and actions

- [13] Am I trapped up?
  Am I running in circles?
  Am I repeating the same and the same all time?
  If yes, I start negations ...
  I have to get out of the loop ...
  Let me use the categories

- [14] What are the opposites
  to all I have done?
  Let me try one by one ...

- [15] Should I try to draw
  some path to the solution?
  Should I decide on a rough line?
  Can I distinguish - maybe decide
  some intermediate points to the solution
  - some intermediate steps?

- [16] Are there any negations:
  not, no, neither, different, unique etc.
  - has the goal only one or two alternatives?

- [17] If yes, let try the contradiction method.
  Is it a large problem statement?
  Let me divide it to smaller problems
  Let me take a small part.

- [18] Let me make reduction to the unit ...
  Let me examine a special case
  Let me examine a more special case

- [19] ...Let me go on step by step
  starting from one unit, to many
  -Let me build little by little

- [20] Do we need some auxiliary elements?
  Is my problem a special case of a more general one?
  So, try to integrate-to complete the expression -do not leave any blank spaces
  -complete the figure, draw one or more lines

-find out symmetries
-Try to generalize your problem
(description will help you in generalizing)
-it might be easier now -make your own synthesis
-combine
-relate
-find analogies
-extract information
-resolve it, develop it
recombine again
-If you feel tired, leave it for some time
Leave it for longer time ......

Now, once more, describe every detail
using short notes. Try your best
to find more aspects, a new aspect ...
-Let us remember some revolutionary examples
concerning *a new aspect.*

1. Thales → new aspect → proof in Mathematics

2. Athens → new aspect → genesis of theatre

3. Aristarchus → new aspect → the planets move around the sun [1]

4. Archimedes → new aspect → calculation of volumes and areas of sphere, parabola etc. about 2000 years before today's calculus was developed.

5. Descartes → new aspect → coordinate geometry

6. Galileo → new aspect → physics (experimenting and thinking)

7. Newton → new aspect → second derivative in Mechanics

8. Faraday → new aspect → generator of electricity [2]

9. Einstein → new aspect → application of Riemann's curvature tensor (general theory of relativity).

... and so on ... So, never give up - just make some breaks ...

The techniques presented in this book are the result of many years of efforts, and sometimes of good luck.

[1]Researchers in the history of science say that Copernicus copied Aristarchus (see: WILL DURANT, *The story of civilization, the renaissance*, Vol. 6, pp. 927).

[2]It was well known that electricity produces magnetic field, but **only** Faraday thought of the inverse aspect : using the magnetic field we can produce electricity ... the basis of today's technology

This initial form of the techniques was tested in USA (where the first author was a Fulbright scholar). Using concepts and methods of Cognitive Psychology, the techniques were transformed into a computer program, and then the computer solved a number of problems on euclidean Geometry and Algebra.

This proved the efficiency of the method.

Raphael's : School of Athens

Platos's Theory of ideas is used by researchers on problem solving. They use the ideas in semantic networks, neccessary for developing computer programs to solve problems.
Plato gestures to the Heavens (world of ideas), while Aristotle gestures to the earth (empirical observations).
An adaptation to our techniques and an extension of the categories from the Logic of Aristotle, help us to approach problems far more efficiently.
This new aspect, is a main idea of this book.

This disagreement between the two great philosophers, has finally been proved useful: Both aspects are found to be necessary for problem solving as they complete each other.
Plato's theory is useful for the semantic networks, while Aristotle's categories make the techniques more efficient.

Instead of an epilogue

A very interesting and different problem

The problem of essay writing

Now, let us deal with a quite different problem, it concerns all students

Problems are not only in Mathematics, or Science. Problems are everywhere, for instance, in **Essay writing**.
Many students are often getting stuck when they are given a topic for essay writing. It is difficult for them to generate ideas about *what to write*.
Well, this is a problem to solve ......
How can we help these students ?

Here is a technique which improves our ability in writing.
It was found after many trials and experiments at the Experimental School of the University of Athens, Greece.
The director of this school, author of this part of the book, retired now, applied a very simple technique, based on Cognitive Psychology (How the mind works, 1956), which really helps students generate ideas in writing.
Let us present this technique using an example.

Assume that in the given topic, an extract of 4 – 5 lines, a key word is *rain*.
According to Cognitive Psychology, there is, in our brain, a parallel connection of *rain* and other knowledge, stored in our memory (see: figure 1).

Figure 1

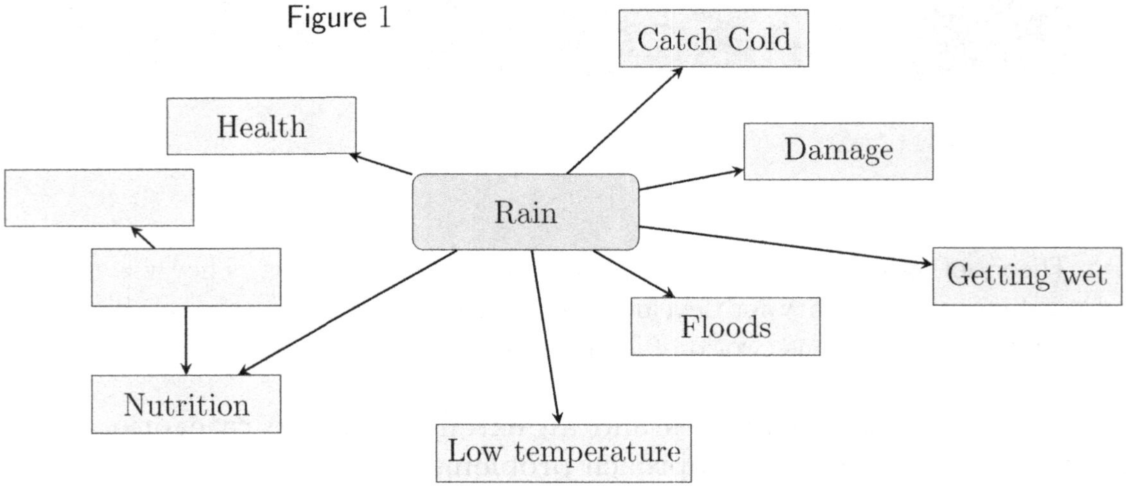

Although there is a lot of knowledge relevant to *rain* in our mind, many students find it difficult to recall some of them. They get stuck.

Look at the following scheme, which makes it very easy to recall knowledge quickly. We do not follow a parallel, but a connection of knowledge in series.
This is a fundamental difference.

Our scheme consists of 3 steps which we describe bellow:
For the first step:

Step 1	**Time**	}	-	Go on little by little in the past ...
	$\Downarrow$	}		
	Today	}	-	Try to remember ... You can recall many things
	$\Downarrow$	}		relative to your topic, for instance here:
	Yesterday	}		
	$\Downarrow$	}		
	Last week	}	-	Floods, storms, damage due to rain and so on
	$\Downarrow$	}		
	Last summer	}		
	$\Downarrow$	}	-	In $1-2$ minutes you can
				write shortly whatever
	A year ago	}		you recall
	$\Downarrow$	}		
	When I was younger	}		

For the second step:

Step 2	**Enviroment**	}	-	Write shortly whatever
				you remember ...
	$\Downarrow$	}		Positive or negative-for example:
	Your home	}	-	
	$\Downarrow$	}		
	Your neighborhood	}	-	An incident with your parents,
				or friends
	$\Downarrow$	}		something you have seen
				in a picture, or in TV
	Your country	}	-	
	$\Downarrow$	}		
	Your school, your sports	}	-	I did not go to school
				because of the storm, etc.
	$\Downarrow$	}	-	In $1-2$ minutes you can
				write shortly whatever
	Friends, parents, compatriots	}		
	$\Downarrow$	}		
	Pictures you have seen	}		
	$\Downarrow$	}		
	Neighboors, schoolmates	}		
	$\Downarrow$	}		
	TV, hobbies	}		
	$\Downarrow$	}		
	Whatever entertains you	}		

For the third step:

Step 3	**Whatever concrens us**	}	-	Write shortly here:
	⇓	}		
	Happiness	}	-	
	We all want be happy	}		Relation of rain with:
	as Aristotels has written	}	-	
	⇓	}	-	Food, no rain, no food
	Here, the following	}	-	economy, no rain, no good economy
	do play a role	}	-	art, I saw a picture with rain
	⇓	}	-	science
	Food, economy, living	}	-	Living, our living depends on food,
	freedom, culture, insurance	}	-	which, depends mainly on rain
	justice, art, science	}	-	culture etc.
	family, friendship	}		
	entertainement, society	}		
	education, health	}		
		}		In $1-2$ minutes you can write
				many things for
		}		any topic.
		}		**Do not forget . . . while writing**
		}		**more ideas are generated**

Note: **You can generate many ideas in 4-5 minutes, for any topic.**

Other methods may also be employed in the case of special topics, e.g: Description (Describe the interior, exterior, its evolution in time etc). Apart from the connection in parallel of knowledge, there is another reason why the generating ideas is hindered. In the case of problem solving, people usually tend to attack the whole problem, and the same they do in composition writing: Pupils are trying to generate ideas all at once. Since this contradicts the laws of the brain functioning (short-term memory etc.) the result is that the brain is blacked out, the most rudimentary laws of its operation being violated. This is precisely the point where the technique we propose, aims: The optimum performance of the brain by means of the right procedure.

My wife Matina helped me in making Step 3. In addition, a well-known play-writer, K. Mourselas, friend of mine, suggested me to add question like : *why did this happen?*, *How did it happen?*.

Using simple concepts of Cognitive Psychology, we can say:

Applying this scheme, more and more nodes in our brain are activated and thereby more information may be retrieved.
Thus the brain is freed of the worry *What shall I write* and it is only devoted to the *How shall I write it.*
It is very easy now to write an essay. For the introduction and the conclusion you can use short notes from Step 3 mainly.
The teachers of the Experimental School applied this technique in class. The results of the application to the various ages, from 8 to 18 years of age, may be described as follows:

- For the very weak students : after enough practice there has been noted a remarkable improvement in generating ideas.

- For the average students: they got easily along with the technique, they did like it and there has been also a remarkable increase in ideas - generation.

Comment:

The well known, for problem solving, or essay writing, *brainstrorming* method (Osborn 1953), for either an individual or for a group of people, consists in generating as many ideas as possible, without criticism.

It welcomes unusual ideas and suggests improvement and combination of ideas. Our technique of short notes has some similarities with the brainstorming method but also an **important difference**.
Our technique shows how someone can easily generate many ideas related to the problem. It helps all students. Experiments have shown that students learn it very easily and they apply it successfully in writing.
The *brainstorming method* asks the students to extremely *press* their mind, so that ideas may be generated. Practice plays actually some role here but, in essay writing, many students cannot find any help using the *brainstorming* method.
The brainstorming method is very good for problems in research, when a group of scientists work on a certain problem.
Note: Researchers in Cognitive Psychology made this comment for the above technique of this book ...... *interesting teaching tool* ... 1983.

Part II

Applications of the techinques to SAT math

By Spyros Kalomitsines & Fotios Travlopanos

Chapter 4

Topic 1 - Arithmetic-Algebra & Geometry

......once upon a time, a king (Ptolemeous A') called in his palace a famous mathematician (Euclid) and asked him: *Tell me quickly what mathematics is. Explain everything to me.*
Then the mathematician answered: *Your majesty, unfortunately there is no royal way to mathematics*, meaning that, in order to learn mathematics you have to work hard ...

However, after today's great progress in Cognitive Psychology (*How the mind works*) and mathematics Education - problem solving techniques, there is an easier way to approach mathematics.

Well, the easiest way to mathematics is:

- (a) To organize everything, adapting and applying knowledge from Cognitive Psychology. With this adaptations and applications you save time and *energy* (as experiments showed it, since 1981).
- (b) To apply, for solving non-routine problems, simple techniques from the first chapters of this book - **short notes, categories, graphs** etc., - many examples are presented.
  So, in less time you will solve more problems, which is quite necessary, to make your mind work faster in these domains. Hence you will get a higher score in the final examinations.

Let us now start presenting SAT math.

1. We have divided everything you have to know in five topics.
2. In every topic you can see about five sections, no more than six.
3. In every section we start by using graphs, formulas, simple symbols and examples. So, we clarify completely everything in the easiest way.
4. For more complicated problems we present some effective, simple techniques - as we do in the first Part of this book. So, we clarify everything in a very simple way using many examples.

5. Finally, multiple choice problems follow, as the ones in the SAT examinations. It is a very good practice. First it is recommended to try to solve the suggested problems by yourself and then you can read the solutions provided by the authors.

Let us repeat something very important:
The more problems you solve, the more practice you get, the higher possible score you will achieve. All these, are saving as more as possible *time* and *energy*, than any other old ways of teaching. We have verified it by performing many experiments. It will not take you too long to go through all topics. It depends on your background in Mathematics.
Do no forget: the more you practice (solving problems) the more the speed of your brain increases, so the higher is the score you will achieve: we can say **Knowledge = Practice**. Here, in this book, we try to make practice easier, saving time and energy ......

4.1 Problems on percentages

The easiest way is:
Group problems together
according to the solution procedure

In this way it is easier to learn and to remember

Grouping problems on percentages
Short mnemonic rules

Group 1: If you are given the **initial price** and the **percentage** then **multiply**

Example:
Initial price 800, **reduction** 12% **per cent ...**

Now, I multiply $800 \times 0.12 = 96$ (reduction)
(Do not forget: $12\% = \frac{12}{100} = 0.12$)
So final price is $800 - 96 = 704$ or, equivalently,
final price $= 800 \times (1 - 0.12) = 800 \times 0.88 = 704$

Summarizing: (mnemonic rules)
Initial price × percentage = reduction
Final price = Initial price − reduction,
Else, equivalently and straight forwardly
Final price = Initial price × (1- reduction)

Note: If we are given an increase percentages, say 12%, then use **mnemonic rule:**

Final price = Initial Price × (1+ percentage)
= Initial Price × $(1 + 0.12)$
= Initial price $\times 1.12$
Generally you can use the formula
Initial Price × (1± percentage) = Final Price

When there is reduction: Initial $\times$ $(1-p)$ = **Final Price**
When there is increase: Initial $\times$ $(1+p)$ = **Final price**
or, if we use the symbols I = **initial,** f = **final, then**
$I \times (1-p) = f$ **in reduction case**
$I \times (1+p) = f$ **in increase case**

Group 2: If you are given the **final price** and the **percentage** of reduction then **divide**

Final price 720, **reduction** was 10%
Find **initial price.**
Now **divide** by $(100\% - 10\%) = 90\% = 0.90$. or by $1 - p = 1 - 0.10 = 0.90$,

$$\frac{720}{0.90} = 800 \text{ initial price}$$

so

$$\frac{\text{Final Price}}{0.90} = \text{Initial price}$$

Summarizing:
Final price $\div$(100% - **Reduction percentage**) = **Initial price**
or, using Algebra
Let x be the **initial price**, then:

$$x \cdot \frac{90}{100} = 720 \Leftrightarrow$$
$$0.90 \cdot x = 720 \Leftrightarrow$$
$$x = 720 \div 0.9 = 800$$

or, use the formula

$$I(1-p) = f \Rightarrow I = \frac{f}{1-p}$$

Attention: you can use either the mnemonic rule, or the formula

$$I(1 \pm p) = f$$

You can solve the formula for I or for p etc. Do whatever seems easier to you.

Some more examples

1) **Group** 1: **goal: final price**

Given:
Initial Price
and percentage } $\rightarrow$ **multiply**

Example 1:

- *Initial Price:* = 600
- *Increase percentage :* $15\% = \frac{15}{100} = 0.15$

- *Final Price:* $600 \times 1.15 = 690$
- **Or - else**
- Increase $= 600 \times 0.15 = 90$
- Final price $= 600 + 90 = 690$

Example 2:

- *Initial Price:* 400
- *Reduction percentage :* $15\% = \frac{15}{100} = 0.15$
- *Final Price:* $400 \times 0.85 = 340$
- **Or - else**
- Reduction $= 400 \times 0.15 = 60$
- Final price $400 - 60 = 340$

2) Group 2: **goal: Initial Price**

Given:
Final Price
and percentage $\}$ → **divide**

Example 1:

- *Finall Price:* $= 560$
- *Increase percentage :* $12\% = \frac{12}{100} = 0.12$
- *Initial Price:* $560 \div 1.12 = 500$

Example 2:

- *Final Price:* $= 160$
- *reduction percentage :* $20\% = \frac{20}{100} = 0.20$
- *Initial Price:* $160 \div 0.80 = 200$

3) Group 3: **goal: Percentage**

Given:
Initial Price
and Final Price $\}$ → **divide**
Mnemonic Rule

Careful!!!!! **Final ÷ Initial**

Example 1:

- *Initial Price:* $= 300$
- *Final Price :* $= 345$
- *Compute:* $345 \div 300 = 1.15$
- *So,* $0.15 =$ *increase percentage* $= 15\%$
- **Or - else** use the formula

$$I(1+p) = F \Rightarrow 300(1+p) = 345 \Rightarrow 1+p = \frac{345}{300} \Rightarrow 1+p = 1.15 \Rightarrow p = 0.15$$

Example 2:

- *Initial Price:* $= 240$
- *Final Price :* $= 216$
- *Compute decrease percentage*
- **Way 1**
- $216 \div 240 = 0.90$ so, decrease percentage $= 0.10 = 10\%$
- **Way 2** Use the formula

$$I(1-p) = f \Rightarrow 240(1-p) = 216 \Rightarrow 1-p = \frac{216}{240} = 0.90 \Rightarrow p = 1 - 0.90 = 0.10$$

General instructions for solving problem

Read Carefully!

- **If you have got stuck with a problem then follow the three steps.**

Step 1

- **Try to remember: Is there any relevant formula? Apply it ...**
  No solution yet?
- **Try to remember: Is there any exact method for similar problems? Apply it ...**
  No solution yet? go to

Step 2

- **Trying to apply formula or exact method**
  (the way that a similar problem has been solved)
  did you notice that something is missing?
  Try to find it.
- **Make clear, explicit all implicit information**

- **Try to spot any subtle point**
  ....
  No solution yet?

- **A more dynamic procedure: in order to help your mind, you should describe all the details using short notes**
  Also use categories for each detail: relation to something, quality, quantity etc.

- **So, you increase the probability that the crucial idea will emerge**

- **Try to surpass yourself ......**
  No solution yet?

- Short notes as it is the writing a children's poem, increases the speed of the brain.
  On the other hand our language was initially developed with short phrases, because this enables people generate more ideas
  Complex phrases were developed later, as we have said in Part 1.

Step 3

We hope you very seldom will be obliged to follow Step 3.
We wish, you will very seldom, need to use this step.

- **Separate the Data & the Goal**
  Do quickly
  Do not write
  just think of these

Data	**Start from the Goal**	**Goal**
Forward procedure	**Do quickly**	**Any previous steps?**
Follow Step 2	**Locate differences between the two columns - left and right**	**Any equivalent step?**
Describe all the details here	**Try to eliminate them, step by step**	
Do not write		
Just think of it		

<u>Compound problems on percentages</u>

Note
Go step by step using exact methods of 1, 2, 3.

Problem (1). *A shop has 25% discount of all items in a sale. George decided to buy a suite and he asked for a further reduction. The reduction for George was 10% and he paid 81. The original price for the suit is*

- *[A]* 90

- *[B]* 100
- *[C]* 110
- *[D]* 120
- *[E]* *none of the above*

Try for a while to solve it, before reading

Solution

Do you know any formula?? or, any exact method?

This question should always be **your first step ...**

The initial price was $81 \div 0.90 = 90$

For the original price we apply again **Group** 2 technique:

$$90 \div (100 - 25) = 90 \div 0.75 = 120$$

Note: Saying **exact** method, we mean for instance a technique for solving equations, or inequalities etc.

Problem (2). *A shop owner increases* 30% *the price of a product, which he has bought for* 200*. Later, he made a reduction on the sale price and the new price is* 221*. The reduction percentage is*

- *[A]* 10%
- *[B]* 12%
- *[C]* 14%
- *[D]* 15%
- *[E]* 18%

Try for a while to solve it, before reading

Solution

Do you know any formula?? or, any exact method?

Application of the technique of Group 1:

$200 \times 0.30 = 60$ so the new price is $200 + 60 = 260$

Now, we have the problem: initial price 260 and the final price is 221

So, by applying **group** 3 technique:

$221 \div 260 = 0.85$ so reduction percentage is 15%

or, we use the formula

$$I(1-p) = f \Rightarrow 1 - p = \frac{f}{I} \Rightarrow p = 1 - \frac{221}{260} = 0.15$$

Problem (3). *The population of a city is* 240.000 *and it increases* 1.2% *per year. In five years the population will be*

- *[A]* 254.000

- *[B]* 254.600
- *[C]* 254.700
- *[D]* 254.749
- *[E]* 255.700

Let us try to apply Group 1 technique:
In one year the popultation will be

$$240.000 \times (1 + \frac{1.2}{100}) = 240.000 \times 1.012$$

So, in 2 years the population will be

$$(240.000 \times 1.012) \times 1.012 = 240.000 \times 1.012^2$$

hence, generally, in five years the population will be

$$240.000 \times 1.012^5 = 254.749.$$

<u>Note</u> **Some complementary simple techniques:**
Fractions:

- **Group 1:** Find $\frac{3}{5}$ of 40 : **Multiply**

$$\frac{3}{5} \times 40 = \frac{120}{5} = 24$$

  Find $\frac{2}{3}$ of $\frac{5}{7}$, **multiply**

$$\frac{2}{3} \times \frac{5}{7} = \frac{10}{21}$$

- **Group 2:** $\frac{2}{5}$ of a number is 80, find the number. **Divide**

$$80 \div \frac{2}{5} = 80 \times \frac{5}{2} = \frac{400}{2} = 200$$

  If $\frac{2}{5}$ of a number is $\frac{3}{7}$, find the number. **Divide**

$$\frac{3}{7} \div \frac{2}{5} = \frac{3}{7} \times \frac{5}{2} = \frac{15}{14}$$

- **Group 3:** What fraction of 50 is 70?.
  Solution

$$\frac{70}{50} = \frac{7}{5}$$

  and verify

$$\frac{7}{5} \times 50 = 7 \times 10 = 70$$

  or, using Algebra

$$50x = 70 \Rightarrow x = \frac{70}{50} = \frac{7}{5}.$$

4.2 Ratio & Proportion

Remember that

- [a] If $\frac{a}{2}=\frac{b}{3}$ then $\frac{a}{2}=\frac{b}{3}=\frac{a+b}{2+3}=\frac{a+b}{5}$
- [b] , If $\frac{a}{2}=\frac{b}{3}=\frac{c}{5}$ then $\frac{a}{2}=\frac{b}{3}=\frac{c}{5}=\frac{a+b+c}{2+3+5}=\frac{a+b+c}{10}$
- [c] If $\frac{a}{b}=\frac{c}{d}$ then $\frac{a+b}{b}=\frac{c+d}{d}$, $\frac{a-b}{b}=\frac{c-d}{d}$, $\frac{a}{b-a}=\frac{c}{d-c}$

Some times in problem solutions we set the ratios equal to some l quantity, i.e:

$$\frac{a}{2}=\frac{b}{3}=l \Rightarrow a=2l,\ b=3l$$

etc.

Example (1). *If*

$$\frac{2a+3b}{3a+4b}=2$$

find the ratio $\frac{b}{a}$ $(a \neq 0)$

Solution: Divide nominator and denominator by a to get:

$$\frac{2a+3b}{3a+4b}=\frac{2+3\frac{b}{a}}{3+4\frac{b}{a}}=2 \overset{set\ \frac{b}{a}=s}{\Rightarrow} 2+3s=2(3+4s) \Rightarrow$$

$$2+3s=6+8s \Rightarrow 5s=-4 \Rightarrow s=-\frac{4}{5}$$

Note: Whenever we divide by a, we mean that $a \neq 0$.

Example (2). *If*

$$\frac{3a^2+5b^2}{4a^2-7b^2}=6$$

find the value of $\frac{b}{a}$.

Solution: Divide nominator and denominator by a^2 to get:

$$\frac{3+5\frac{b^2}{a^2}}{4-7\frac{b^2}{a^2}}=6 \overset{set\ \frac{b}{a}=s}{\Rightarrow} \frac{3+5s^2}{4-7s^2}=6 \Rightarrow$$

$$3+5s^2=24-42s^2 \Rightarrow s^2=\frac{21}{47} \Rightarrow s=\pm\sqrt{\frac{21}{47}}$$

Problem (1). *If* $\frac{2a+3b}{4a+5b}=3$ *then*

$$\frac{2a^2+3b^2}{5a^2+7b^2}$$

is equal to

- *[A]* $\dfrac{133}{355}$
- *[B]* $\dfrac{140}{355}$
- *[C]* $\dfrac{147}{355}$
- *[D]* $\dfrac{150}{355}$
- *[E]* *none of the above*

Think for a while before reading on

Solution

Apply the exact technique-method of example 1, divide by a and set $\frac{b}{a} = s$

$$\frac{2a+3b}{4a+5b} = 3 \Rightarrow \frac{2+3s}{4+5s} = 3 \Rightarrow s = \frac{-5}{6}$$

hence

$$\frac{2a^2+3b^2}{5a^2+7b^2} = \frac{2+3s^2}{5+7s^2} = \frac{2+3\times\frac{25}{36}}{5+7\cdot\frac{25}{36}} = \frac{147}{355}$$

Problem (2). *If $a^2 - 3ab - 4b^2 = 0$ and a, b are positive numbers then $\frac{a}{b}$ is equal to*

- *[A]* 1
- *[B]* 2
- *[C]* 3
- *[D]* 4
- *[E]* 5

Think for a while before reading on

Solution *Apply the exact method*

$$\left(\frac{a}{b}\right)^2 - 3\frac{a}{b} - 4 = 0 \overset{set\ \frac{a}{b}=s}{\Leftrightarrow} s^2 - 3s - 4 = 0$$

with roots 4, -1 and since a, b are positive we get $\frac{a}{b} = 4$.

Problem (3). *If the ratio $\hat{A} : \hat{B} : \hat{C}$ of the angles $\hat{A}$, $\hat{B}$, $\hat{C}$ of the triangle $\triangle ABC$ is $1:2:3$ then the ratio of its corresponding sides $a:b:c$ is equal to*

- *[A]* $1:\sqrt{3}:2$
- *[B]* $2:\sqrt{3}:1$
- *[C]* $1:2:3$
- *[D]* $\sqrt{3}:2:1$

- *[E]* $1 : \sqrt{3} : \sqrt{2}$

Think for a while before reading on

Solution *Describe - Use the formula :*

$$\frac{\hat{A}}{1} = \frac{\hat{B}}{2} = \frac{\hat{C}}{3} = \frac{\hat{A}+\hat{B}+\hat{C}}{1+2+3} = \frac{180^o}{6} = 30^o$$

hence

$$\hat{A} = 30^o, \ \hat{B} = 60^o, \ \hat{C} = 90^o$$

and now use the sine law

$$\frac{a}{\sin A} = \frac{b}{\sin B} = \frac{c}{\sin C} \Rightarrow$$
$$\frac{a}{\sin 30} = \frac{b}{\sin 60} = \frac{c}{\sin 90} \Rightarrow$$
$$\frac{a}{\frac{1}{2}} = \frac{b}{\frac{\sqrt{3}}{2}} = \frac{c}{\frac{2}{2}} \Rightarrow$$
$$a : b : c = 1 : \sqrt{3} : 2$$

So $[A]$ is the correct answer.

Problem (4). *If the ratio $\hat{A} : \hat{B} : \hat{C}$ of the measures of the angles $\hat{A}$, $\hat{B}$, $\hat{C}$ of the triangle $\triangle ABC$ is $2 : 3 : 5$ then the smallest angle is*

- *[A]* 18°
- *[B]* 36°
- *[C]* 54°
- *[D]* 60°
- *[E]* 72°

Think for a while before reading on

Solution *Let us apply the exact method*

$$\frac{\hat{A}}{2} = \frac{\hat{B}}{3} = \frac{\hat{C}}{5} = \frac{\hat{A}+\hat{B}+\hat{C}}{2+3+5} = \frac{180^o}{10} = 18^o$$

hence

$$\frac{\hat{A}}{2} = 18^o \Rightarrow \hat{A} = 36^o,$$
$$\frac{\hat{B}}{3} = 18^o \Rightarrow \hat{B} = 54^o,$$
$$\frac{\hat{C}}{5} = 18^o \Rightarrow \hat{C} = 90^o,$$

4.2.1 Examples & Techniques for problems on directly, or, inversely proportional values

Problem (1). *If x is directly proportional to y and $x = 12$ when $y = 15$, then the value of x when $y = 20$ is*

- *[A]* 14
- *[B]* 16
- *[C]* 18
- *[D]* 20
- *[E]* *none of the above*

Solution:

- **Way** 1 **Technique:** Consider the ratio

  values of x and y *before*

  $$\frac{12}{x} = \frac{15}{20}$$

  values of x and y *after*

  and cross-multiply

  $$15x = 240 \Rightarrow x = \frac{240}{15} = 16$$

  the correct answer is $[B]$.

- **Way** 2 **Technique:** It is

  $$x = ky \Rightarrow 12 = 15k \Rightarrow k = \frac{12}{15} = \frac{4}{5}$$

  hence if $y = 20$ then

  $$x = \frac{4}{5} \times 20 = 16.$$

Problem (2). *If x is inversely proportional to y and $x = 10$ when $y = 18$ then the value of y, when $x = 6$, is*

- *[A]* 8
- *[B]* 10.8
- *[C]* 16
- *[D]* 24
- *[E]* 30

Solution:

- **Way** 1 **- Technique:** Consider the ratio

x - value *before* y - value *after*

$$\frac{10}{6} = \frac{y}{18}$$

x - value *after* y - value *before*

Now, cross multiply to get:

$$6y = 180 \Rightarrow y = 30$$

- **Way** 2 It is

$$x = \frac{k}{y} \Rightarrow 10 = \frac{k}{18} \Rightarrow k = 180$$

hence, if $x = 6$ then

$$6 = \frac{180}{y} \Rightarrow y = \frac{180}{6} = 30.$$

Problem (3). *If x is directly proportional to y and y is inversely proportional to z and $x = 8$ when $y = 6$, $z = 10$ then the value of xy, when $z = 20$ is equal to:*

- *[A]* 12
- *[B]* 20
- *[C]* 24
- *[D]* 36
- *[E]* 48

Solution:

- **Way** 1**:** Consider the ratios:

for directly proportional x, y:

values *before*

$$\frac{8}{x} = \frac{6}{y}$$

values *after*

for inversely proportional y, z:

before values *after*

$$\frac{6}{y} = \frac{20}{10}$$

after values *before*

Now, cross multiply

$$8y = 6x \text{ and } 60 = 20y \Rightarrow y = 3 \overset{8y=6x}{\Rightarrow}$$
$$24 = 6x \Rightarrow x = 4 \Rightarrow xy = 12.$$

Correct answer is $[A]$.

- **Way** 2: Using direct proportionality

$$x = ky \Rightarrow 8 = 6k \Rightarrow k = \frac{4}{3}$$

Now, using inverse proportionality

$$y = \frac{m}{z} \Rightarrow 6 = \frac{m}{10} \Rightarrow m = 60$$

therefore if

$$z = 20 \Rightarrow y = \frac{60}{20} = 3$$

hence

$$x = \frac{4}{3} \times 3 = 4$$

and thus

$$xy = 4 \cdot 3 = 12.$$

4.3 Inverse, odd-even & Combining functions

Inverse functions

Quick, standard technique: Given a function $y = f(x)$ and a domain where f is defined and its inverse does exist on it, then you have to try to resolve the equation

$$y = f(x)$$

with respect to x so that we deduce an equation of the form

$$x = f^{-1}(y).$$

Example (1). *If $f(x) = ax + b$, find the inverse $f^{-1}(x)$.*

Solution: There are two very easy and equivalent ways:

Way 1: Set $f(x) = y$ and interchange x and y. Solve with respect to y:

$$y = ax + b \Rightarrow x = ay + b$$

Solve for y:

$$y = \frac{x-b}{a}$$

hence

$$f^{-1}(x) = \frac{x-b}{a}$$

Way 2: Set

$$f(x) = y \Rightarrow y = ax + b$$

solve for x:

$$x = \frac{y-b}{a}$$

and **interchange** the role of x and y:

$$y = \frac{x-b}{a} \Rightarrow f^{-1}(x) = \frac{x-b}{a}$$

Follow any way you prefer.

Example (2). *If $f(x) = 3x + 5$ then $f^{-1}(x)$ is*

- *A* $\frac{5-x}{3}$
- *B* $\frac{5+x}{3}$
- *C* $x + \frac{5}{3}$
- *D* $\frac{x-5}{3}$
- *E* $\frac{x+3}{5}$

Solution:

It is $y = 3x + 5$, interchange

$$x = 3y + 5$$

and solve for y:

$$y = \frac{x-5}{3}$$

hence

$$f^{-1}(x) = \frac{x-5}{3}$$

Example (3). • *(i) If $f(x) = 2x^3 + x + 1$ find $f^{-1}(4)$*

- *(ii) If $f(x) = x^3 + 10$ find $f^{-1}(2)$*

Solution: (i) It is

$$f(1) = 4 \Rightarrow f^{-1}(4) = 1$$

(ii) Also

$$f(-2) = (-2)^3 + 10 = -8 + 10 = 2 \Rightarrow f^{-1}(2) = -2$$

Example (4). *If $f(x) = x^2 + 3x - 1$ find all values of $f^{-1}(2)$*

Solution: Let us use the classic symbols:

$$y = f(x), \; x = f^{-1}(y)$$

Since we are looking for $f^{-1}(2)$ we have

$$x = f^{-1}(2), \; y = 2$$

Therefore

$$2 = x^2 + 3x - 1 \Rightarrow x^2 - 3x - 3 = 0$$

hence

$$x = \frac{-3 \pm \sqrt{21}}{3}$$

so the values of $f^{-1}(2)$ are

$$\frac{-3+\sqrt{21}}{3}, \frac{-3-\sqrt{21}}{3}$$

Example (5). *If $f(x) = 5\sqrt{2x}$, $x \geq 0$ find $f^{-1}(x)$.*

Solution - Technique: Set $y = f(x)$ solve the equation $y = f(x)$ with respect to x. It is:

$$y = 5\sqrt{2x} \Leftrightarrow y^2 = 50x \Rightarrow x = \frac{y^2}{50}$$

hence

$$f^{-1}(x) = \frac{x^2}{50}, \quad x \geq 0.$$

Example (6). *If $f(x) = x^2 + 3x - 1$ find the positive values of **(i)** $f^{-1}(0)$ and **(ii)** $f^{-1}(2)$.*

Solution: As in the previous examples

$$f(x) = 0, \ f^{-1}(0) = x$$

so

$$x^2 - 3x - 1 = 0 \Rightarrow x = \frac{-3 \pm \sqrt{13}}{2}$$

and the positive value corresponds to

$$x = \frac{-3+\sqrt{13}}{2} \sim 0.3$$

Similarly working

$$x^2 + 3x - 1 = 2 \Rightarrow x = \frac{-3 \pm \sqrt{21}}{2}$$

hence the required value of

$$x = \frac{-3+\sqrt{21}}{2} \sim 0.8$$

Problem (1). *The inverse function of $f(x) = 2x^3 - 8$ is*

- *[A]* $-\sqrt[3]{\frac{x}{2}+4}$
- *[B]* $\sqrt[3]{\frac{x}{2}+4}$
- *[C]* $\sqrt[3]{\frac{x}{2}-4}$
- *[D]* $-\sqrt[3]{\frac{x}{2}+8}$

- *[E] none of the above*

Solution: Follow the suggested technique, let $y = 2x^3 - 8$ and solve it in x:

$$y = 2x^3 - 8 \Rightarrow 2x^3 = y + 8 \Rightarrow x^3 = \frac{y}{2} + 4 \Rightarrow x = \sqrt[3]{\frac{y}{2} + 4}$$

hence $f^{-1}(x) = \sqrt[3]{\frac{x}{2} + 4}$ and the correct answer is $[B]$.
Problem (2). *If $f(x) = x^2 + 9$, $x < 0$ then its inverse is*

- *[A]* $-\sqrt{x - 9}$
- *[B]* $\sqrt{x - 9}$
- *[C]* $\sqrt{x + 9}$
- *[D]* $-\sqrt{x + 9}$
- *[E]* *none of the above*

Solution: Use standard technique solving $y = f(x)$ in x:

$$y = x^2 + 9 \Rightarrow x^2 = y - 9 \overset{y=x^2+9>9}{\Rightarrow} |x| = \sqrt{y - 9} \overset{x \leq 0}{\Rightarrow} x = -\sqrt{y - 9}$$

hence

$$f^{-1}(x) = -\sqrt{x - 9}$$

Correct answer is $[A]$.
Problem (3). *If $f(x) = x^2 - 4$ and $x \geq 0$ then $f^{-1}(x)$ is*

- *[A]* $-\sqrt{x + 4}$
- *[B]* $-\sqrt{x + 2}$
- *[C]* $\pm\sqrt{x + 2}$
- *[D]* $\sqrt{x + 4}$
- *[E]* $x + 4$

Solution: Use standard technique :

$$y = x^2 - 4 \Rightarrow x^2 = y + 4 \Rightarrow x = \pm\sqrt{y + 4}, \text{ and } x \geq 0 \Rightarrow f^{-1}(x) = \sqrt{x + 4}$$

hence $[D]$ is the correct answer.

4.3.1 Odd - Even Functions

By definition a function is **odd** if and only if

$$f(-x) = -f(x)$$

e.g: $f(x) = x^3$
and is said to be **even** if and only if

$$f(-x) = f(x)$$

e.g: $f(x) = x^2$.

Remark (1). *The graph of an odd function* ***is symmetric with respect to the origin*** *of the coordinates,*
The graph of an even function ***is symmetric with respect to the vertical coordinate axis*** $yy' : \ x = 0$.

Problem (1). *Find which of the below functions are odd*

$$I: \ y = x, \ II: \ y = x^2, \ III: \ y = x^3, \ IV: \ y = \sqrt{x^8}, \ V: \ y = \sqrt{|x|}$$

- *[A]* I *and* III *only*
- *[B]* I *and* V
- *[C]* II *and* V
- *[D]* II *and* III
- *[E]* III *and* V

Solution: Let us remember that : if $f(x) = f(-x)$ then f is even and if $f(-x) = -f(x)$ then f is even.
Start checking one by one:

$$f(x) = x \Rightarrow f(-x) = -x = -f(x)$$

hence $y = x$ is **odd**.

$$f(x) = x^2 = (-x)^2 \Rightarrow f(x) = f(-x)$$

hence $y = x^2$ is **even**.

$$f(x) = x^3 = -(-x)^3 \Rightarrow f(-x) = -f(x)$$

hence $y = x^3$ is **odd**.

$$f(x) = \sqrt{x^8} = \sqrt{(-x)^8} \Rightarrow f(x) = f(-x)$$

hence $y = \sqrt{x^8}$ is **even** and finally

$$f(x) = \sqrt{|x|} = \sqrt{|-x|} \Rightarrow f(x) = f(-x)$$

hence $y = \sqrt{|x|}$ is **even**.
Correct answer is $[A]$.

Problem (2). *Which of the following functions are neither odd nor even*

$$I: \ f(x) = x^3 + x^2 + 1, \ II: \ f(x) = x^2 + x - 1, \ III: \ f(x) = x^3 + x^2 + 3,$$
$$IV: \ f(x) = x^4 + x^2 - 1, \ V: \ f(x) = x^3 + 5x$$

- *[A]* I, II *and* III
- *[B]* I *and* II
- *[C]* II *and* IV

- *[D] I, II and IV*
- *[E] II and V*

Solution: Start by checking I:

$$f(x) = x^3 + x^2 + 1 \Rightarrow f(-x) = -x^3 + x^2 + 1$$

hence I is neither odd nor even.
For II:

$$f(x) = x^2 + x - 1 \Rightarrow f(-x) = x^2 - x - 1$$

hence II is neither odd nor even.
For III:

$$f(x) = x^3 + x^2 + 3 \Rightarrow f(-x) = -x^3 + x^2 + 3$$

hence III is neither odd nor even.
Since $IV : f(x) = x^4 + x^2 - 1$ obviously is an even function and $V : f(x) = x^3 + 5x$ the $f(-x) = -f(x)$ is an odd function.
So, the correct answer is $[A]$.

4.3.2 Combining functions & inverse functions

We start with some examples
Example (1). *If $f(x) = 2x + 3$ and $g(x) = 3x + 5$ compute $f(g(x))$ and $g(f(x))$*
Solution:

- For $f(g(x))$ we have:

$$f(g(x)) = f(3x + 5) = 2(3x + 5) + 3 = 6x + 13$$

- and for $g(f(x))$:

$$g(f(x)) = g(2x + 3) = 3(2x + 3) + 5 = 6x + 14.$$

Example (2). *If $f(x) = x^2 + x - 1$ and $g(x) = 3x + 4$ then compute $f(g(3))$:*
Solution: It is

$$f(g(3)) = f(13) = 13^2 + 13 - 1 = 169 + 13 - 1 = 181$$

Example (3). *If $f(x) = 3x+5$, $g(x) = x^2-x+1$ and $h(x) = 2x-1$ then compute $f(g(h(x)))$*
Solution: an easy way, start with:

$$g(h(x)) = g(2x - 1) = (2x - 1)^2 - (2x - 1) + 1 = 4x^2 - 6x + 3$$

and thus

$$f(g(h(x))) = f(4x^2 - 6x + 3) = 3(4x^2 - 6x + 3) + 5 = 12x^2 - 18x + 14.$$

We go on with some problems:

Problem (1). *If $f(x) = x^3 + x$, $g(x) = \sin x^2$ which of the following is odd?*

$$\begin{aligned} \text{I}: &\ (f \circ g)(x) = f(g(x)), \\ \text{II}: &\ (g \circ f)(x) = g(f(x)), \\ \text{III}: &\ f \cdot g(x) = f(x) \cdot g(x) \end{aligned}$$

- *[A]* *only I*
- *[B]* *only II*
- *[C]* *only III*
- *[D]* *only I and II*
- *[E]* *none of the above*

Solution: It is

$$\begin{aligned} f(g(x)) &= f(\sin x^2) = sin^3 x^2 + sin x^2, \text{ even} \\ f(g(-x)) &= f(sin(-x)^2) = f(sin x^2) = f(g(x)), \text{ even} \\ g(f(x)) &= g(x^3 + x) = sin(x^3 + x)^2 + x^3 + x, \text{ neither odd, nor even} \end{aligned}$$

and the correct answer is $[C]$.

Problem (2). *If $f(x) = 3x - 4$, $g(x) = x - 7$ and $h(x) = 7x - 3$ then $f(g(h(2)))$ is equal to*

- *[A]* 4
- *[B]* 8
- *[C]* 11
- *[D]* 12
- *[E]* 16

Think for a while before reading on

Solution: It is $h(2) = 14 - 3 = 11$ so

$$f(g(h(2))) = f(g(11)) = f(11 - 7) = f(4) = 8.$$

So, the correct answer is $[B]$.

Problem (3). *If $g(x) = x + 1$ and $f(g(x)) = 3x^2 + 6x + 2$ then $f(x)$ is*

- *[A]* $x^2 - 3$
- *[B]* $x^2 - 1$
- *[C]* $x^2 + 3$
- *[D]* $3x^2 - 1$
- *[E]* $3x^2$

Solution: Exact method:
Step 1:

$$g(x) = x + 1 \Rightarrow x = g(x) - 1$$

Step 2:

$$f(g(x)) = 3(g(x) - 1)^2 + 6(g(x) - 1) + 2 = 3g^2(x) - 1$$

hence

$$f(x) = 3x^2 - 1.$$

Correct answer is $[D]$.

Problem (4). *If $g(x) = x + 1$ and $f(g(x)) = 2x + 5$ then $f(x)$ is*

- *[A]* $2x + 1$
- *[B]* $2x + 2$
- *[C]* $x + 5$
- *[D]* $2x + 3$
- *[E]* $2x + 5$

Solution: Let us use the data:

$$g(x) = x + 1 \Rightarrow x = g(x) - 1 \Rightarrow$$
$$f(g(x)) = 2(g(x) - 1) + 5 = 2g(x) + 3$$

hence $f(x) = 2x + 3$.
Correct answer is $[D]$.

Problem (5). *If $f(x) = 2x + 1$, $g(x) = 2x$, $h(x) = 3x$ then $g^{-1}(f(h(2)))$ is*

- *[A]* 3
- *[B]* 4
- *[C]* 5
- *[D]* 6
- *[E]* 6.5

Solution: Simple technique, we start from the inside function:

$$h(2) = 6 \Rightarrow f(h(2)) = f(6) = 13 \Rightarrow g^{-1}(f(h(2))) = g^{-1}(13)$$

where

$$g^{-1}(x) = \frac{x}{2} \Rightarrow g^{-1}(13) = \frac{13}{2} = 6.5.$$

Correct answer is $[E]$.

Problem (6). *If* $f(x) = \sqrt[3]{x^5} - 1$ *then* $f^{-1}(x)$ *is*

- *[A]* $\sqrt[3]{x^5} + 1$
- *[B]* $\sqrt[5]{(x+1)^3}$
- *[C]* $\sqrt[3]{x^5 + 1}$
- *[D]* $\frac{1}{\sqrt[5]{x^3}-1}$
- *[E]* $\frac{1}{\sqrt[5]{x^3}+1}$

Solution: Apply standard technique solving $y = f(x)$ in x, it is:

$$y = \sqrt[3]{x^5} - 1 \Rightarrow \sqrt[3]{x^5} = y + 1 \Rightarrow$$
$$x^5 = (y+1)^3 \Rightarrow x = \sqrt[5]{(y+1)^3}$$

hence $f^{-1}(x) = \sqrt[5]{(x+1)^3}$.
Correct answer is $[B]$.

Problem (7). *If* $f(x) = \sqrt{6-x}$ *and* $g(x) = x^2 + x$, *the domain of* $f \circ g$ *is*

- *[A]* $[-3, -2]$
- *[B]* $[-3, 2]$
- *[C]* $[-2, 3]$
- *[D]* $[2, 3]$
- *[E]* $[2, 6]$

Solution: It is

$$(f \circ g)(x) = f(x^2 + x) = \sqrt{6 - x^2 - x}$$

is defined for $x \in \mathbb{R}$ such that

$$6 - x^2 - x \geq 0 \Leftrightarrow -(x+3)(x-2) \geq 0 \Leftrightarrow (x+3)(2-x) \geq 0$$

hence

$$x \in [-3, 2]$$

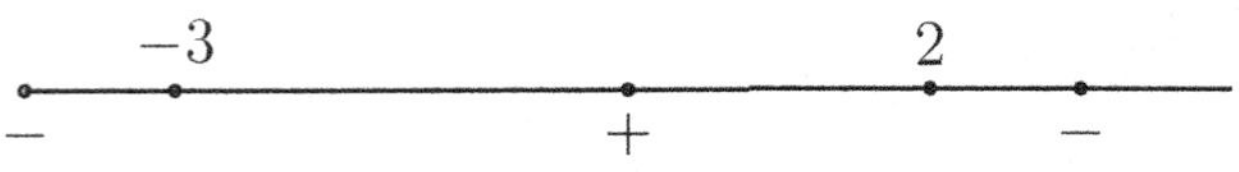

Correct answer is $[B]$.

Problem (8). *Given that*

$$f(x) = \frac{x-1}{x+3}$$

and

$$g(x) = f(\frac{x+1}{2})$$

the domain of g *is*

- *[A]* $\mathbb{R}$
- *[B]* $x \in \mathbb{R}^+$
- *[C]* $x \neq -7$
- *[D]* $x \neq -3$
- *[E]* $x \neq -1$

Solution: It is

$$g(x) = f(\frac{x+1}{2}) = \frac{\frac{x+1}{2} - 1}{\frac{x+1}{2} + 3} = \frac{x-1}{x+7}$$

Correct answer is $[C]$. which is defined if and only if

$$x + 7 \neq 0 \Leftrightarrow x \neq -7.$$

4.4 Absolute value & Inequalities

The **absolute value** of a real number x is defined as

$$|x| = x, \text{ if } x \geq 0 \text{ or } |x| = -x, \ x < 0.$$

Example:

$$|4| = 4, \ \ |-4| = -(-4) = +4$$

Triangular inequality

$$||x| - |y|| \leq |x + y| \leq |x| + |y|$$

Be careful!!!

$$\sqrt{x^2} = |x|$$

Example (1). *Solve the equation*

$$|x - 1| = 2x + 1$$

Solution:

- **Way 1:** It must be

$$2x + 1 \geq 0 \Leftrightarrow x \geq \frac{-1}{2} \ \ \textbf{(i)}$$

If $x - 1 \geq 0$ then $|x - 1| = x - 1$ so the equality becomes

$$x - 1 = 2x + 1 \Rightarrow x = -2 \ \textbf{(ii)}$$

and **(i)**, **(ii)** contradict each other having no common solution.
If $x - 1 < 0$ then $|x - 1| = 1 - x$ and the equality becomes

$$1 - x = 2x + 1 \Rightarrow -3x = 0 \Rightarrow x = 0$$

and $x = 0 > -\frac{1}{2}$ is the **accepted** solution.

- **Way 2:** It must be $x > \frac{-1}{2}$ and we square

$$(x-1)^2 = (2x+1)^2 \Rightarrow 3x^2 + 6x = 0 \Rightarrow 3x(x+2) = 0 \Rightarrow x+2 = 0 \text{ or } x = 0$$

and thus $x = 0$ is the accepted solution, whereas the solution $x = -2$ is rejected.

Example (2). *Solve the following inequalities*

- $|x| < 4$ *It is:*

$$|x| < 4 \Leftrightarrow -4 < x < 4$$

- $|x| > 4$ *It is:*

$$|x| > 4 \Leftrightarrow x < -4 \text{ or } x > 4$$

- $|2x+1| < 3$ *It is:*

$$|2x+1| < 3 \Leftrightarrow -3 < 2x+1 < 3 \Leftrightarrow -4 < 2x < 2 \Leftrightarrow -2 < x < 1$$

- $|x+3| > 5$ *It is:*

$$|x+3| > 5 \Leftrightarrow x+3 < -5 \text{ or } x+3 > 5 \Leftrightarrow x < -8 \text{ or } x > 2$$

Remark. ***Careful!!:*** *the above techniques should also be considered as showing us techniques for solving similar problems -* ***Keep them in mind!!***

Example (3). *Solve the inequality*

$$|x-1| \leq |2x+1|$$

Solution: Read carefully the solution.
We should consider it as the application of a simple technique
(we can do this because both sides of the inequality are non negative.) We square both sides to get:

$$(x-1)^2 < (2x+1)^2 \Leftrightarrow (x-1)^2 - (2x+1)^2 < 0 \Leftrightarrow 3x(-x-2) < 0$$

The two roots are $x = -2$, $x = 0$ and is easy to check that the sign of

$$p(x) = 3x(-x-2)$$

is graphically described by the following

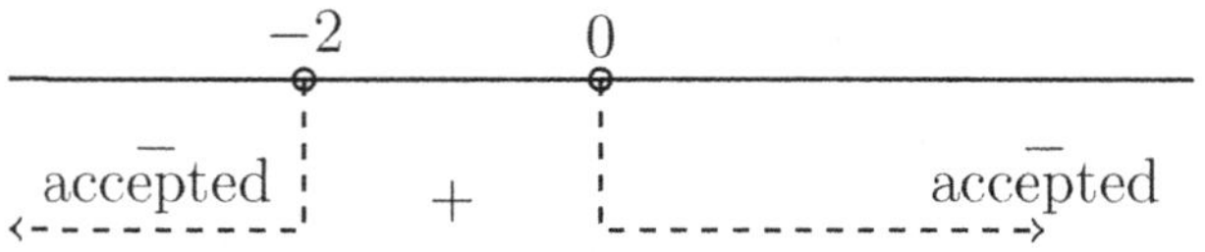

Note: Here there are three intervals. If for a random value of x , say for $x = 2$, the inequality is true, then for all values $x > 0$ it is true. Now, for the other intervals , we alternate +, −. This is a standard simple technique. Finally, we get

$$x < -2 \text{ or } x > 0.$$

Read some more examples

Example (4).

- *(i) Solve the equality*

$$\frac{1}{x} = \frac{3}{2}$$

and then

- *(ii) Solve the inequality*

$$\frac{1}{x} < \frac{3}{2}$$

Solution: To solve the equality **cross - multiply** to get:

$$3x = 2 \Rightarrow x = \frac{3}{2}$$

For the inequality $\frac{1}{x} < \frac{3}{2}$ **Be Careful!! We cannot cross-multiply** here ...
Instead, we apply the following exact method (*this is the easiest way*)
Step 1: $\frac{1}{x} - \frac{3}{2} < 0$, the first side should become one fraction

$$\Rightarrow \frac{2-3x}{2x} < 0$$

Step 2: The root of the nominator is $\frac{2}{3}$
The root of the denominator is 0
We write these roots on a line

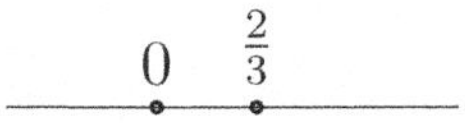

There are three intervals:
We consider any number in the first interval, $x < 0$, say $x = -1$. Then,

$$\frac{2-3(-1)}{2(-1)} = \frac{-5}{2} < 0$$

and we can say that the inequality is true for all values $x < 0$.
Then, we alternate the sign for the other intervals and we accept

$$x < 0, \text{ or } x > \frac{3}{2}$$

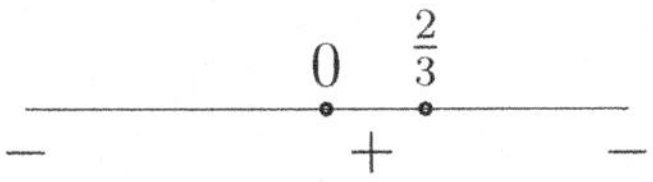

The solution is $x < 0$ or $x > \frac{2}{3}$.
This is a very good mnemonic rule: Let us repeat:
We try any number x, say $x < 0$, and we see that $\frac{2-3x}{2x} < 0$, we write $-$ under this interval. Then, we do not have to do any more trials, we just alternate $-$, $+$ etc for the other intervals.
In the final fraction there should not be any multiple roots in nominator or denominator. The fraction should also be simplified. This is the easiest way to solve inequalities.
Read more examples.

Example (5). *Solve the inequality*

$$\frac{2}{x-1} < \frac{3}{x-2}$$

Solution:

- **Step 1:** We make it one fraction

$$\frac{2}{x-1} - \frac{3}{x-2} < 0 \Leftrightarrow \frac{2(x-2)-3(x-1)}{(x-1)(x-2)} < 0 \Leftrightarrow \frac{-x-1}{(x-1)(x-2)} < 0$$

- **Step 2:** Root of nominator is -1,
  Roots of denominator are 1 and 2 and draw them on a straight line

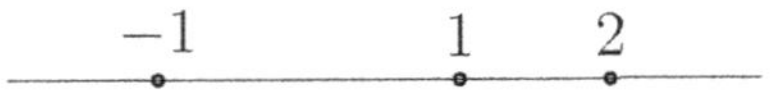

If $x > 2$, say $x = 3$, then the fraction is negative. So, alternatively we have:

Finally, the solution is $-1 < x < 1$, or, $x > 2$.

Note: We repeat here that this technique is valid if there are only simple roots, not for instance for the polynomial inequality $(x-1)^2(x+2) < 0$.

Example (6). *Solve the inequality*

$$T(x) = \frac{(x-1)(x+3)(x-4)}{(x-5)(x+7)} > 0$$

The roots are $\{-7, -3, 1, 4, 5\}$ *and we represent them on a line*

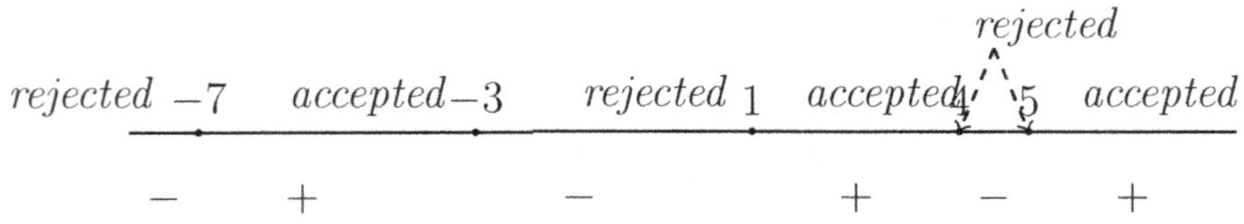

Consider a number between -3 *and* 1, *say* $x = 0$. *Then:*

$$T(0) = \frac{-1 \cdot 3 \cdot (-4)}{(-5)7} = \frac{12}{-35} < 0, \text{ rejected}$$

because we want $T(x) > 0$.
Alternate *to get the solution*

$$-7 < x < -3, \text{ or, } 1 < x < 4 \text{ or } x > 5$$

Example (7). *Solve the inequality*

$$S(x) = \frac{1}{x+1} - 3 < 0$$

Solution:

$$\frac{1}{x+1} - 3 < 0 \Leftrightarrow \frac{-3x-2}{x+1} < 0$$

The roots are $\{-1, \frac{-2}{3}\}$ *and represent them on a line*

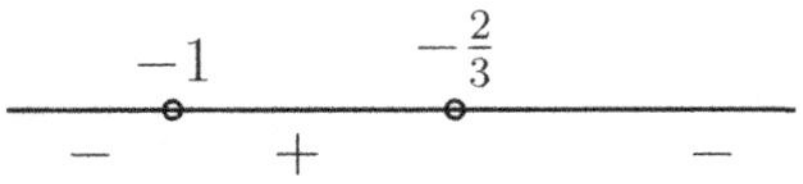

Test for $0 \in (-\frac{2}{3}, \infty)$*:*

$$S(0) = -2 < 0 \text{ accepted}$$

because we want $S(x) < 0$ *and* ***Alternate*** *to obtain:*

$$S(x) < 0 \Leftrightarrow x < -1 \text{ or } x > -\frac{2}{3}$$

Problem (1). *The solution of*

$$\frac{x-3}{(x+2)(x+4)} > 0$$

is

- *[A]* $-4 < x < -2$
- *[B]* $-2 < x < 3$
- *[C]* $x > 3$
- *[D]* $-4 < x < -2$ *or* $3 < x$
- *[E]* $x < -4$ *or* $x > 3$

Try for a while to solve it . . .

Solution: *Apply the exact method for inequalities of the previous examples.*

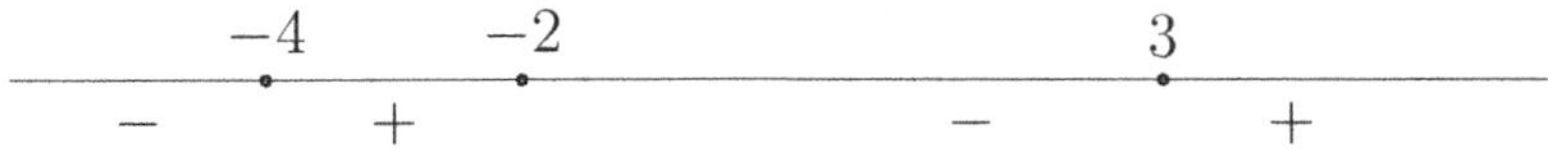

and the solution is $-4 < x < -2$, or, $x > 3$ and the correct answer is $[D]$.

Problem (2). *The solution of*

$$\frac{x^2-2x+1}{(x-1)(x+2)} < 0$$

is

- *[A]* $x < -2$
- *[B]* $-2 < x < 1$
- *[C]* $x > 1$
- *[D]* $x < -2$ *or* $x > 1$
- *[E]* *none of the above*

Try for a while to solve it. . .

Solution:
Apply the exact method

$$\frac{x^2-2x+1}{(x-1)(x+2)}<0 \Leftrightarrow \frac{(x-1)^2}{(x-1)(x+2)}<0 \Leftrightarrow \frac{x-1}{x+2}<0$$

The fraction should be simplified

hence the solution is $-2 < x < +1$.

Problem (3). *The solution of*

$$\frac{x^3+3x^2+3x+1}{x+1}<0$$

is

- *[A]* $x > -1$
- *[B]* $x < -1$
- *[C]* $x > -1$ *or* $x = 0$
- *[D]* $x < -2$ *or* $x > 1$
- *[E]* *no solution*

Try for a while to solve it...
Solution:
Try to apply the exact method. Any implicit information? Any subtle point?

$$x^3+3x^2+3x+1=(x+1)^3$$

hence, for $x \neq -1$ it becomes:

$$\frac{x^3+3x^2+3x+1}{x+1}<0 \Leftrightarrow \frac{(x+1)^3}{x+1}<0 \Leftrightarrow (x+1)^2<0$$

which is **impossible!!!** and the correct answer is $[E]$.

Problem (4). *The solution of*

$$|2x+1|<1$$

is

- *[A]* $x < -1$
- *[B]* $x < -1$ *or* $x > 0$
- *[C]* $-1 < x < 0$
- *[D]* $x > 0$
- *[E]* *none of the above*

Try for a while to solve it...

Solution:

Any subtle point? Since both sides are positive, we square

$$4x^2+4x+1<1 \Leftrightarrow 4x^2+4x<0 \Leftrightarrow 4x(x+1)<0$$

Follow the exact method

$$\begin{array}{ccccc} & -1 & & 0 & \\ + & & - & & + \end{array}$$

so, $-1<x<0$ is the solution. Correct answer is $[C]$.

Problem (5). *The solution of*

$$|2x-1|<x$$

is

- *[A]* $x<\frac{1}{3}$
- *[B]* $x>\frac{1}{3}$
- *[C]* $0<x<\frac{1}{3}$
- *[D]* $\frac{1}{3}<x<1$
- *[E]* *none of the above*

Try for a while to solve it...

Solution:

Let us try to spot any subtle point

Describe everything in short notes

If $x<0$ the inequality results to be **impossible**

If $x\geq 0$ then we can square bothe sides

$$4x^2-4x+1<x^2 \Leftrightarrow 3x^2-4x+1<0 \Leftrightarrow (x-1)(x-\frac{1}{3})<0$$

Hence (*exact method*)

$$\begin{array}{ccccc} & \frac{1}{3} & & 1 & \\ + & & - & & + \end{array}$$

and the correct answer is $[D]$.

Problem (6). *If*

$$\sqrt{(a+1)^2}=\sqrt{a^2}$$

then

- *[A]* $a<\frac{-1}{2}$
- *[B]* $\frac{-1}{2}<a<0$
- *[C]* $0<a<\frac{1}{2}$
- *[D]* $a>\frac{1}{2}$

- *[E] none of the above*

Solution: Recall $\sqrt{x^2} = |x|$ so

$$\sqrt{(a+1)^2} = \sqrt{a^2} \Rightarrow |a+1| = |a| \Rightarrow$$
$$a = 1 + a \text{ or } a + 1 = -a$$

hence

$$1 = 0 \text{ or } a = \frac{-1}{2}$$

and the only accepted solution is $a = \frac{-1}{2}$, i.e: the correct answer is $[E]$.

Problem (7). *If*

$$(x-y)^2 - 3|y-x| - 28 = 0$$

then the value of $|x-y|$ *is*

- *[A]* 3
- *[B]* 7
- *[C]* -7
- *[D]* 28
- *[E] none of the above*

Solution: **Subtle point?**

$$|x-y| = |y-x|, \ |x|^2 = x^2$$

so by setting $|x-y| = \omega \geq 0$ we get:

$$\omega^2 - 3\omega - 28 = 0 \Rightarrow \omega_{1,2} = \frac{3 \pm \sqrt{81}}{2} = \frac{-3 \pm 9}{2}$$

hence $\omega_1 = 7$, $\omega_2 = -4$ with $\omega_2 = -4$ non - acceptable, so the solution we ask for is $\omega_1 = +7$. Correct answer $[B]$.

Problem (8). *If*

$$\frac{1}{x} < 2 < \frac{1}{x-1}$$

then

- *[A]* $x < 0$ *or* $x > \frac{1}{2}$
- *[B]* $0 < x < \frac{1}{2}$
- *[C]* $x > \frac{1}{2}$
- *[D]* $1 < x < \frac{3}{2}$

- *[E] none of the above*

Solution: Where is the subtle point?
Let us solve **separately** each inequality:

$$\frac{1}{x} < 2 \Rightarrow \frac{1-2x}{x} < 0 \Rightarrow x < 0 \text{ or } x > \frac{1}{2}$$

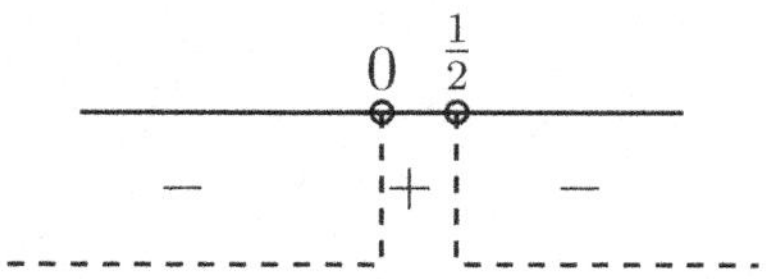

Also

$$2 < \frac{1}{x-1} \Rightarrow \frac{3-2x}{x-1} > 0 \Rightarrow 1 < x < \frac{3}{2}$$

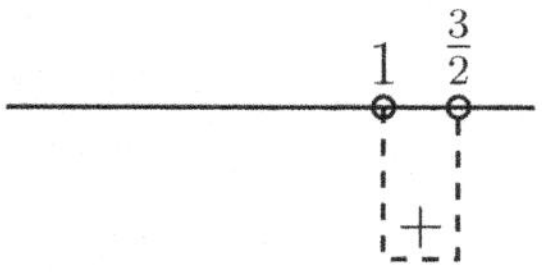

hence common interval of the two inequalities is $(1, \frac{3}{2})$. Correct answer is $[D]$.

Problem (9). *If*

$$2x^2 + 6 > 7x$$

then

- *[A]* $x < \frac{3}{2}$ *or* $x > 2$
- *[B]* $\frac{3}{2} < x < 2$
- *[C]* $x < 2$
- *[D]* $x > \frac{3}{2}$
- *[E] none of the above*

Solution: Apply the usual technique:

$$2x^2 + 6 > 7x \Leftrightarrow 2x^2 - 7x + 6 > 0 \Rightarrow x < \frac{3}{2} \text{ or } x > 2$$

where

$$x_1 = \frac{7-\sqrt{1}}{4} = \frac{3}{2}, \quad x_2 = \frac{7+\sqrt{1}}{4} = 2$$

Correct answer $[A]$.

An equivalent way to solve an inequality of the form

$$ax^2 + bx + c > 0, \quad \text{or} \quad ax^2 + bx + c < 0$$

- If there are **two** zeros x_1, x_2 and $a > 0$ then

  $a > 0$

  $+ \quad x_1 \quad - \quad x_2 \quad +$

  So $ax^2 + bx + c > 0$ is true if and only if $x < x_1$ or $x > x_2$ and
  $ax^2 + bx + c < 0$ is true if and only if $x_1 < x < x_2$
  If $a < 0$ then

  $a < 0$

  $- \quad x_1 \quad + \quad x_2 \quad -$

  so $ax^2 + bx + c > 0$ is true if only if $x_1 < x < x_2$ and
  $ax^2 + bx + c < 0$ if and only if $x < x_1$ or $x > x_2$.

- If the discriminant is $\Delta = b^2 - 4ac < 0$ then, there are no real roots and if $a > 0$ then $ax^2 + bx + c > 0$ for any $x \in \mathbb{R}$ and if $a < 0$ then $ax^2 + bx + c < 0$ for any $x \in \mathbb{R}$

- If $\Delta = b^2 - 4ac = 0$ then there is a root of multiplicity 2 and

  $$ax^2 + bx + c = a(x + \frac{b}{2a})^2$$

  hence,
  if $a > 0$ then $ax^2 + bx + c = a(x + \frac{b}{2a})^2 \geq 0$ for any $x \in \mathbb{R}$ and
  if $a < 0$ then $ax^2 + bx + c = a(x + \frac{b}{2a})^2 \leq 0$ holds true for all $x \in \mathbb{R}$.

Example (8).

- *Solve the inequality*

  $$x^2 - 3x + 2 > 0,$$

  the zeros are $x_1 = 1$, $x_2 = 2$ and $a = 1 > 0$ so the inequality holds true for

  $$x < 1 \text{ or } x > 2$$

  $+ \quad 1 \; - \; 2 \; +$

- *Solve the inequality*

  $$x^2 - 5x + 6 < 0,$$

  the zeros are $x_1 = 2$, $x_2 = 3$ and $a = 1 > 0$ so the inequality holds true for

  $$2 < x < 3$$

  $+ \quad 2 \; - \; 3 \; +$

- *Solve the inequality*

$$-x^2 + 4x - 3 > 0,$$

the zeros are $x_1 = 1$, $x_2 = 3$ *and* $a = -1 < 0$ *so the inequality holds true for*

$$1 < x < 3$$

− 1 + 3 −

- *Solve the inequality*

$$-x^2 + 7x - 6 < 0,$$

the zeros are $x_1 = 1$, $x_2 = 6$ *and* $a = -1 < 0$ *so the inequality holds true for*

$$x < 1 \text{ or } x > 6$$

− 1 + 6 −

Example (9). • *Solve the inequality*

$$x^2 + 2x + 5 < 0,$$

it is

$$\Delta = 4 - 20 = -16 < 0, \quad a = 1 > 0$$

so it is ***impossible***.

- *Solve the inequality*

$$x^2 - 7x + 15 > 0$$

it is

$$\Delta = 49 - 60 = -11 < 0, \quad a = 1 > 0$$

so it is true for any $x \in \mathbb{R}$

- *Solve the inequality*

$$-x^2 + 8x - 20 < 0$$

it is

$$\Delta = 64 - 80 = -16 < 0, \quad a = -1 < 0$$

so it is true for any $x \in \mathbb{R}$

- *Solve the inequality*

$$x^2 - 4x + 4 \geq 0$$

it is

$$\Delta = 0, \quad a = 1 > 0, \quad x_1 = x_2 = 2$$

hence

$$x^2 - 4x + 4 = (x - 2)^2 \geq 0$$

for any $x \in \mathbb{R}$.

Problem (10). *The solution of*

$$\frac{x^2-3x}{4-x}>0$$

is

- *[A]* $x<0$ or $x>4$
- *[B]* $x<0$ or $3<x<4$
- *[C]* $x<0$
- *[D]* $x>4$
- *[E]* *none of the above*

Solution: Apply the usual technique, for inequalities:
Draw a line and denote the roots of nominator and denominator on the axis
Roots of nominator

$$x^2-3x=0 \Leftrightarrow x(x-3)=0 \Leftrightarrow x=0 \text{ or } x=3$$

Roots of denominator

$$4-x=0 \Leftrightarrow x=4$$

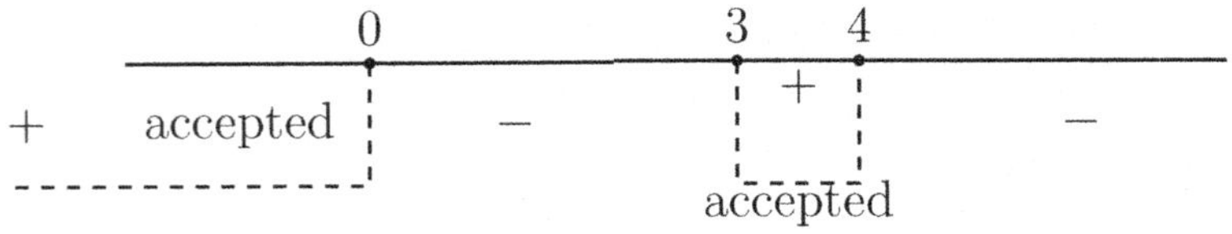

hence

$$x<0 \text{ or } 3<x<4$$

and the correct answer is $[B]$.

Problem (11). *The solution of the inequality*

$$x-4+\frac{5}{x}<0$$

is

- *[A]* $x<0$
- *[B]* $-4<x<5$
- *[C]* $x>0$
- *[D]* $0<x<4$
- *[E]* $x>5$

Solution: It is

$$x-4+\frac{5}{x}<0 \Leftrightarrow \frac{x^2-4x+5}{x}<0 \Leftrightarrow \frac{(x-2)^2+1}{x}<0 \Leftrightarrow x<0$$

or, the discriminant of $x^2 - 4x + 5$ is $\Delta = 16 - 20 < 0$ so

$$x^2 - 4x + 5 > 0$$

for all $x \in \mathbb{R}$ and the solution is $x < 0$.
Correct answer is $[A]$.

Problem (12). *If*

$$x^3 + 3x^2 + 6x > 0$$

then, the values of x are

- *[A]* $x < 0$
- *[B]* $x > 0$
- *[C]* $x \geq 0$
- *[D]* $x \leq 0$
- *[E]* *none of the above*

Solution:
Think for a while before you read on
Apply exact method

$$x^3 + 3x^2 + 6x > 0 \Leftrightarrow x(x^2 + 3x + 6) > 0$$

Observe: the discriminant of $x^2 + 3x + 6$ is

$$\Delta = 3^2 - 4 \cdot 6 = 9 - 24 < 0$$

hence

$$x^2 + 3x + 6 > 0$$

for any x so the inequality holds true for $x > 0$.

Problem (13). *If*

$$(2x + 3)(x^3 + x^2 + x) \geq 0$$

the values of x are

- *[A]* $x \leq -\frac{3}{2}$
- *[B]* $x > 0$
- *[C]* $-\frac{3}{2} < x < 0$
- *[D]* $x \leq -\frac{3}{2}$ *or* $x \geq 0$
- *[E]* *none of the above*

Solution:
Think for a while before you read on
Apply exact method - factorization:

$$(2x+3)(x^3+x^2+x) \geq 0 \Leftrightarrow (2x+3)x(x^2+x+1) \geq 0$$

where the **discriminant** of

$$x^2+x+1$$

is $\Delta = 1-4 = -3 < 0$ hence

$$x^2+x+1 > 0$$

for all x and the initial inequality becomes

$$(2x+3)(x^3+x^2+x) \geq 0 \Leftrightarrow x(2x+3) \geq 0$$

The roots of $q(x) = x(2x+3)$ are $0, \frac{-3}{2}$ and thus

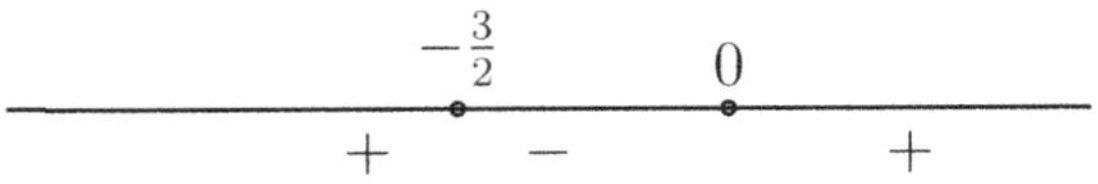

the correct answer is $[D]$.

Problem (14). *If*

$$-(x-1)^2(x^2-x+1)(x^2+3) \leq 0$$

then, the values of x are

- *[A]* $x \leq 0$
- *[B]* $x \geq 0$
- *[C]* $\mathbb{R}$
- *[D]* $\emptyset$ ***(impossible)***
- *[E]* *none of the above*

Solution:
Think for a while before you read on
Observe:

$$(x-1)^2 \geq 0,$$
$$x^2-x+1 \geq 0, \text{ since } \Delta = -3 < 0,$$
$$x^2+3 \geq 3 > 0$$

hence

$$-(x-1)^2(x^2-x+1)(x^2+3) \leq 0$$

for all $x \in \mathbb{R}$ and the correct answer is $[C]$.

Problem (15). *If*

$$\frac{(x-1)^3(x+2)}{x^2-2x+1} < 0$$

then, the values of x are

- *[A]* $x < -2$
- *[B]* $x < 1$
- *[C]* $x > -2$
- *[D]* $-2 < x < 1$
- *[E]* $x > 1$

Solution: Think for a while before you read on
Observe that for $x \neq 1$:

$$x^2 - 2x + 1 = (x-1)^2 > 0, \text{ for } x \neq 1$$

therefore

$$\frac{(x-1)^3(x+2)}{x^2-2x+1} < 0 \Leftrightarrow (x-1)^3(x+2) < 0$$

The **roots** of $q(x) = (x-1)^3(x+2)$ are -2, 1, so:

and the correct answer is $[D]$.

4.5 Geometry & Formulas

4.5.1 A summary of geometric formulas

Start with a few formulae from 2- dimensional Euclidean Geometry:

- **Orthogonal parallelogram**

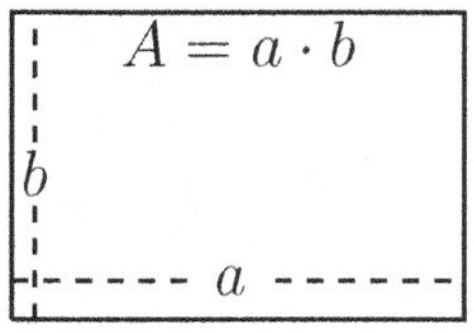

- **Parallelogram**

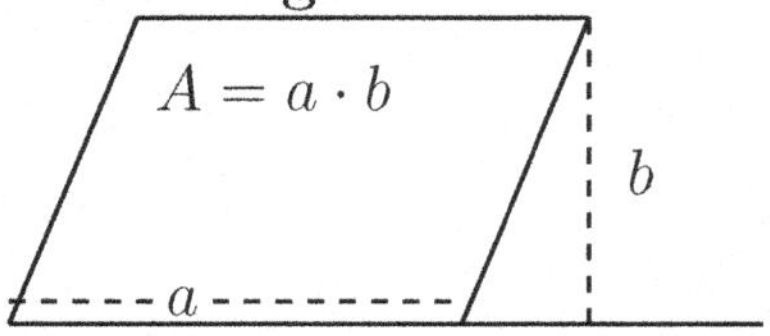

- **Triangle** of sides a, b, c

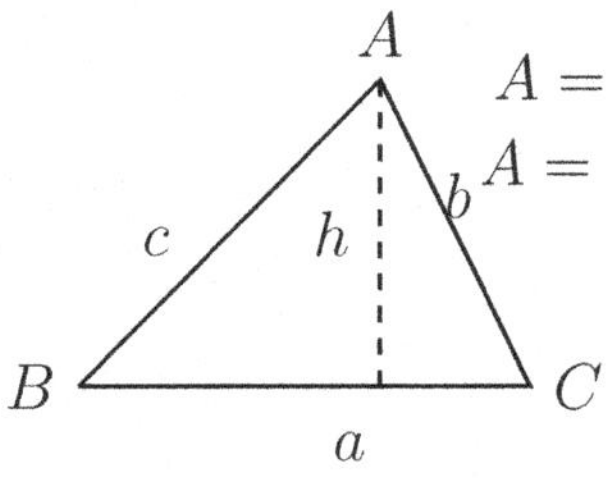

$A = \frac{1}{2}a \cdot h = \frac{1}{2}b \cdot c \cdot \sin \hat{A}$

$A = \sqrt{s(s-a)(s-b)(s-c)}$ Heron's Formula

$s = \frac{a+b+c}{2}$ semiperimeter

- **Trapezium**

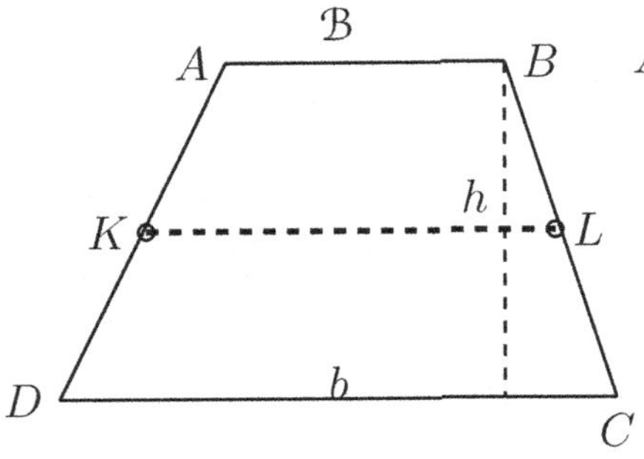

$A = \frac{\mathcal{B}+b}{2} \cdot h, \quad KL = \frac{\mathcal{B}+b}{2}$ the semi-median

- **Rhombus**

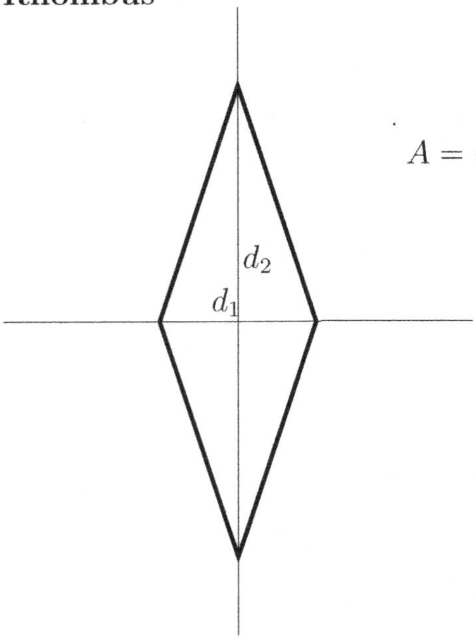

$A = \frac{1}{2} d_1 \cdot d_2$

- **Circle**

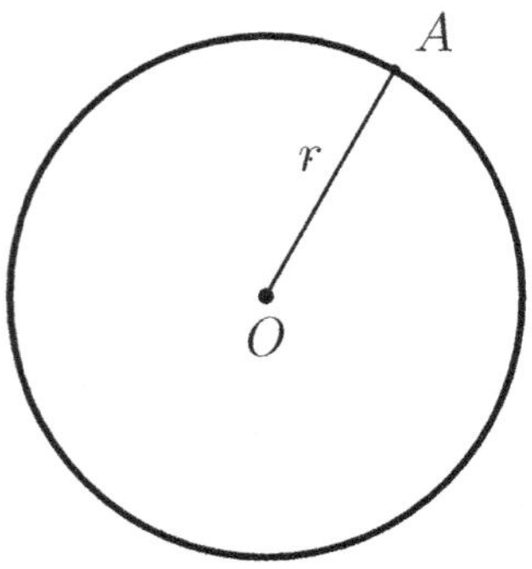

$L = 2\pi r$ - circle perimeter

Area $= \pi r^2$

We go on with some useful formulae from Euclidean Geometry of the 3-dimensional space:

- **Pyramid**

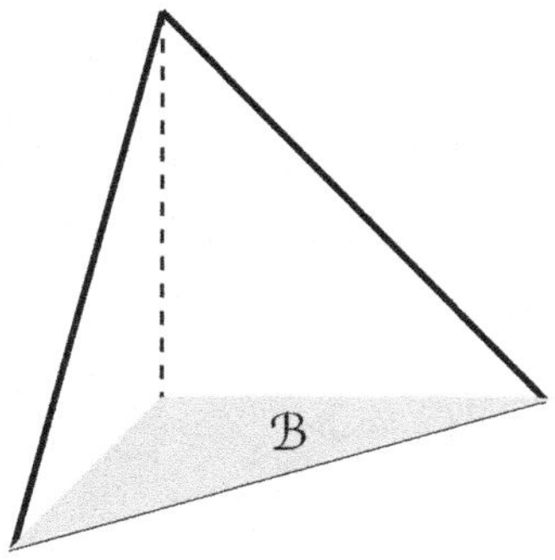

- Volume $= \frac{1}{3}\mathcal{B}h$
- $\mathcal{B}$ = Area of base
- h = height

- **Cylinder**

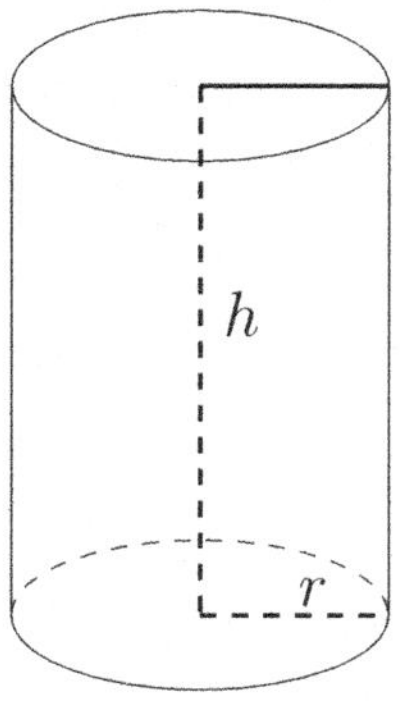

$V = \pi r^2 h$ - volume

$A_l = 2\pi r h$ -lateral area

$A_t = 2\pi r^2 + 2\pi r h$ - total area

r - radius of base, h - height

- **Cone**

$V = \frac{1}{3}\pi r^2 h$ - volume

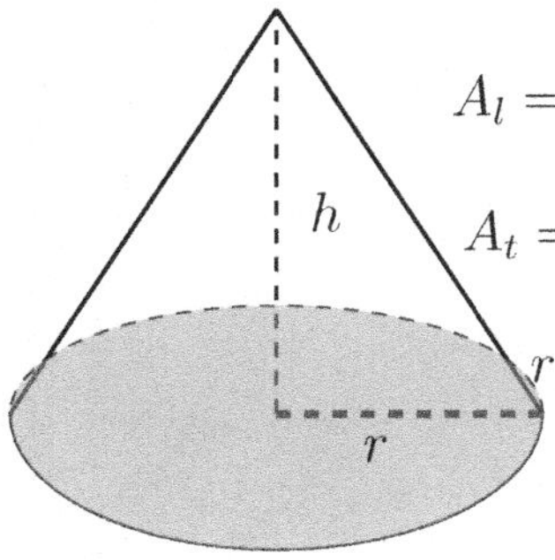

$A_l = \pi r\sqrt{r^2 + h^2}$ -lateral area

$A_t = \pi r\sqrt{r^2 + h^2} + \pi r^2$ -total area

r - radius of base, h- height

- **Sphere**

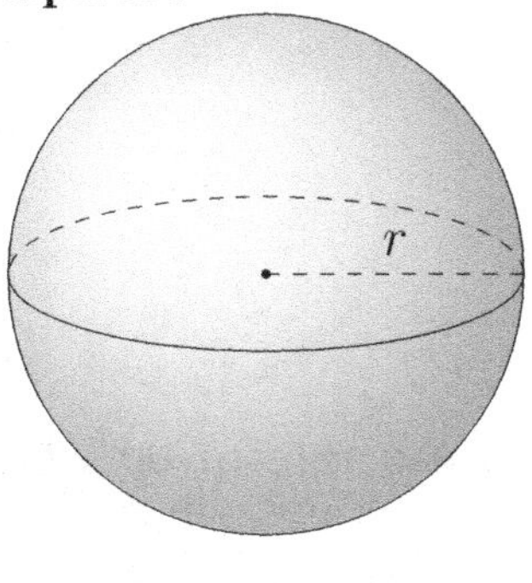

$V = \frac{4}{3}\pi r^3$ -volume

$A_s = 4\pi r^2$ -Surface Area

r -radius

Internal & external bisectors

Basic theorems on bisectors

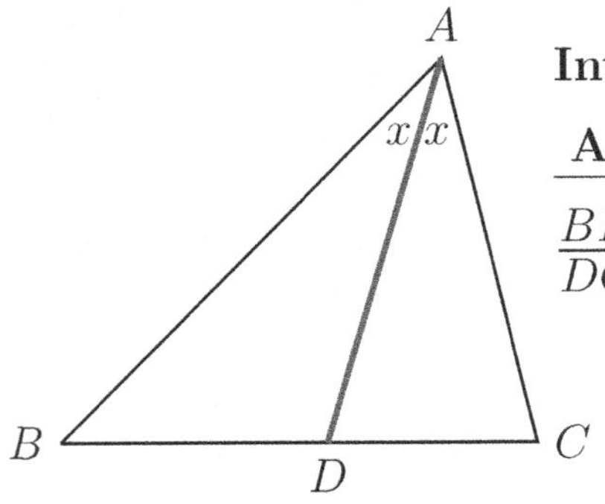

Internal Bisector $AD: B\hat{A}D = D\hat{A}C = x$

Angle Bisector theorem

$$\frac{BD}{DC} = \frac{AB}{AC}$$

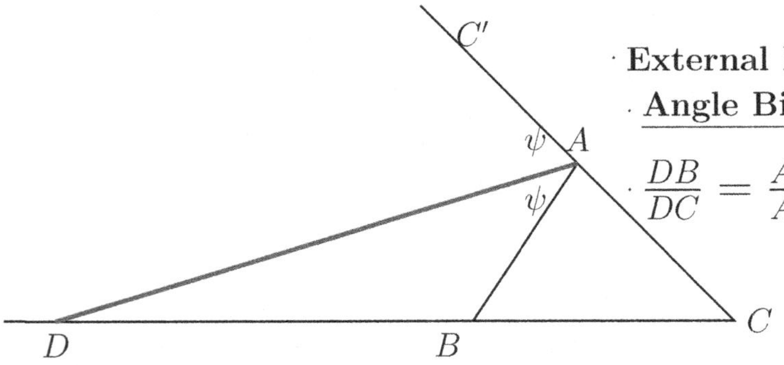

· **External Bisector:** $AD: C'\hat{A}D = D\hat{A}B = \psi$

· **Angle Bisector Theorem**

$$\cdot\ \frac{DB}{DC} = \frac{AB}{AC}$$

Angles inscribed in a circle

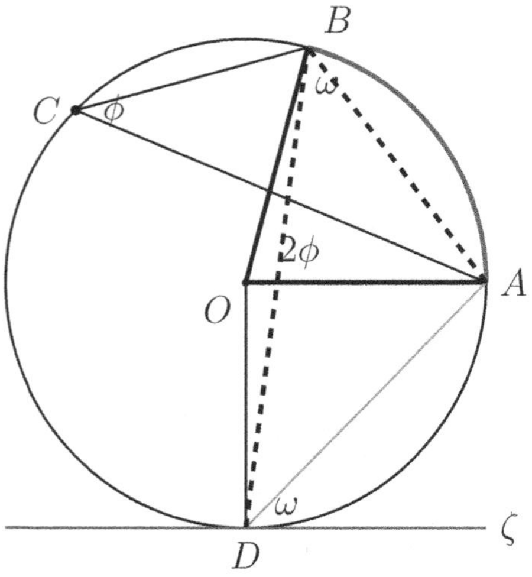

- $B\hat{C}A = \phi$
- $B\hat{O}A = 2\phi = 2B\hat{C}A$
- $D\zeta \perp OD$ i.e: $D\zeta$ tangent to the circle at D then:
- $\omega = A\hat{D}\zeta = D\hat{B}A = \frac{1}{2}D\hat{O}A$

As an immediate consequence we have that:
for a point on a circle the point *views* the diameter by an angle of 90° degrees and vice-versa :

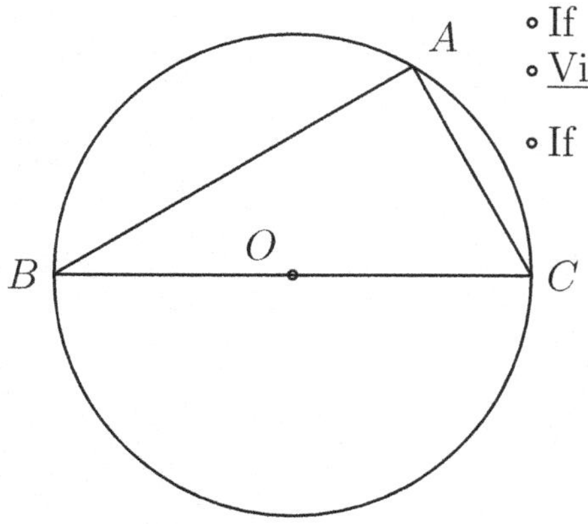

- Direct
- If BOC is a diameter - $B\hat{O}C = \pi$ then $B\hat{A}C = \frac{\pi}{2}$
- Vice-versa
- If $B\hat{A}C = \frac{\pi}{2}$ then BC is a diameter - $B\hat{O}C = \pi$

4.5.2 Geometry Examples & Problems

Example (1). *Given the isosceles trapezium $ABCD$, in the figure below compute AD:*

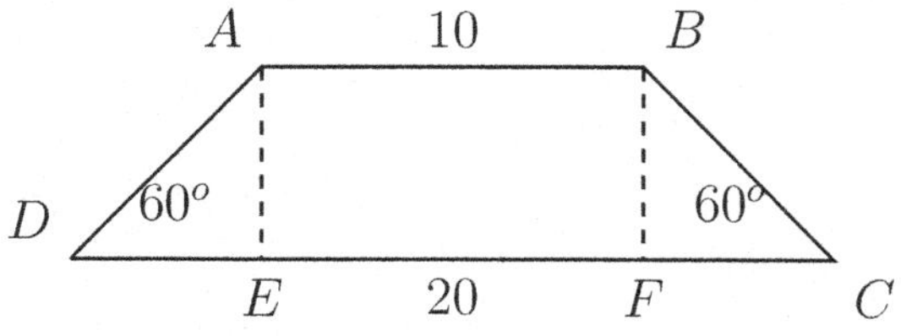

- $AD = BC$
- $AB = 10,\ DC = 20$
- $A\hat{D}E = B\hat{C}F = 60^o$

It is

$$DE = FC = \frac{DC - EF}{2} = \frac{DC - AB}{2} = \frac{20 - 10}{2} = 5$$

and since

$$D\hat{A}E = F\hat{B}C = 30^o$$

it follows that

$$AD = BC = 2DE = 2FC = 10$$

Example (2). *In the quadrilateral below find x:*

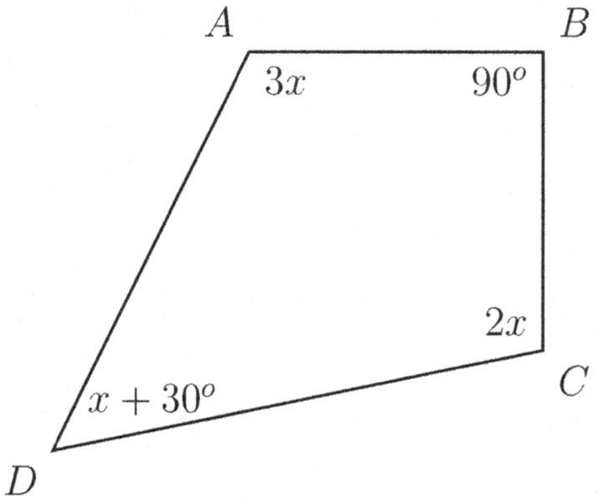

It is

$$3x + 90 + 2x + x + 30 = 360 \Rightarrow 6x + 120 = 360 \Rightarrow$$

$$6x = 240 \Rightarrow x = \frac{240}{6} = 40^o$$

Example (3). *Given a* ***rectangle***

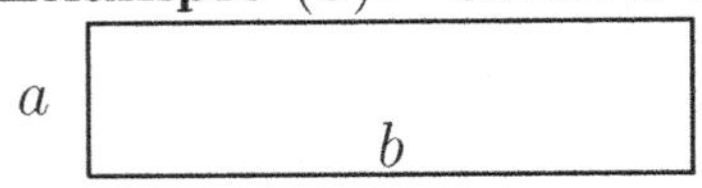

its area is given by $A = a \cdot b$.
For a ***parallelogram***

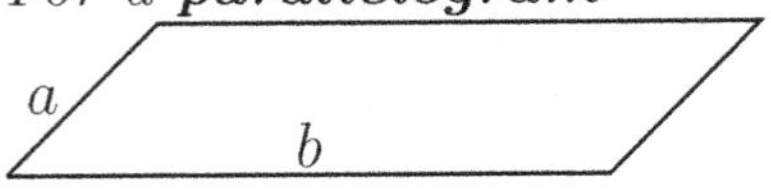

how can we find its area? ***Draw auxiliary lines - height:***

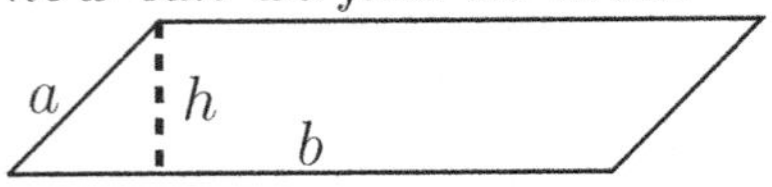

hence Area $= b \cdot h$.
For a ***triangle*** $\triangle ABC$ *draw auxiliary* $AD = h \perp BC$

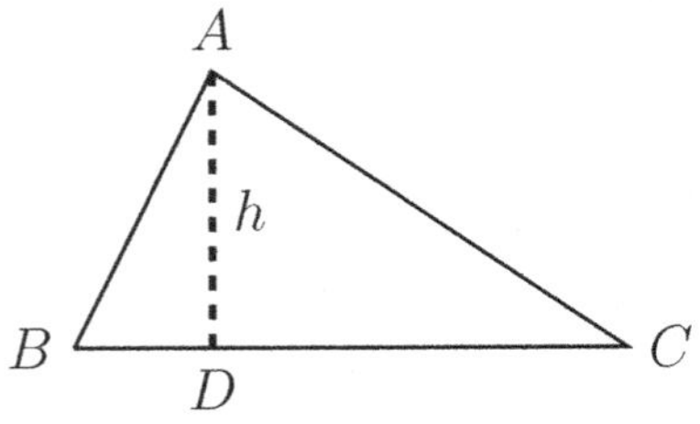

then Area $= \frac{1}{2}BC \cdot h$.
For a ***trapezium***

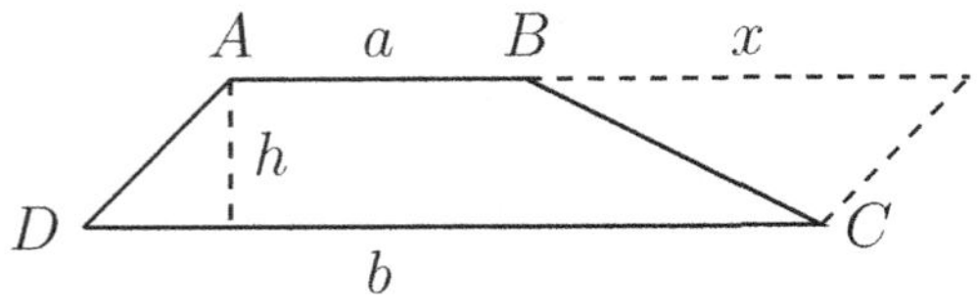

Area $= h \cdot \frac{b+a}{2}$

Proof: Area $= b \cdot h - \frac{1}{2}x \cdot h = h\frac{2b-x}{2} = h\frac{b+a}{2}$

Problem (1). *Let* $\triangle ABC$ *be an isosceles triangle with* $AB = AC = 10$ *and height* $AM = 8$. *Then, its area is:*

- *[A]* 40
- *[B]* 46
- *[C]* 48
- *[D]* 50
- *[E] none of the above*

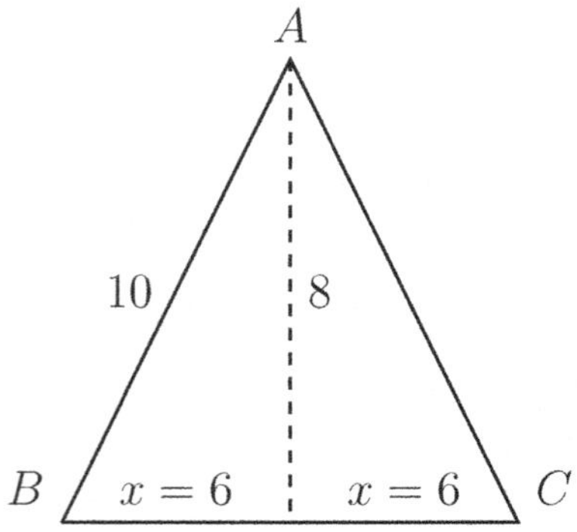

Solution: It is

$$x^2 = 10^2 - 8^2 = 36 \Rightarrow x = 6$$

hence its area A is

$$A = \frac{1}{2} 2x \cdot h = x \cdot h = 6 \cdot 8 = 48.$$

Problem (2). *If $\triangle ABC$ is inscribed in a circle, AB is a diameter, $AB = 10$ and $AC = 8$ then the area of the triangle is*

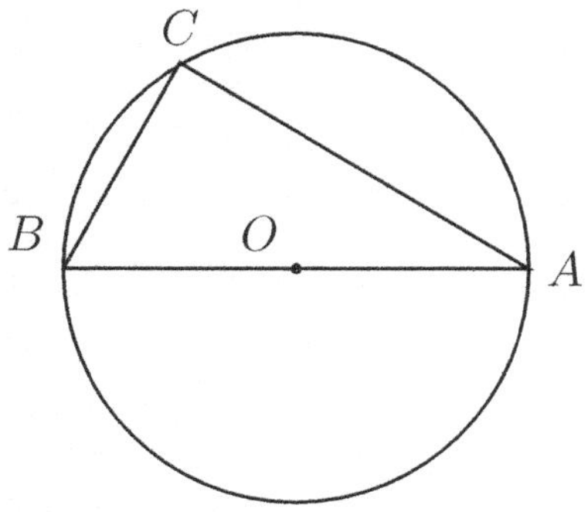

- *[A]* 24
- *[B]* 40
- *[C]* 46
- *[D]* 48
- *[E]* 50

Try to solve it before reading on.
Solution: Since $\hat{C} = 90^o$, using Pythagorean theorem:

$$CB^2 = AB^2 - AC^2 = 10^2 - 8^2 = 36 \Rightarrow CB = 6$$

hence Area $= \frac{1}{2} AC \cdot CB = \frac{1}{2} 6 \times 8 = 24$.
Correct answer is $[A]$.

Problem (3). *If the perimeter of a rectangle is $28m$ and the length of a diagonal is $10m$ its area is*

- *[A]* 40
- *[B]* 48
- *[C]* 52
- *[D]* 56
- *[E]* 60

Try to solve it before go on reading

Solution: If x, y the sides of the rectangle and δ its diagonal then:

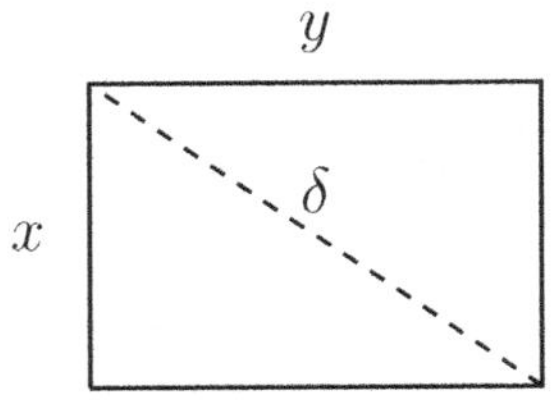

$$\left\{\begin{array}{c} x+y=14 \\ x^2+y^2=10^2 \end{array}\right\} \overset{y=14-x}{\Rightarrow} x^2+(14-x)^2=10^2 \Rightarrow x_{1,2}=\frac{14\pm\sqrt{4}}{2}$$

hence $x=8$, $y=6$ or $x=6$, $y=8$ and thus the area is $A=6\times 8=48$.
Correct answer is $[B]$.

Problem (4). *Let the triangle* $\triangle ABC$ *such that* $\hat{A}=90^o$, Area $=30$ *and* $BC=13$ *then* $AB+AC$ *is equal to*

- *[A]* 17
- *[B]* 28
- *[C]* 20
- *[D]* 25
- *[E] none of the above*

Try to solve it before reading on.

Solution: Let $AB=x$, $AC=y$ then:

$$\left\{\begin{array}{c} \frac{1}{2}xy=30 \\ x^2+y^2=13^2 \end{array}\right\} \Rightarrow \left\{\begin{array}{c} xy=60 \\ x^2+y^2=169 \end{array}\right\} \Rightarrow$$
$$(x+y)^2=x^2+y^2+2xy=169+120=289 \Rightarrow x+y=17.$$

The correct answer is $[A]$.

Problem (5). *If* $AB=9$, $DE=DC=4$, $BC=8$ *and* $\hat{B}=\hat{C}=\hat{D}=90^o$

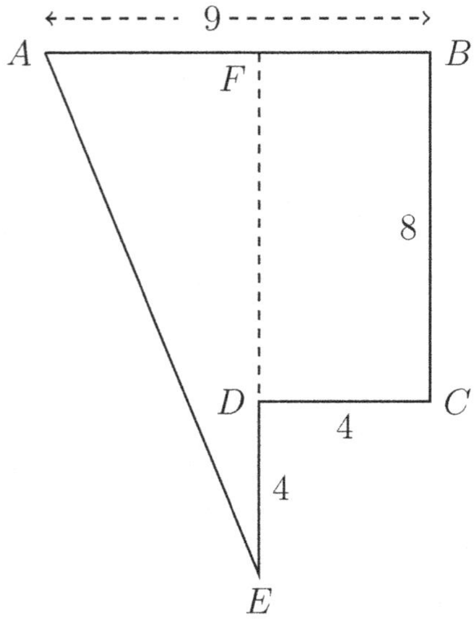

then AE *is equal to*

- *[A]* 10
- *[B]* 11
- *[C]* 12
- *[D]* 13
- *[E] none of the above*

Try to solve it before reading on.

Solution: By construction $FB = DC = 4$, draw auxiliary line

$$FD \perp AB \parallel DC$$

then

$$FD = BC = 8$$

so

$$FE = FD + DE = 8 + 4 = 12$$

and

$$AF = AB - FB = 9 - 4 = 5$$

with $A\hat{F}E = 90^o$, hence

$$AE^2 = AF^2 + FE^2 = 5^2 + 12^2 = 169 \Rightarrow AE = 13.$$

Correct answer is $[D]$.

Problem (6). *A circle with radius $R = 1$ is inscribed in a triangle $\triangle ABC$ with $AB = AC$ and height $AD = 4$, then the area of $\triangle ABC$ is*

- *[A]* $2\sqrt{2}$ *or* 4.5
- *[B]* $3\sqrt{2}$ *or* 4.8
- *[C]* $4\sqrt{2}$ *or* 5.2
- *[D]* $6\sqrt{2}$ *or* 5.4
- *[E]* 5.6 *or* 6

Think for a while before you go on reading

Have you got stuck?

Solution: Denote with ρ the radius of the inscribed circle

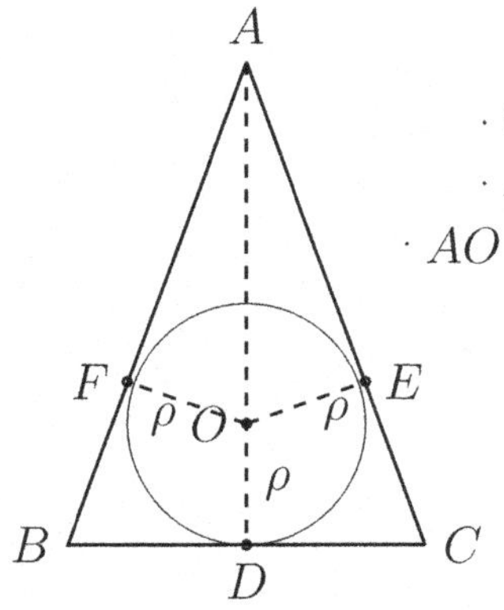

· $OF = OE = OD = \rho = 1$

· $AD = 4$, $OD = 1$ so $AO = 3$

· $AO = 3$, $OE = \rho = 1$, $A\hat{E}O = 90^o$ so $AE = \sqrt{OE^2 - OE^2} = \sqrt{8} = 2\sqrt{2}$

Describe everything - step by step
Draw a rough graph-use simple symbols
By the above figure we have

$$OE = OF = OD = 1, \ AD = 4, \ AO = 3, \ AE = 2\sqrt{2}$$

where

$$OE \perp AC, \ OF \perp AB, \ OD \perp BC$$

We have all the necessary information for the quadrilateral $AFOE$
What is the subtle point we miss?
We need some extra information on

$$DC = EC = \frac{BC}{2} = BD = BF$$

Use **similarity of triangles!**
From $\triangle AEO \sim \triangle ADC$ with:
$O\hat{A}E$ common angle, $A\hat{O}E = A\hat{C}D$, $A\hat{E}O = A\hat{D}C = 90^o$. We get:

$$\frac{DC}{OE} = \frac{AD}{AE} \Rightarrow \frac{DC}{1} = \frac{4}{2\sqrt{2}} \Rightarrow DC = \frac{2}{\sqrt{2}} = \sqrt{2}$$

hence $BC = 2DC = 2\sqrt{2}$ and thus

$$A_{\triangle ABC} = \frac{1}{2}BC \cdot AD = \frac{1}{2}2\sqrt{24} = 4\sqrt{2} \sim 5.66.$$

Correct answer is $[E]$.

Problem (7). *A rectangle is inscribed in a circle. If $AB = 8$, $BC = 6$ then, the area of the circle outside the rectangle is*

- *[A]* 30
- *[B]* 30.5
- *[C]* 31
- *[D]* 32
- *[E]* 32.5

Think for a while before you go on reading
Solution: Draw auxiliary lines $OK \perp AB$, $AO = OB = r$, where r the radius of the circle.

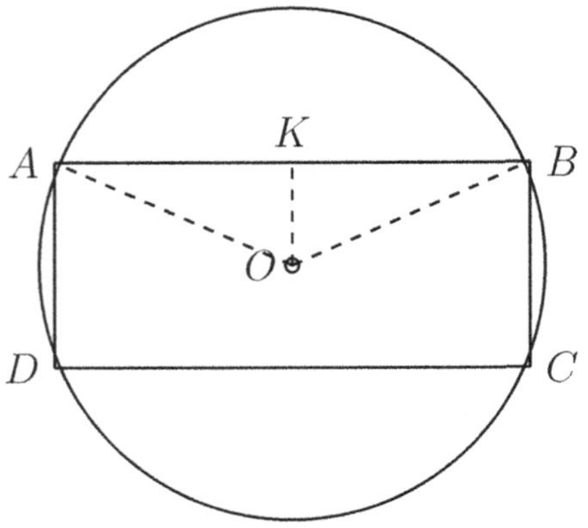

∘ $AB = 8$
∘ $BC = 6$
∘ $OK = \frac{BC}{2} = 3, \quad AK = KB = \frac{AB}{2} = 5$

Since $OK = 4$ then

$$r^2 = OB^2 = OK^2 + KB^2 = 3^2 + 4^2 = 25 \Rightarrow r = 5$$

hence the area A_{ext} of the circle outside of the rectangle is

$$A_{ext} = \pi r^2 - \text{Area}_{ABCD} = 25\pi - 48 \sim 30.5$$

Correct answer is $[B]$.

Problem (8). *$\triangle ABC$ is an equilateral triangle inscribed in a circle with radius $R = 4$. Then, the area of the triangle is*

- *[A]* $4\sqrt{3}$
- *[B]* $8\sqrt{3}$
- *[C]* $10\sqrt{3}$
- *[D]* $12\sqrt{3}$
- *[E]* $16\sqrt{3}$

Think for a while before you go on reading

Solution: It is $O\hat{B}K = \frac{\hat{B}}{2} = \frac{60^o}{2} = 30^o$ and $O\hat{K}B = 90^o$

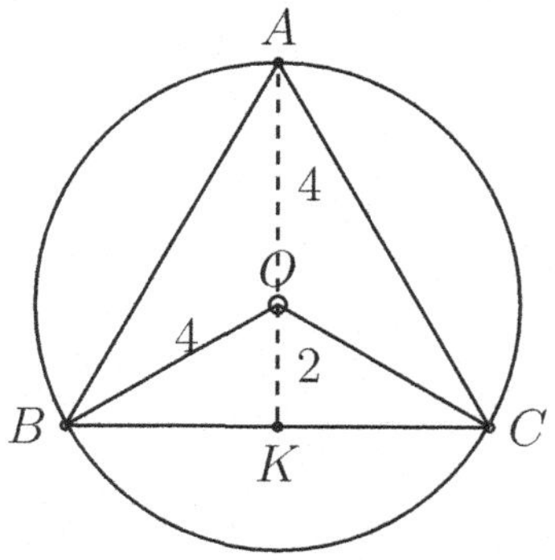

so $OK = \frac{OB}{2} = 2$
hence

$$BK^2 = OB^2 - OK^2 = 12 \Rightarrow BK = \sqrt{12} = 2\sqrt{3}$$

and thus

$$a = BC = 2BK = 4\sqrt{3}$$

therefore

$$\mathrm{A}_{\triangle ABC} = \frac{1}{2}BC \cdot AD = 12\sqrt{3}.$$

Correct answer is $[D]$.

Problem (9). *If $\triangle ABC$ is an orthogonal triangle with $AM = 5$, M midpoint of BC and $AC = 8$ then its height AD is equal to*

- *[A]* 4
- *[B]* 4.2

- *[C]* 4.4
- *[D]* 4.6
- *[E]* 4.8

Think for a while to solve it before you go on

Solution: Observe that: since $AM = 5$ and AM median of ABC, $\hat{A} = 90^0$ then $BC = 2AM = 10$

Further using the similarity

$$\triangle ADC \sim \triangle ABC$$

we get

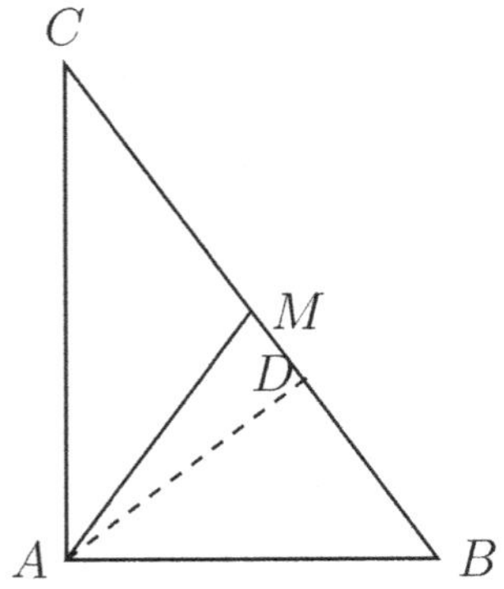

Data	Goal
$AM = 5$	$AD = ?$
$AC = 8$	
$\hat{A} = 90^o$	

What's the subtle point?

$\hat{A} = 90^o \Rightarrow 2AM = BC = 10$

$\triangle ADC \sim \triangle ABC \Rightarrow \frac{AD}{AB} = \frac{AC}{BC}$

$$AB^2 = BC^2 - AC^2 = 100 - 64 = 36 \Rightarrow AB = 6$$

hence

$$AD \cdot BC = AB \cdot AC \Rightarrow AD = \frac{AB \cdot AC}{BC} = 4.8.$$

Way 2: Use the area formula

$$\text{Area}_{\triangle ABC} = \frac{1}{2}AB \cdot AC = \frac{1}{2}BC \cdot AD \Rightarrow AD \cdot BC = AB \cdot AC$$

and thus

$$AD = \frac{AB \cdot AC}{BC} = \frac{6 \cdot 8}{10} = 4.8.$$

Correct answer is $[E]$.

Problem (10). *Given a trapezium $ABCD$, $AB \parallel CD$, $\hat{B} = \hat{C} = 90^o$, $MN = 15$ (the median), $AC = 10$ and area $= 120$. Then AB is equal to*

- *[A]* 4
- *[B]* 6
- *[C]* 8
- *[D]* 9
- *[E] none of the above*

Think for a while to solve it before you go on reading

Solution:

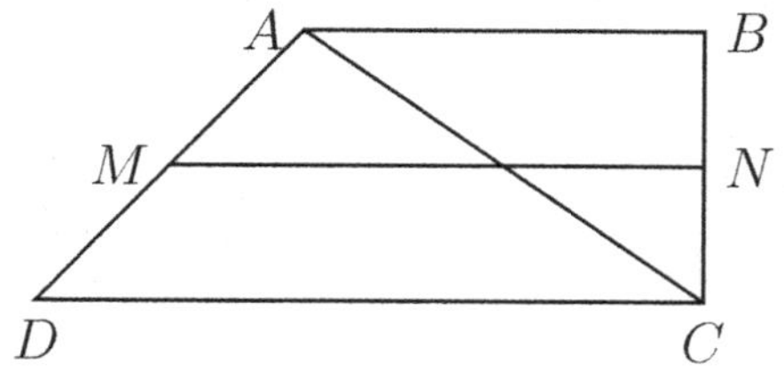

$MN = 15$
$AC = 10$
$\hat{B} = \hat{C} = 90^o$
$\text{Area}_{ABCD} = 120$

Sublte point?

$MN \cdot BC = \text{Area}_{ABCD}$

so

$$120 = 15 \cdot BC \Rightarrow BC = 8$$

and from the orthogonal triangle $\triangle ABC$, $\hat{B} = 90^o$ we get

$$AB^2 = AC^2 - BC^2 \Rightarrow AB = \sqrt{AC^2 - BC^2} = \sqrt{10^2 - 8^2} = \sqrt{36} = 6.$$

Correct answer is $[B]$.

Problem (11). *The volume of a cube inscribed in a sphere is* $1m^3$. *Then,*

(i) : *the surface of the sphere is:*

- *[A]* $\frac{3}{2}\pi$
- *[B]* 3π
- *[C]* 4π
- *[D]* 6π
- *[E] none of the above*

(ii) : *the volume of the sphere is*

- *[A]* $\frac{\pi\sqrt{3}}{4}\pi$
- *[B]* $\frac{\pi\sqrt{3}}{2}$
- *[C]* $\frac{4}{3}\pi$
- *[D]* 2π
- *[E]* 4π

(iii) : *the volume between the sphere and the cube is*

- *[A]* $\frac{\pi\sqrt{3}-4}{4}$
- *[B]* $\frac{\pi\sqrt{3}-3}{4}$
- *[C]* $\frac{\pi\sqrt{3}-4}{2}$
- *[D]* $\frac{\pi\sqrt{3}-2}{2}$
- *[E]* $\frac{\pi\sqrt{3}+2}{2}$

Solution:

Try for a while to solve before you go on reading

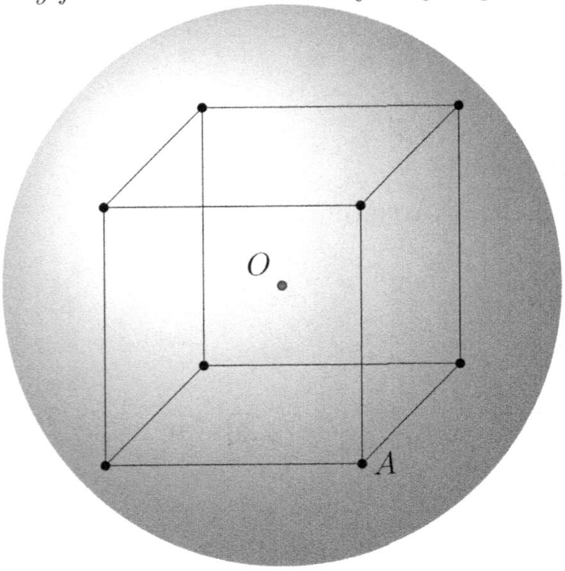

(i) : The volume of the cube is $V_{cube} = a^3$ so the side of the cube is $a = 1$.
If $d = 2R$ is the double of the radius of the sphere, then:

$$3a^2 = d^2 \Rightarrow d^2 = 3 \Rightarrow d = \sqrt{3} \Rightarrow R = \frac{\sqrt{3}}{2}$$

hence, the surface of the sphere is

$$S_{sphere} = 4\pi R^2 = 4\pi\frac{3}{4} \Rightarrow S_{sphere} = 3\pi.$$

(ii) : The volume of a sphere is

$$V_{sphere} = \frac{4}{3}\pi R^3 = \frac{\pi\sqrt{3}}{2}.$$

(iii) : The volume between the sphere and the cube is

$$V_{S\setminus C} = \frac{\pi\sqrt{3}}{2} - 1 = \frac{\pi\sqrt{3} - 2}{2}.$$

Problem (12). *An equilateral triangle, with side $a = 1$, is inscribed in a circle, then the area of the circle is*

- *[A]* $\frac{\pi}{6}$
- *[B]* $\frac{\pi}{5}$
- *[C]* $\frac{\pi}{4}$
- *[D]* $\frac{\pi}{3}$
- *[E] none of the above*

Try to solve it before you go on reading.

Solution:

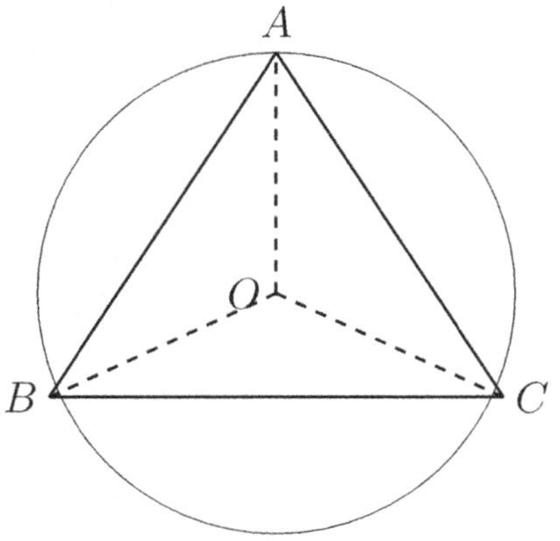

$\cdot AB = BC = AC = a = 1$
$\cdot OA = OB = OC = \rho$

The height is (using Pythagorean Theorem)

$$h = a\frac{\sqrt{3}}{2}$$

Since O is the center of gravity it follows that

$$R = \frac{2}{3}a\frac{\sqrt{3}}{2} = a\frac{\sqrt{3}}{3}$$

If R is the radius of the circle and a the side of the equilateral triangle, then:

$$R = \frac{\sqrt{3}}{3}a = \frac{\sqrt{3}}{3}$$

hence the area of the circle

$$A_{\text{circle}} = \pi R^2 = \frac{\pi}{3}.$$

Correct answer is $[D]$.

Problem (13). *A cone is inscribed in a sphere with radius $R = 1$. The height of the cone is $\frac{4}{3}$. Then, the volume of the cone is*

- *[A]* $\dfrac{16\pi}{81}$
- *[B]* $\dfrac{32\pi}{81}$
- *[C]* $\dfrac{64\pi}{81}$
- *[D]* 2π
- *[E]* 4π

Solution:

Try to solve it before you go on reading.

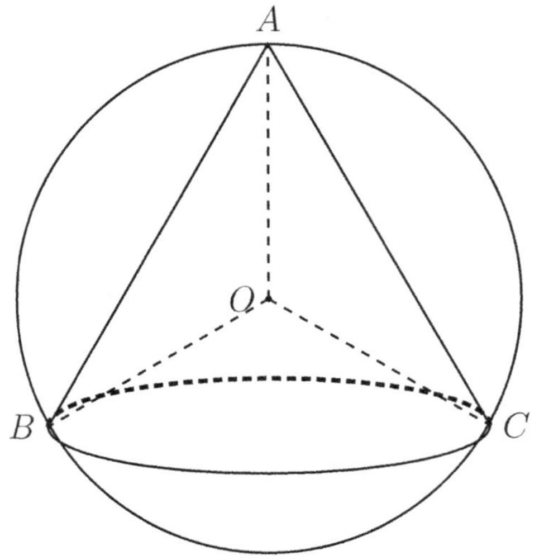

Consider the triangle

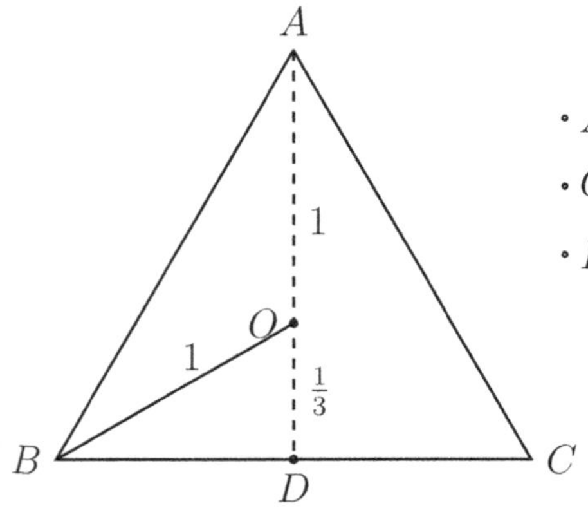

- $AO = OB = 1$
- $OD = AD - AO = \frac{4}{3} - 1 = \frac{1}{3}$
- BD is the radius of the circle - basis of the cone

Subtle point? we need the radius BD of the circle - basis of the cone.
From the orthogonal triangle $\triangle BDO$, $\hat{D} = 90^o$ we get:

$$BD^2 = BO^2 - OD^2 = 1^2 - (\frac{1}{3})^2 = \frac{8}{9} \Rightarrow BD = \frac{2\sqrt{2}}{3}$$

hence

$$V_{\text{cone}} = \frac{1}{3}\pi BD^2 \cdot AD = \frac{1}{3} \cdot \pi \cdot \frac{8}{9} \cdot \frac{4}{3} = \frac{32\pi}{81}$$

Correct answer is $[B]$.

Problem (14). *A sphere of radius $R = 1$ is inscribed in a cone of height $h = 3$. Then, the area of the cone A_{cone} is*

- *[A]* 7π
- *[B]* 7.5π
- *[C]* 8.5π
- *[D]* 9π
- *[E]* $\frac{82}{9}\pi$

Try to solve it before you go on reading.

Solution:

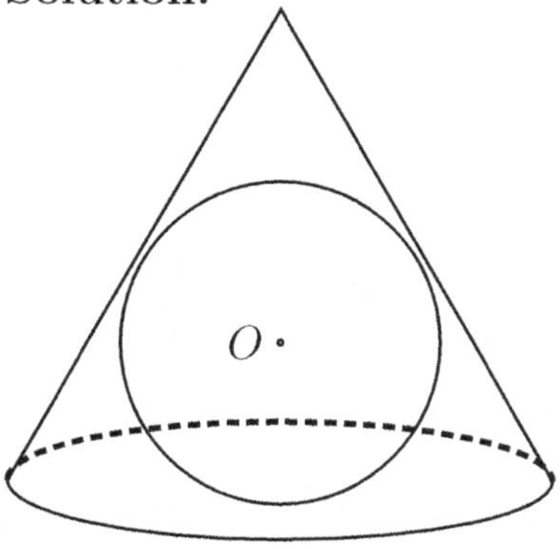

Consider the triangle and the inscribed circle

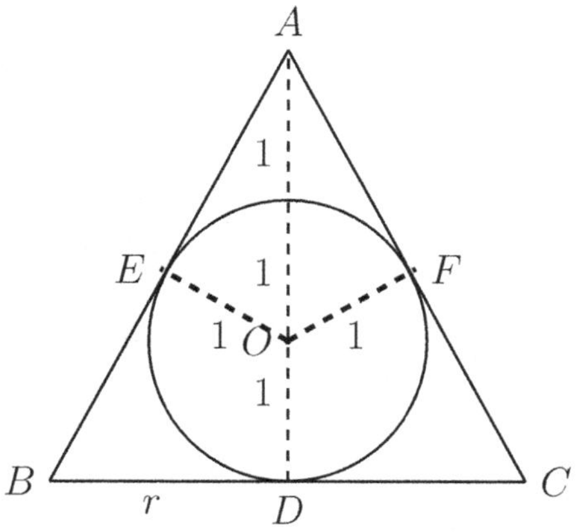

- $\hat{D} = \hat{E} = \hat{F} = 90^o$
- $EO = OF = OD = 1$
- $AO = 2, \quad AD = h = 3$
- Let $\hat{A}_1 = E\hat{A}O = \phi$

Subtle point? $r = BD = DC$ is the radius of the circle - basis of the cone.
Use **trigonometry** to relate $r = BD$ with $AD = 3$ in the orthogonal triangle $\triangle ADB$.
The $\tan\phi$ relates AD and BD.
Use orthogonal triangle $\triangle AEO$ to compute:

$$\sin\phi = \frac{EO}{AO} = \frac{1}{2}$$

so

$$\cos\phi = \sqrt{1 - \frac{1}{4}} = \frac{\sqrt{3}}{2}$$

and thus

$$\tan\phi = \frac{\sin\phi}{\cos\phi} = \frac{1}{\sqrt{3}} = \frac{\sqrt{3}}{3}$$

hence

$$\tan\phi = \frac{BD}{AD} = \frac{r}{AD} \Rightarrow r = \tan\phi \cdot AD = 3\frac{\sqrt{3}}{3} = \sqrt{3}$$

Now, use formula for the total area of the cone

$$A_{\text{total}} = \pi r\sqrt{r^2 + h^2} + \pi r^2 \overset{h=3,\ r=\sqrt{3}}{=} 9\pi$$

Else:

$$AE = AF = \sqrt{4 - 1} = \sqrt{3} \text{ Pythagorean Theorem}$$

Now, using similar triangles

$$\triangle AEO \sim \triangle ADB \Rightarrow \frac{\sqrt{3}}{1} = \frac{3}{BD} \Rightarrow BD = \sqrt{3}$$

so

$$AB = \sqrt{3^2 + (\sqrt{3})^2} = \sqrt{12} = 2\sqrt{3}$$

hence

$$\text{Area} = \pi(\sqrt{3})^2 + \pi\sqrt{3}2\sqrt{3} = 3\pi + 6\pi = 9\pi.$$

Correct answer is $[D]$.

Problem (15). *The measures of the angles $\hat{A}, \hat{B}, \hat{C}, \hat{D}$ of a quadrilateral $ABCD$ are in ratio $1 : 2 : 3 : 4$. The greatest of the angles is*

- *[A]* 108°
- *[B]* 112°
- *[C]* 144°
- *[D]* 150°
- *[E]* 160°

Try for a while to solve it before you go on reading

Solution: Use the formula

$$\frac{\hat{A}}{1} = \frac{\hat{B}}{2} = \frac{\hat{C}}{3} = \frac{\hat{D}}{4} = \frac{\hat{A} + \hat{B} + \hat{C} + \hat{D}}{1+2+3+4} = \frac{360^o}{10} = 36^o$$

hence the angle

$$\hat{D} = 36^o \times 4 = 144^o$$

is the greatest angle.

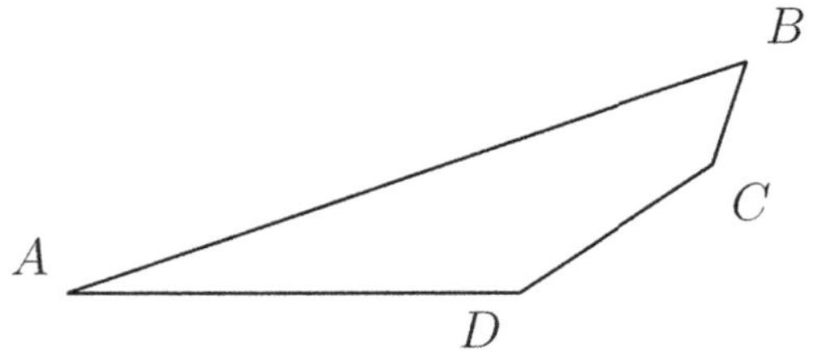

Correct answer is $[C]$.

Problem (16). *If $DE \parallel BC$ in the triangle $\triangle ABC$ and $\hat{A} = 120^o$, $\hat{D} = A\hat{D}E = x^o$ and $\hat{C} = x^o - 20^o$ then x^o is equal to*

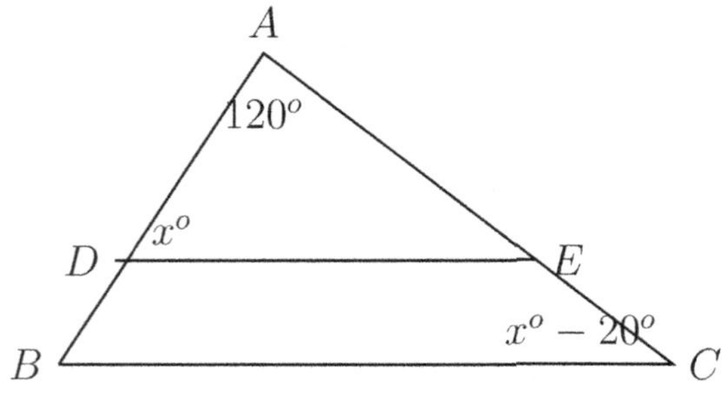

- *[A]* 35°
- *[B]* 40°
- *[C]* 45°
- *[D]* 50°
- *[E]* 55°

Solution: Subtle point?
Parallelism $DE \parallel BC$ hence

$$\hat{E} = A\hat{C}B = x^o - 20^o$$

and thus

$$120^o + x^o + x^o - 20^o = 180^o \Rightarrow 2x^o = 80^o \Rightarrow x^o = 40^o$$

Correct answer is $[B]$.

Problem (17). *Given the orthogonal triangle $\triangle ABC$, $\hat{A} = 90^o$, $AD \perp BC$, $AB = 6$, $AC = 8$ then $DC = ?$*

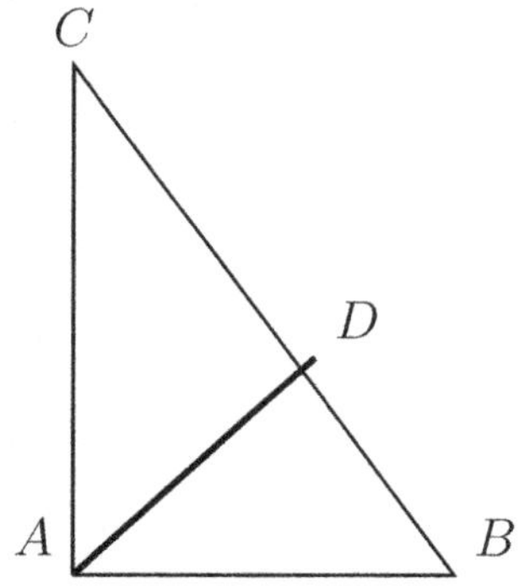

- *[A]* 3.6
- *[B]* 4.5
- *[C]* 6.4
- *[D]* 6.8
- *[E]* 6.9

Solution: Where is the subtle point?
Pythagoras allows us to compute

$$BC = \sqrt{AB^2 + AC^2} = \sqrt{6^2 + 8^2} = \sqrt{100} = 10$$

and from standard relations for the orthogonal triangles we get:

$$AC^2 = BC \cdot DC \Rightarrow 8^2 = 10CD \Rightarrow CD = 6.4$$

Correct answer is $[C]$.

Chapter 5

Topic 2 - Functions

5.1 Linear Functions

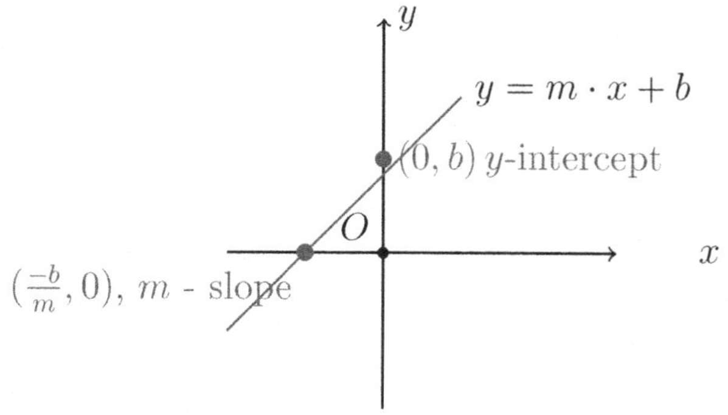

Let $y = mx + b$ be the equation of a **straight line**,

- $m =$ **slope** (or **gradient**)
- $b =$ y**-intercept**

Example. *Let* $y = 2 \cdot x + 3$, *then*

- *if* $x = 0$, $y = 3$ *(*y*-intercept)*
- *if* $y = 0$, $x = -\frac{3}{2}$

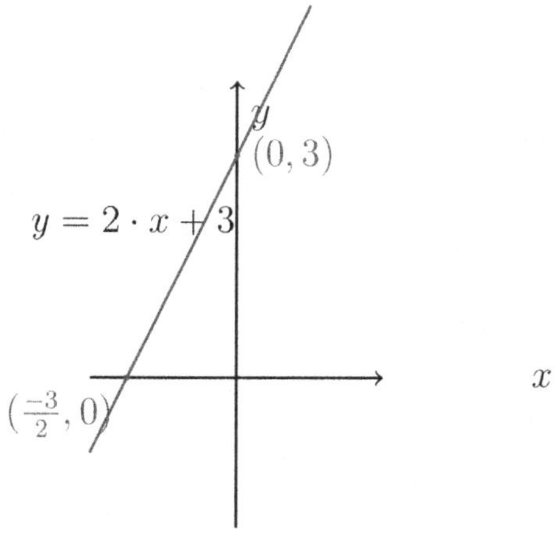

Instead of $y = mx + b$ we can also write $f(x) = mx + b$.

5.1.1 General form of the equation

General form is given by:

$$Ax + By + C = 0$$

or, equivalently, solving with respect to y, by:

$$y = -\frac{A}{B}x - \frac{C}{B}, \quad \text{slope} = \frac{-A}{B}, \quad B \neq 0.$$

Example. *Let* $2x + 3y + 6 = 0$ *then:*

$$y = \frac{-2}{3}x - 2$$

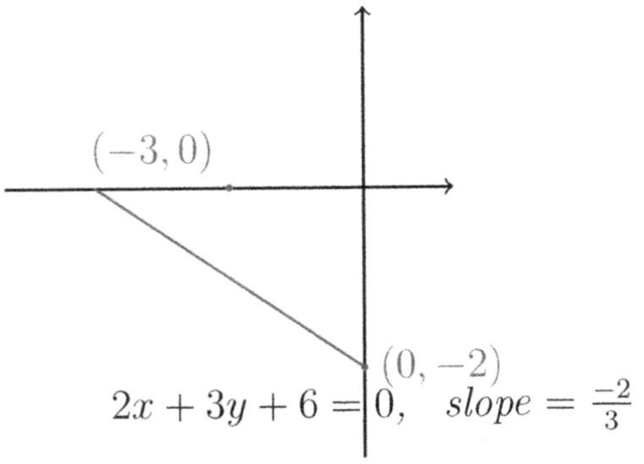

5.1.2 Slope formula

If two points $(x_1, y_1), (x_2, y_2)$, with $x_1 \neq x_2$, are given:

$$m = \frac{y_2 - y_1}{x_2 - x_1} = \frac{y_1 - y_2}{x_1 - x_2}$$

or

$$m = \tan\theta$$

5.1.3 Parallel lines

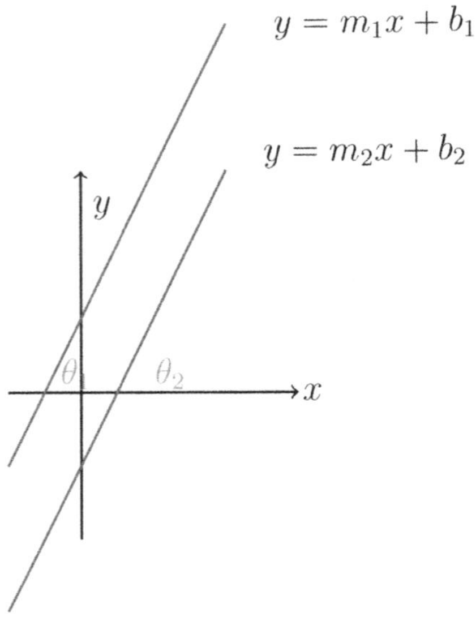

Parallel lines → equal slopes i.e: $m_1 = m_2$.

5.1.4 Perpendicular lines

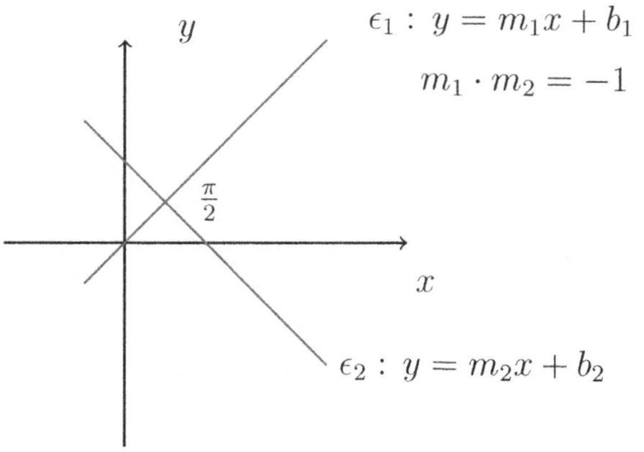

Perpendicularity of straight lines:

$$\epsilon_1 \perp \epsilon_2 \Leftrightarrow m_1 \cdot m_2 = -1$$

5.1.5 Midpoint & Distance of points

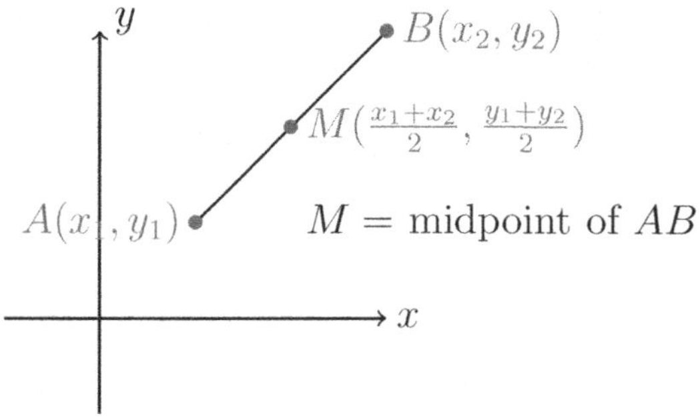

Let $P(x_1, y_1)$, $Q(x_2, y_2)$ points on the plane, then:

- *Midpoint* M of AB: $M = (\frac{x_1+x_2}{2}, \frac{y_1+y_2}{2})$
- *Distance* of A, B: $d(A, B) = \sqrt{(x_1 - x_2)^2 + (y_1 - y_2)^2}$

5.1.6 Distance of a point from a straight line

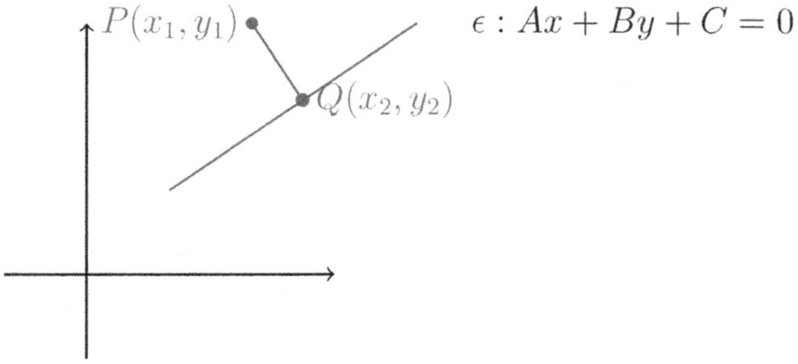

Given $P(x_1, y_1)$ its distance from $\epsilon : Ax + By + C = 0$ is given by

$$d(P, \epsilon) = \frac{|Ax_1 + By_1 + C|}{\sqrt{A^2 + B^2}}$$

5.1.7 Angle

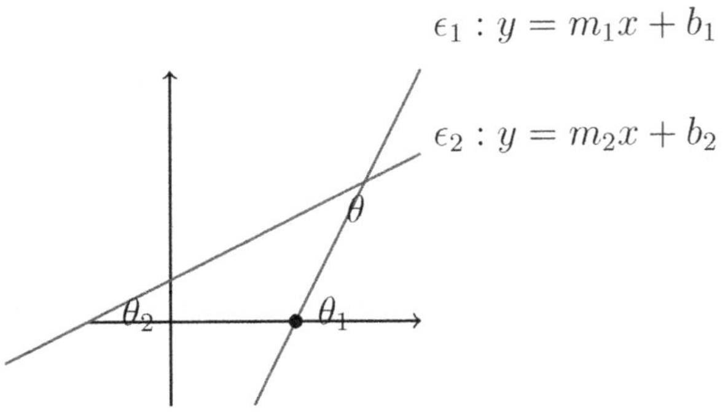

Let $\epsilon_1 : y = m_1x + b_1$, $\epsilon_2 : y = m_2x + b_2$ be two straight lines, then :

$$\theta = \theta_1 - \theta_2$$

and

$$\tan\theta = \frac{m_1 - m_2}{1 + m_1 \cdot m_2}$$

5.2 Examples of Linear functions & graphs

Let us start with an example

Example (1). *Let*

$$f(x) = 2x + 1, \text{ or } y = 2x + 1$$

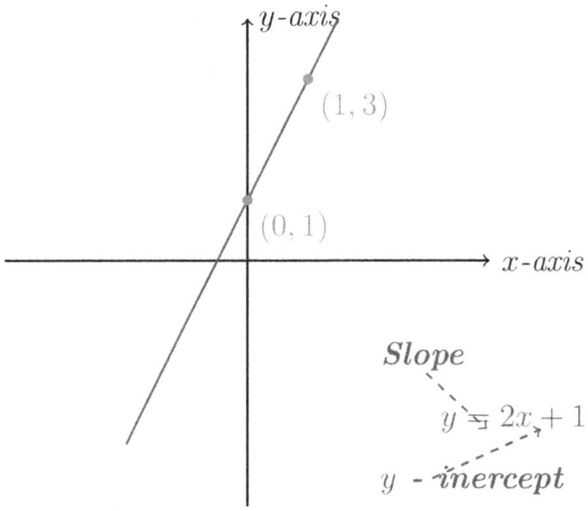

Graph:

Example. *Let*

$$2y + 3x - 5 = 0$$

solve with respect to y:

$$y = -\frac{3}{2}x + \frac{5}{2}, \quad \textbf{Slope} = \frac{-3}{2}, \quad \textbf{y} - \textbf{intercept} = \frac{5}{2}$$

Graph:

- *for* $x = 0$, $y = \frac{5}{2}$ *(y-intercept)*
- *for* $x = 1$, $y = 1$

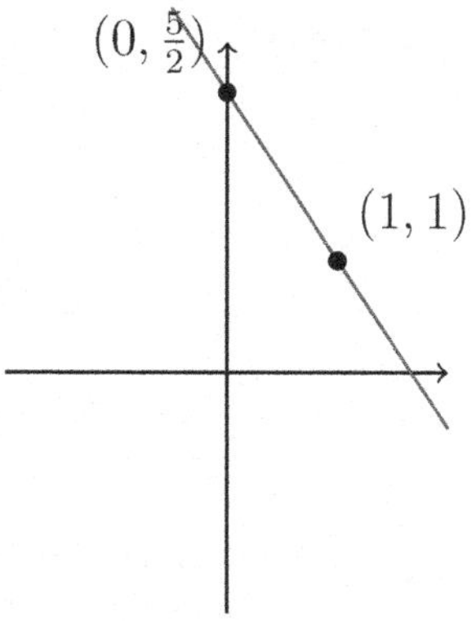

Note: The general form is

$$y = mx + b, \text{ or } f(x) = mx + b$$

where $m \to$ Slope and $b \to y -$ intercept.
If the generic equation

$$Ax + By + C = 0$$

is solved for y, then:

$$y = -\frac{A}{B}x - \frac{C}{B}$$

where $-\frac{A}{B}$ - Slope and $-\frac{C}{B}$ the y- intercept.
Example (2). *Let*

$$2x + 5y - 7 = 0$$

then

$$y = -\frac{2}{5}x + \frac{7}{5}$$

with

$$\text{Slope} = -\frac{2}{5} \ \& \ \text{y} - \text{intercept} = \frac{7}{5}$$

Let us repeat now:

Slope - formula: If (x_1, y_1), (x_2, y_2) are two points on a straight line then, the slope m is given by:

$$m = \frac{y_2 - y_1}{x_2 - x_1} = \frac{y_1 - y_2}{x_1 - x_2}$$

Note: The above formula for m has a meaning if and only if $x_1 \neq x_2$ i.e: **if and only if the two points do not belong to the same vertical line.**

Careful! If $(2, 3)$ is a point of the graph of a linear function f, then we usually write $f(2) = 3$, that is for $x = 2$ it is $y = 3$.
Generically, if (a, b) is a point of the graph, then $f(a) = b$.

Another useful formula: If m is the slope of a straight line and (x_1, y_1) is a point of it, then the equation of the linear function is

$$y - y_1 = m(x - x_1),$$

which is equivalent to

$$m = \frac{y - y_1}{x - x_1}$$

Example (3). *A straight line has a slope of* 2 *and* $(3, 5)$ *is one point on it. The equation of this straight line is*

$$y - 5 = 2(x - 3)$$

or, equivalently,

$$y = 2x - 1$$

Parallel lines & Perpendicular lines

Let the equations of two straight lines be

$$\epsilon : y = m_1 x + b_1, \quad \zeta : y = m_2 x + b_2,$$

- If they are **parallel** then $m_1 = m_2$
- If they are **perpendicular** then $m_1 \cdot m_2 = -1$

Example (4).

1. *Let* $\left.\begin{array}{l} y = 2x + 3 \\ y = 2x - 5 \end{array}\right\},$ *then the graphs are parallel lines since they have equal slopes* $m_1 = 2 = m_2 = 2.$

2. *Let* $\left.\begin{array}{l} y = 5x + 2 \\ y = \frac{-1}{5}x + 7 \end{array}\right\},$ *It is* $5 \cdot \frac{-1}{5} = -1$ *hence the graphs are perpendicular straight lines.*

   ,

3. *Let* $\begin{array}{ll} 2x + 3y - 5 = & 0 \\ -3x + 2y + 7 = & 0 \end{array}$ *solving with respect to y:* $\begin{array}{l} y = \frac{-2}{3}x + \frac{5}{3} \\ y = \frac{3}{2}x - \frac{7}{2} \end{array}$ *so* $\begin{array}{l} m_1 = \frac{-2}{3} \\ m_2 = \frac{3}{2} \end{array}$ *hence* $m_1 \cdot m_2 = -1$ *and thus the straight lines under consideration are perpendicular.*

The **distance** between two points $P(x_1, y_1)$, $Q(x_2, y_2)$ is given by

$$d(P, Q) = \sqrt{(x_1 - x_2)^2 + (y_1 - y_2)^2}$$

The distance between $P(2, 3)$, $Q(6, 7)$ is

$$d(P, Q) = \sqrt{(2 - 6)^2 + (3 - 7)^2} = \sqrt{2 \cdot 4^2} = \sqrt{32} = \sqrt{2^5} = \sqrt{2 \cdot 16} = 4\sqrt{2}$$

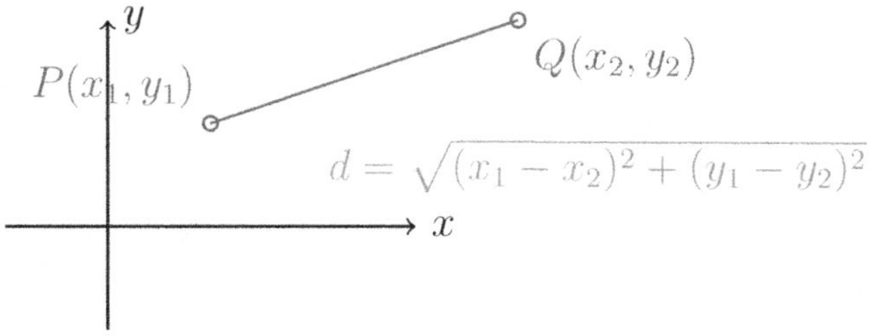

The angle θ between the straight lines

$$y = m_1 x + b_1, \quad y = m_2 x + b_2$$

is given by

$$\tan\theta = \frac{m_1 - m_2}{1 + m_1 \cdot m_2}$$

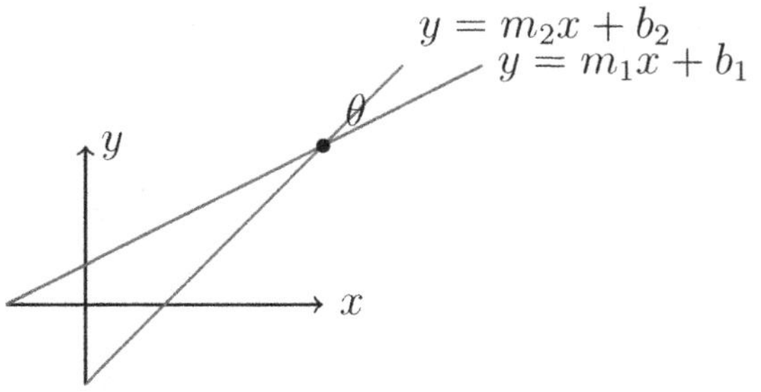

Example (5). *Let*

$$y = 3x - 1, \; y = x + 1$$

then

$$\tan\theta = \frac{3-1}{1+3\cdot 1} = \frac{2}{4} = \frac{1}{2}.$$

The **distance** between a point $P(x_1, y_1)$ and a straight line $\epsilon : Ax + By + C = 0$ is given by

$$d(P, \epsilon) = \frac{|Ax_1 + By_1 + C|}{\sqrt{A^2 + B^2}}$$

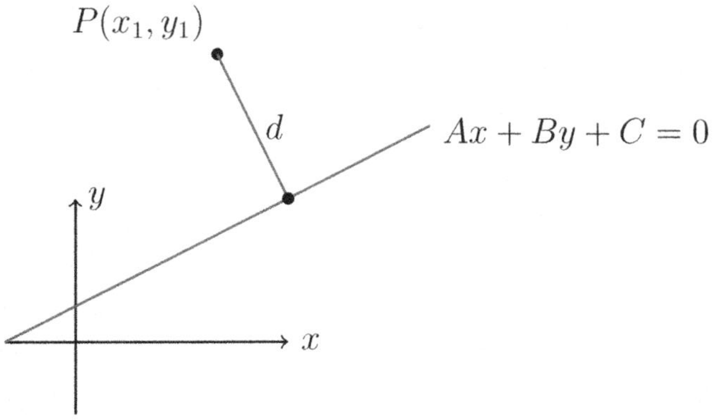

Example (6). *The distance between the origin* $O(0, 0)$ *and the straight line* $\epsilon : 2x + 3y - 5 = 0$ *is*

$$d(\epsilon, O) = \frac{|0 + 0 - 5|}{\sqrt{2^2 + 3^2}} = \frac{5}{\sqrt{13}}.$$

The **Midpoint** coordinates (x_M, y_M) of the straight line segment defined by the points $A(x_1, y_1)$, $B(x_2, y_2)$ are

$$x_M = \frac{x_1 + x_2}{2}, \quad y_M = \frac{y_1 + y_2}{2}$$

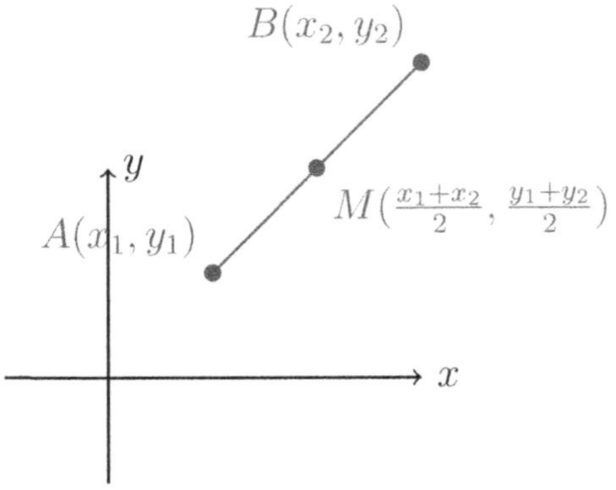

Example (7). *The midpoint between* $(2, 3)$ *and* $(4, 5)$ *is*

$$M(\frac{2+4}{2}, \frac{3+5}{2}) \equiv M(3, 4)$$

Note: Do not forget : the equation

$$y = mx + b$$

can be written

$$f(x) = mx + b$$

Example (8). *Let*

$$f(x) = 2x + 1,$$

for $x = 2$ *we have* $f(2) = 2 \cdot 2 + 1 = 5$, *so* $(2, 5)$ *is a point of the graph of* $f(x) = 2x + 1$, *or,* $y = 2x + 1$.

5.2.1 Multiple Choice Problems

Let us repeat here our technique

Step 1

Try your best

Use formulas - quickly

No solution yet?

Is there any similarity to another problem known to you?

Follow the same technique

......

No solution yet?

Step 2 - quickly ...

Emphasis on the goal

What is missing?

Any backward step from the goal?

Make full use of the data ......

No solution yet?

Step 3

Data
Make clear, explicit, all the details - all the subtle points: describe the details, quickly, using short notes.
Think using short notes ...
Try to find **more aspects** for each detail.
Think of former steps from the data

Try to eliminate the difference between goal and data
In this way, you will very soon hit on the crucial idea ...
Continue with self-confidence
Self - confidence is very important here
Yes, you can solve it ...
No solution yet? ...
Try your best to surpass yourself
No solution yet?
Try for a while the technique of getting out of loops ...
No solution yet?
Start from the beginning Step 3.

Goal
Any previous steps?
Any equivalent steps?
Backward procedure

We start with some problems where we practise in applications of formulas. Then, we continue with problems where we apply Step 2 of our technique.

Problem (1). *If the graphs of the straight lines with equations $ax+2y-3=0$ and $y-x+1=0$ are perpendicular then a is*

- *[A]* -2
- *[B]* -1
- *[C]* 0
- *[D]* 1
- *[E]* 2

Think for a while before you go on reading
Solution: The slope of
$$ax + 2y - 3 = 0$$
is $m_1 = \frac{-a}{2}$ and the slope of
$$y = x - 1$$
is $m_2 = 1$, hence perpendicularity condition $m_1 \cdot m_2 = -1$ yields
$$\frac{-a}{2} \cdot 1 = -1 \Rightarrow a = 2.$$
So, correct answer is $[E]$.

Remark. ***Vertical lines*** *of the form $x = k$ have* ***no slope****.*

Problem (2). *If the slope of a straight line with equation $3y - (2k+1)x - 5 = 0$ is 4 then k is equal to*

- *[A]* 2
- *[B]* 3
- *[C]* 5
- *[D]* 5.5
- *[E]* 8

Think for a while before go on reading
Solution: Solve with respect to y, it is:

$$y = \frac{2k+1}{3}x + \frac{5}{3},$$

Slope is $m = \frac{2k+1}{3}$ hence

$$\frac{2k+1}{3} = 4 \Rightarrow 2k+1 = 12 \Rightarrow k = 5.5$$

Correct answer is $[D]$.

Problem (3). *The slope of the median AM of a triangle with vertices $A(a,5)$, $B(0,4)$ and $C(6,2)$ is 2. Then the value a is*

- *[A]* 1
- *[B]* 2
- *[C]* 3
- *[D]* 4
- *[E]* 5

Think for a while before you go on reading
Solution:
The midpoint M of BC has coordinates $M(x_M, y_M) = M(\frac{0+6}{2}, \frac{4+2}{2}) = M(3,3)$. The slope m_{AM} of AM is given by

$$m_{AM} = \frac{y_A - y_M}{x_A - x_M} = \frac{5-3}{a-3} \overset{m_{AM}=2}{\Rightarrow} \frac{2}{a-3} = 2 \Rightarrow a = 4.$$

Correct answer is $[D]$.

Problem (4). *The slope of the line AB is*

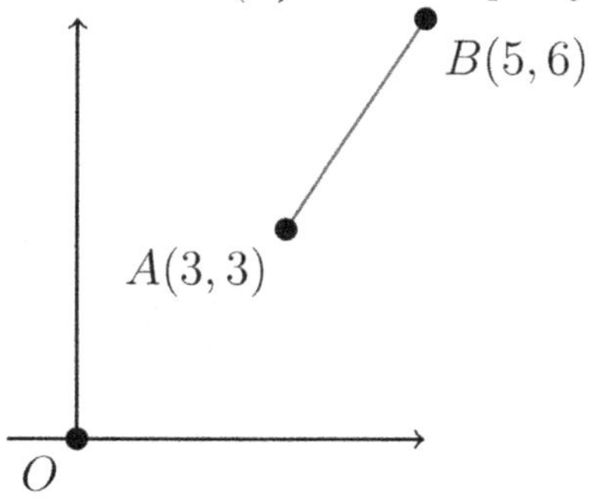

- *[A]* $\frac{-2}{3}$
- *[B]* $\frac{3}{2}$
- *[C]* 2
- *[D]* 3
- *[E]* 4

Answer: Use the formula

$$m = \frac{y_2 - y_1}{x_2 - x_1},$$

so

$$m = \frac{6-3}{5-3} = \frac{3}{2}$$

and the correct answer is $[B]$.

Problem (5). *If the slope of the straight line is* 4 *then the value of* x *is*

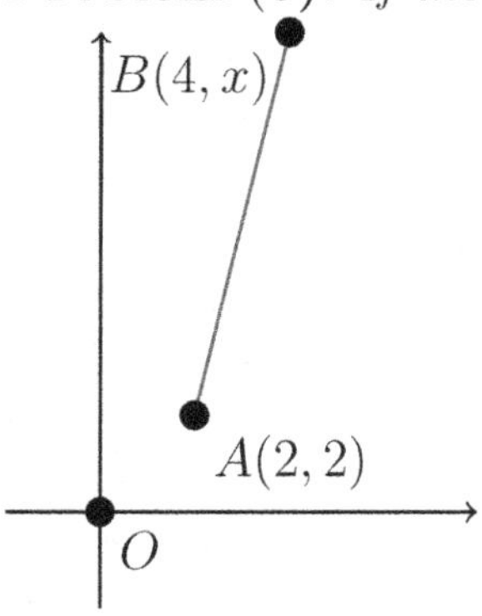

- *[A]* 2
- *[B]* 4
- *[C]* 6
- *[D]* 8
- *[E]* 10

Answer: Use the formula

$$m = \frac{y_2 - y_1}{x_2 - x_1} = \frac{x-2}{4-2} = 4$$

Solve with respect to x:

$$x - 2 = 8 \Rightarrow x = 10$$

and the correct answer is $[E]$.

Problem (6). *The slope of a straight line perpendicular to* $\epsilon : 2x - 3y + 1 = 0$ *is*

- *[A]* -2
- *[B]* $\frac{-3}{2}$
- *[C]* $\frac{-2}{3}$
- *[D]* 1
- *[E]* $\frac{3}{2}$

Solution: The formula for perpendicular straight lines is

$$m_1 \cdot m_2 = -1$$

and since the slope of ϵ is $m_1 = \frac{2}{3}$ it follows that

$$m_2 = \frac{-1}{m_1} = \frac{-3}{2}$$

the correct answer is $[B]$.

Problem (7). *If* $M(x, y)$ *is the midpoint of* AB *then* (x, y) *is:*

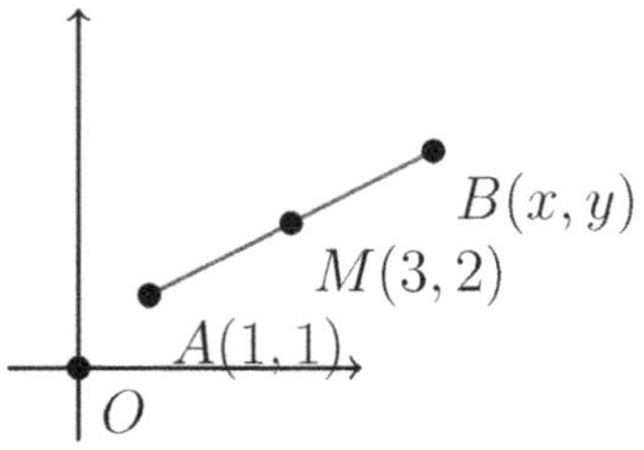

- *[A]* $(-3, 5)$
- *[B]* $(3, 5)$
- *[C]* $(3, 3)$
- *[D]* $(5, 2)$

- *[E]* $(5, 3)$

Solution: The formula for the midpoint is

$$\left(\frac{x_1 + x_2}{2}, \frac{y_1 + y_2}{2}\right),$$

hence

$$3 = \frac{x+1}{2} \Rightarrow x + 1 = 6 \Rightarrow x = 5,$$
$$2 = \frac{1+y}{2} \Rightarrow y + 1 = 4 \Rightarrow y = 3$$

therefore the correct answer is [E] $((x, y) = (5, 3))$.

Problem (8). *The slope of the median*

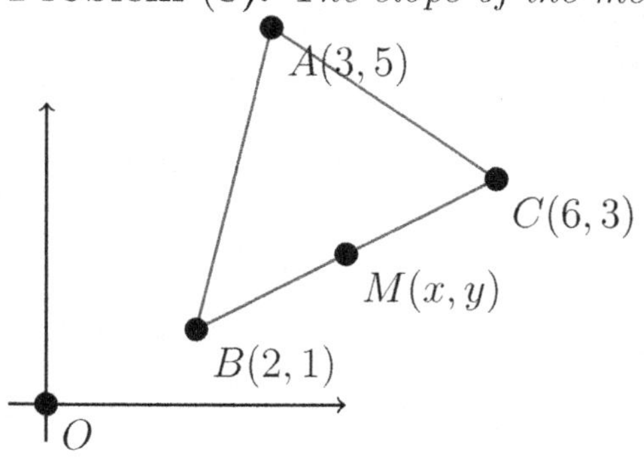

is

- *[A]* -3
- *[B]* -2
- *[C]* -1
- *[D]* 2
- *[E]* 3

Solution: Formula for the slope

$$m = \frac{y_2 - y_1}{x_2 - x_1}$$

Missing ? Coordinates of M
Formula ?

$$M\left(\frac{x_1 + x_2}{2}, \frac{y_1 + y_2}{2}\right) = \left(\frac{2+6}{2}, \frac{1+3}{2}\right) = (4, 2)$$

hence

$$m = \frac{y_2 - y_1}{x_2 - x_1} = \frac{5-2}{3-4} = -3$$

so the correct answer is [A].

Problem (9). *The slope of the graph of a linear function is* -3 *and* $f(2) = 3$. *The value of* $f(1)$ *is*

- *[A]* -6
- *[B]* -3
- *[C]* 0
- *[D]* 3
- *[E]* 6

Solution - procedure:

(**Emphasis on the goal**)
$f(1) = ?$ formula?
$f(x) = mx + b$ Use of data:
So $f(1) = m \cdot 1 + b = -3 + b$
Missing? The value of b.
Have I used all the data?
All the aspects of the data?
$f(2) = 3$ hence
$2(-3) + b = 3$
or
$b = 3 + 6 = 9$
Finally $f(1) = -3 + 9 = 6$
and the correct answer is $[E]$.

Another Way:

$f(1) = ?$ formula?
$f(x) = mx + b$ or $y = mx + b$
Have a look at all aspects

$$m = \frac{y_2 - y_1}{x_2 - x_1}$$

Can we apply it?
Let us use all the data
$m = -3$
$x_1 = 1,\ y_1 = f(1)$
from $f(2) = 3$ we deduce:
$x_2 = 2,\ y_2 = 3$
hence

$$-3 = \frac{3 - f(1)}{2 - 1}$$

or

$$-3 = 3 - f(1)$$

hence $f(1) = 6$.
Correct answer is $[E]$.

Problem (10). *For a linear function we are given that $f(1) = 4$, $f(x) = 8$ and the slope of the graph is 2. Then, the value of x is:*

- *[A]* -4
- *[B]* -3
- *[C]* 1
- *[D]* 2
- *[E]* 3

Solution - procedure:

Emphasis on the goal

$f(x) = 8$

Let us use all the aspects of $f(x)$

$f(x)$ linear so

$f(x) = mx + b$ therefore

since $f(x) = 8$ then

$8 = 2x + b$

Missing? b

Use all the data

$f(1) = 4 \rightarrow 2 \cdot 1 + b = 4$

hence $b = 2$ and thus

$8 = 2x + 2$

therefore $x = 3$

and the correct answer is $[D]$.

Another way

Slope $= \frac{y_2 - y_1}{x_2 - x_1}$ or

$2 = \frac{8-4}{x-1}$

solve with respect to x:

$2(x-1) = 4$

$2x - 2 = 4$

$x = 3.$

Correct answer is $[E]$.

Problem (11). *The equation of the height of the triangle $\triangle AOB$ is*

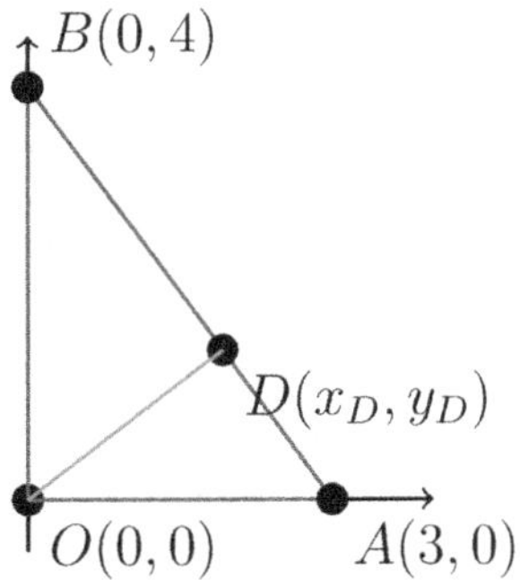

- *[A]* $y = -3x$
- *[B]* $y = -3x + 2$
- *[C]* $y = -x$

- *[D]* $y = \frac{3}{4}x$
- *[E]* $y = 3x$

Solution - procedure:

Emphasis on the goal

The equation is of the form $y = mx + b$ Missing?

m and b ...

Use the data:

b is the y- intercept

hence $b = 0$

$m = ?$

Another aspect?

$OD \perp AB$

hence $m \cdot m_{AB} = -1$

Missing ? m_{AB}

Formula? $m_{AB} = \frac{y_2 - y_1}{x_2 - x_1} = \frac{4-0}{0-3} = -\frac{4}{3}$

so $m \cdot (\frac{-4}{3}) = -1$ or $m = \frac{3}{4}$

The equation is $y = \frac{3}{4}x$

Correct answer is $[D]$.

Problem (12). *If for a linear function it is* $f(2) = 10$, $f(3) = 13$ *then the value of* $f(4)$ *is*

- *[A]* 8
- *[B]* 10
- *[C]* 12
- *[D]* 14
- *[E]* 16

Solution - procedure:

Emphasis on the goal

$f(4) = ?$

f linear function $\to f(x) = mx + b$

$f(4) = 4m + b$ Missing ? m and b

$f(2) = 10 \to 10 = 2m + b$

$f(3) = 13 \to 13 = 3m + b$

System - subtract

$m = 3$ and thus $10 = 6 + b$ so $b = 4$

therefore $f(4) = 4 \cdot 3 + 4 = 16$

the correct answer is $[E]$.

Another way

$f(2) = 10 \to (2, 10)$ is a point on the graph

$f(3) = 13 \to (3, 13)$ is a point on the graph

so

$$m = \frac{y_2 - y_1}{x_2 - x_1} = \frac{3}{1} = 3$$

an using the formula $y - y_1 = m(x - x_1)$ we get
$y - 10 = 3(x - 2)$ or
$y = 3x + 4$
hence
$f(4) = 3 \cdot 4 + 4 = 16.$
Correct answer is $[E]$.

Problem (13). *The coordinates of the point which lies midway between the origin and the line*

$$x + 2y - 4 = 0$$

are

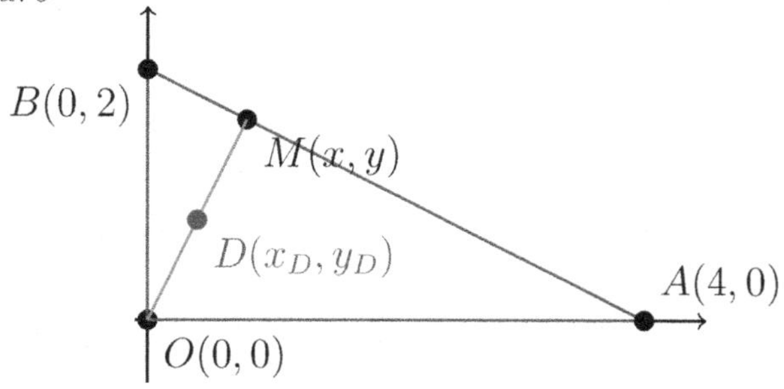

- *[A]* $(-\frac{4}{5}, \frac{8}{5})$
- *[B]* $(-\frac{2}{5}, \frac{4}{5})$
- *[C]* $(\frac{2}{5}, -\frac{4}{5})$
- *[D]* $(\frac{2}{5}, \frac{4}{5})$
- *[E]* $(\frac{4}{5}, \frac{8}{5})$

Solution - procedure:

Equation $\epsilon_{AB} : x + 2y - 4 = 0$ is written
$y = \frac{-1}{2}x + 2$
so $m_{\epsilon_{AB}} = \frac{-1}{2}$ but
and $m_{OD} \cdot m_{AB} = -1$ hence
$m_{OD} = 2$ and
since OD passes through the origin
$\zeta_{OD} : y = 2x$
$M(x, y)$ has coordinates satisfying
the system $y = 2x,\ y = \frac{-1}{2}x + 2$ i.e:
$x = \frac{4}{5}$ and $y = \frac{8}{5}$
Find midpoint $D(x_D, y_D)$ of OM using
the formula $x_D = \frac{x_1+x_2}{2}$, $y_D = \frac{y_1+y_2}{2}$
so

$$x_D = \frac{\frac{4}{5} + 0}{2} = \frac{4}{10} = \frac{2}{5},\ y_D = \frac{\frac{8}{5} + 0}{2} = \frac{8}{10} = \frac{4}{5},$$

hence the correct answer is $[D]$.

Problem (14). *If the points $A(2,3)$, $B(4,6)$ and $C(a,8)$ lie on the same line then , the value of a is*

- *[A]* $\dfrac{14}{3}$
- *[B]* 6
- *[C]* 5
- *[D]* $\dfrac{16}{3}$
- *[E]* $\dfrac{17}{3}$

Solution-procedure: **Subtle point?**
Slope of the straight line segments AB, BC must be constant, i.e:

$$\frac{6-3}{4-2} = \frac{8-6}{a-4} \Rightarrow 3a-12=4 \Rightarrow 3a=16 \Rightarrow a=\frac{16}{3}.$$

Correct answer is $[D]$.

5.3 Quadratic functions

Let us have a look at the graph:
Let $f(x) = ax^2+bx+c$ or $y = ax^2+bx+c$ then

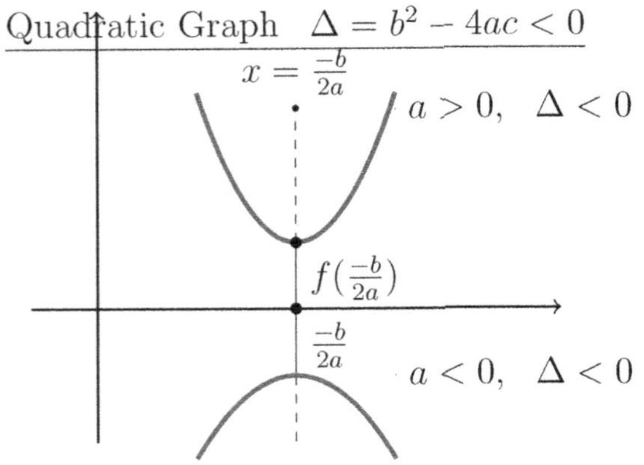

- **Axis of symmetry:** $x=\frac{-b}{2a}$
- **Roots:** $x_{1,2}=\frac{-b\pm\sqrt{b^2-4ac}}{2a}$ if $b^2-4ac\geq 0$
- **Sum of roots:** $x_1+x_2=\frac{-b}{a}$
- **Product of roots:** $x_1\cdot x_2=\frac{c}{a}$
- If $b^2-4ac<0$ there are two complex - conjugate roots $x_{1,2}=\frac{-b\pm i\sqrt{4ac-b^2}}{2a}$
- If $b^2-4ac=0$ there is one - double root $x_1=x_2=\frac{-b}{2a}$ and the function is written $f(x)=a(x+\frac{b}{2a})^2$

If $a > 0$ and $\Delta = b^2 - 4ac < 0$ then $ax^2 + bx + c > 0$ for all $x \in \mathbb{R}$,
If $a < 0$ and $\Delta = b^2 - 4ac < 0$ then $ax^2 + bx + c < 0$ for all $x \in \mathbb{R}$.

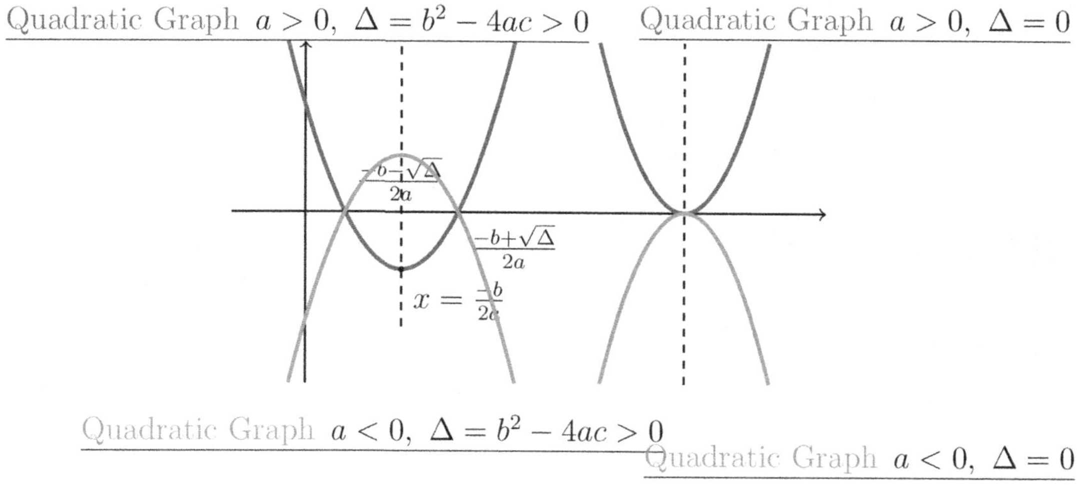

Quadratic Inequalities

- If $\Delta = b^2 - 4ac > 0$ and $x_1 = \frac{-b-\sqrt{\Delta}}{2a} < \frac{-b+\sqrt{\Delta}}{2a} = x_2$ then $f(x)$ has the same sign with a in $(-\infty, x_1) \cup (x_2, +\infty)$ and the opposite sign of a in the interval (x_1, x_2).
- If $\Delta = b^2 - 4ac < 0$ then $f(x) = a\left\{(x + \frac{b}{2a})^2 + \frac{4ac-b^2}{4a^2}\right\}$ has always the same sign with a.
- If $\Delta = 0$ then $f(x) = a(x + \frac{b}{2a})^2$ has the same sign with a.

Read carefully the following examples as well as the section on inequalities - Topic 1.
Examples (1). *Solve the following inequalities:*

- *[(i)]* $x^2 + 5x + 6 > 0$
  Roots of $x^2 + 5x + 6 = 0$*:* $x_1 = -2,\ x_2 = -3,\ \Delta = 1 > 0,\quad a = 1 > 0$ *hence the solution is*
  $$x < -3 \text{ or } x > -2$$
- *[(ii)]* $-x^2 + 3x - 2 > 0$*:*
  Roots of $-x^2 + 3x - 2 = 0$*:* $x_1 = 1,\ x_2 = 2,\ \Delta = 1 > 0,\quad a = -1 < 0$ *hence the solution is*
  $$1 < x < 2$$
- *[(iii)]* $x^2 + x + 1 > 0$
  Roots of $x^2 + x + 1 = 0$*:* $\Delta = -3 < 0$ *so there are no real roots and* $a = 1 > 0$ *hence the solution is all* $\mathbb{R}$.

Examples (2). *Solve the following inequalities:*

- *[(i)]* $-x^2 + 2x - 7 < 0$*:*
  Roots of $-x^2 + 2x - 7 = 0$*:* $\Delta = 4 - 28 = -14 < 0$ *so there are no real roots and since* $a = -1 < 0$ *the solution is all* $\mathbb{R}$.

- *[(ii)]* $x^2 + 2x + 2 < 0$:
  Roots of $x^2 + 2x + 2 = 0$: $\Delta = -4 < 0$ *so there are no real roots and since* $a = 1 > 0$ *we have that*

$$x^2 + 2x + 2 = \{x^2 + 2x + 1\} + 1 = (x + 1)^2 + 1 \geq 1$$

  for all $x \in \mathbb{R}$ *and the inequality is* ***impossible***.

5.3.1 Problems on quadratics

Problem (1). *If*

$$x^2 - 4x + k = 0, \;\; k \in \mathbb{R}$$

has roots x_1, x_2 *such that*

$$x_1^3 + x_2^3 = -128$$

then k *is equal to*

- *[A]* 14
- *[B]* 16
- *[C]* 18
- *[D]* 20
- *[E] none of the above*

Solution:
Think for a while before you read on
Use formula

$$x_1^3 + x_2^3 = (x_1 + x_2)(x_1^2 - x_1x_2 + x_2^2), \tag{1}$$

or

$$x_1^3 + x_2^3 = (x_1 + x_2)^3 - 3x_1x_2(x_1 + x_2), \tag{2}$$

where

$$x_1 + x_2 = -\frac{-4}{1} = 4,$$
$$x_1x_2 = \frac{k}{1} = k$$

and

$$x_1^2 + x_2^2 = (x_1 + x_2)^2 - 2x_1x_2$$

hence, using the second formula we have:

$$-128 = 4^3 - 3 \cdot 4k \Rightarrow 12k = 192 \Rightarrow k = 16$$

Correct answer is $[B]$.

Problem (2). *If the equation* $2x^2 - kx + k - 1 = 0$ *has no real roots then the values of* k *are*

- *[A]* $k < 4 - 2\sqrt{2}$

- *[B]* $4-2\sqrt{2}<k<4+2\sqrt{2}$
- *[C]* $k<4-\sqrt{2}$ *or* $k>4+\sqrt{2}$
- *[D]* $k>4+2\sqrt{2}$
- *[E] none of the above*

Solution – Think for a while before you go on reading: Use formula for $\Delta<0$:

$$\Delta<0 \Leftrightarrow (-k)^2-4\cdot 2\cdot (k-1)<0 \Leftrightarrow k^2-8k+8<0$$

Roots of $k^2-8k+8=0$ are

$$k_{1,2}=\frac{8\pm\sqrt{64-32}}{2}=\frac{8\pm 4\sqrt{2}}{2}=4\pm 2\sqrt{2}$$

hence

$$4-2\sqrt{2}<k<4+2\sqrt{2}$$

and the correct answer is [B].
Else, use the

Problem (3). *If the equation* $x^2-2mx+m+1=0$ *has real roots then, the values of* m *are*

- *[A]* $m\le\frac{1-\sqrt{5}}{2}$
- *[B]* $\frac{1-\sqrt{5}}{2}<m<\frac{1+\sqrt{5}}{2}$
- *[C]* $m<\frac{1-\sqrt{5}}{2}$ *or* $m>\frac{1+\sqrt{5}}{2}$
- *[D]* $m>\frac{1+\sqrt{5}}{2}$
- *[E]* $m\le\frac{1-\sqrt{5}}{2}$ *or* $m\ge\frac{1+\sqrt{5}}{2}$

Solution – Think for a while before you go on reading Use formula for $\Delta\ge 0$, it is:

$$\Delta\ge 0 \Leftrightarrow 4m^2-4(m+1)\ge 0 \Leftrightarrow m^2-m-1\ge 0$$

Roots of $m^2-m-1=0$ are

$$m_{1,2}=\frac{1\pm\sqrt{5}}{2}$$

hence

$$m\le\frac{1-\sqrt{5}}{2} \text{ or } m\ge\frac{1+\sqrt{5}}{2}.$$

Correct answer is [E].

Problem (4). *If r_1, r_2 are the roots of $x^2 - 4mx + m + 1 = 0$ and $r_1 = 3r_2$ then m is*

- *[A]* $\dfrac{-1-\sqrt{13}}{6}$
- *[B]* $\dfrac{1-\sqrt{13}}{6}$
- *[C]* $\dfrac{1+\sqrt{13}}{2}$
- *[D]* $\dfrac{1-\sqrt{13}}{2}$ *or* $\dfrac{1+\sqrt{13}}{2}$
- *[E] none of the above*

Solution – Think for a while before you go on reading
Write down all the relevant formula:

$$\left\{\begin{array}{cc} r_1 + r_2 & = 4m \\ r_1 \cdot r_2 & = m+1 \\ r_1 & = 3r_2 \end{array}\right\} \overset{\text{eliminate step by step}}{\Longrightarrow} 4r_2 = 4m \Rightarrow r_2 = m \overset{r_1 r_2 = m+1}{\Longrightarrow}$$

$$r_1 = \frac{m+1}{m} \overset{x^2-4mx+m+1=0}{\Longrightarrow} 3m^2 - m - 1 = 0 \Rightarrow m = \frac{1 \pm \sqrt{13}}{2}$$

Correct answer is $[D]$.

Problem (5). *The x - coordinate of the vertex of the graph of $y = (2x-3)^2 + x - 4$ is*

- *[A]* $\dfrac{-11}{8}$
- *[B]* $\dfrac{-10}{8}$
- *[C]* $\dfrac{10}{8}$
- *[D]* $\dfrac{11}{8}$
- *[E]* $\dfrac{3}{2}$

Solution – Think for a while before you go on reading
Use the formula $x = \frac{-b}{2a}$, hence $x = -\frac{-11}{8} = \frac{11}{8}$.
Correct answer is $[D]$.

Problem (6). *If both roots of the equation $ax^2 + bx + c = 0$ are reduced each one by k, then, in the equation $ax^2 + mx + d = 0$ with these new roots, m is*

- *[A]* $b - 2ka$
- *[B]* $b - ka$
- *[C]* $b + ka$

- *[D]* $b + 2ka$
- *[E] none of the above*

Solution – Think for a while before you go on reading
The new roots of $ax^2+mx+d=0$ are r_1-k, r_2-k where r_1, r_2 roots of $ax^2+bx+c=0$.So

$$\frac{-m}{a} = r_1 - k + r_2 - k = r_1 + r_2 - 2k = \frac{-b}{a} - 2k = \frac{-b-2ka}{a} \Rightarrow$$
$$\frac{-m}{a} = \frac{-b-2ka}{a} \Rightarrow m = b + 2ka$$

Correct answer is $[D]$.

Problem (7). *If x is a real number then $\frac{x^2-12}{2x-7}$ cannot lie between*

- *[A] -3 and 4*
- *[B] -3 and 0*
- *[C] 0 and 4*
- *[D] 2 and 3*
- *[E] 3 and 4*

Solution – Think for a while before you go on reading
Any subtle point?
Any implicit information?
Yes, the values of $\frac{x^2-12}{2x-7}$, therefore let us write $y = \frac{x^2-12}{2x-7}$ and see it under as a new quadratic with parameter y and x real:

$$x^2 - 2yx + (7y-12) = 0 \overset{x\in\mathbb{R}}{\Rightarrow} \Delta(y) \geq 0 \Rightarrow$$
$$y^2 - 7y + 12 \geq 0 \Rightarrow y \in (-\infty, 3] \cup [4, \infty)$$

Roots of $y^2 - 7y + 12 = 0$ are $y_1 = 3$, $y_2 = 4$ and thus

$$y \leq 3 \text{ or } y \geq 4$$

Correct answer is $[E]$.

Problem (8). *The greatest value of m for which $x^2 + 10x + m$ has real roots is*

- *[A]* -5
- *[B]* 5
- *[C]* 10
- *[D]* 25
- *[E] none of the above*

Solution
Think for a while before you go on reading Use that

$$\Delta \geq 0 \Leftrightarrow 10^2 - 4m^2 \geq 0 \Leftrightarrow 25 - m^2 \geq 0 \Leftrightarrow |m| \leq 5 \Leftrightarrow m \in [-5, +5]$$

Else,

$$m^2 = 25 \Leftrightarrow m = \pm 5$$

then use
Correct answer is $[B]$.

Problem (9). *If $x^2 - ax + b^2 = 0$ has equal roots then*

- *[A]* $a = -2b$
- *[B]* $a = 2b$
- *[C]* $|a| = 2b$
- *[D]* $a = 2b$ *or* $a = -2b$
- *[E] none of the above*

Solution – Think for a while before you go on reading
Use the fact that equal roots means

$$\Delta = 0 \Leftrightarrow a^2 - 4b^2 = 0 \Leftrightarrow (a - b) \cdot (a + 2b) = 0$$

and the correct answer is $[D]$.

Problem (10). *If $3x^2 - 4x + 2m^2 = 0$ has equal roots then m is*

- *[A]* $\frac{-2}{3}$
- *[B]* $-\sqrt{\frac{2}{3}}$ *or* $\sqrt{\frac{3}{2}}$
- *[C]* $\sqrt{\frac{2}{3}}$
- *[D]* $\sqrt{\frac{2}{3}}$
- *[E]* $\frac{2}{3}$

Solution – Think for a while before you go on reading
Equal roots means

$$\Delta = 0 \Leftrightarrow 16 - 24m^2 = 0 \Leftrightarrow |m| = \sqrt{\frac{2}{3}} \Leftrightarrow m = \pm\sqrt{\frac{2}{3}}$$

Correct answer is $[B]$.

Problem (11). *If $a \neq 0$ is one root of $ax^2 + mx - 3am = 0$ then m is*

- *[A]* a
- *[B]* $\frac{a}{2}$
- *[C]* $2a$
- *[D]* a^2
- *[E]* $\frac{a^2}{4}$

Solution – Think for a while before you go on reading:
Since a is a root then

$$a^3 - ma - 3am = 0 \Rightarrow a^2 - 4m = 0 \Rightarrow m = \frac{a^2}{4}$$

Correct answer is $[E]$.

Problem (12). *If* 1 *is a root of* $x^4 - 5x^2 + a = 0$ *then the other roots are*

- *[A]* $-4, -2, -1$
- *[B]* $-4, -2, 2$
- *[C]* $-2, -1, 2$
- *[D]* $-2, -1, 4$
- *[E]* *none of the above*

Solution – Think for a while before you go on reading
Since a is a root

$$1 - 5 + a = 0 \Rightarrow a = 4$$

hence

$$x^4 - 5x^2 + 4 = 0 \Rightarrow x^2 = 1, \; x^2 = 4$$

so

$$x = \pm 1, \; x = \pm 2.$$

Correct answer is $[C]$.

Problem (13). *If* r_1, r_2 *are the roots of* $x^2 - 5x + 3 = 0$ *then* $r_1^3 + r_2^3$ *is*

- *[A]* 40
- *[B]* 64
- *[C]* 70
- *[D]* 80
- *[E]* 90

Solution – Think for a while before you go on reading
Using standard identities

$$r_1^3 + r_2^3 = (r_1 + r_2)(r_1^2 - r_1 r_2 + r_2^2)$$

where

$$r_1 + r_2 = -\frac{-5}{1} = 5,\ r_1 \cdot r_2 = \frac{4}{1} = 3,$$
$$r_1^2 + r_2^2 = (r_1 + r_2)^2 - 2r_1 \cdot r_2 = 5^2 - 2 \cdot 3 = 19$$

hence

$$r_1^3 + r_2^3 = 5 \cdot 16 = 80$$

Correct answer is $[D]$.

Problem (14). *If r_1, r_2 are the roots of $x^2 - 3x + 1 = 0$ then $r_1^4 + r_2^4$ is*

- *[A]* 30
- *[B]* 40
- *[C]* 47
- *[D]* 48
- *[E]* 50

Solution – Think for a while before you go on reading
Since

$$r_1^4 + r_2^4 = (r_1^2 + r_2^2)^2 - 2r_1^2 \cdot r_2^2 = \left\{(r_1 + r_2)^2 - 2r_1 \cdot r_2\right\}^2 - 2(r_1 \cdot r_2)^2$$

where

$$r_1 + r_2 = -\frac{(-3)}{1} = 3,\ r_1 \cdot r_2 = \frac{1}{1} = 1$$

hence $r_1^4 + r_2^4 = 47$.
Correct answer is $[C]$.

Problem (15). *If the equation $3x^2 - 4x + 4a^2$ has two equal roots then a is*

- *[A]* $\frac{-1}{2}$ *or* $\frac{-1}{4}$
- *[B]* $\frac{-1}{4}$ *or* $\frac{-1}{3}$
- *[C]* $\frac{1}{4}$ *or* $\frac{-1}{2}$
- *[D]* $\frac{-1}{2}$ *or* $\frac{1}{\sqrt{3}}$

- *[E]* $\frac{-1}{\sqrt{3}}$ *or* $\frac{1}{\sqrt{3}}$

Solution – Think for a while before you go on reading
Equal roots means

$$\Delta = 0 \Leftrightarrow 16 - 48a^2 = 0 \Leftrightarrow a^2 = \frac{1}{3} \Leftrightarrow a \in \{-\frac{1}{\sqrt{3}}, \frac{1}{\sqrt{3}}\}$$

the correct answer is $[E]$.

Problem (16). *If the equation $3x^2 + (a+1)x + 3 = 0$ has two distinct real roots the the values of a are*

- *[A]* $a < -7$
- *[B]* $a < -5$
- *[C]* $a < -7$ *or* $a > 5$
- *[D]* $a \leq -7$ *or* $a \geq 5$
- *[E]* $-7 < a < 5$

Solution – Think for a while before you go on reading
Distinct real roots means

$$\Delta > 0 \Leftrightarrow (a+1)^2 - 6^2 > 0 \Leftrightarrow (a-5)(a+7) > 0 \Leftrightarrow a \in (-\infty, -7) \cup (5, \infty)$$

$\cdot\, a < -7$ or $a > 5$

-7 $+$ 5 $+$

Correct answer is $[C]$.

Problem (17). *If r_1, r_2 are the roots of $x^2 + x - 1 = 0$ then $\frac{1}{r_1}, \frac{1}{r_2}$ are the roots*

- *[A]* $x^2 + x + 1 = 0$
- *[B]* $x^2 - x + 1 = 0$
- *[C]* $x^2 - x - 1 = 0$
- *[D]* $x^2 + x - 1 = 0$
- *[E]* $x^2 + x - 2 = 0$

Solution – Think for a while before you go on reading
Use formula

$$\frac{1}{r_1} + \frac{1}{r_2} = \frac{r_1 + r_2}{r_1 \cdot r_2} = \frac{-1}{-1} = 1,$$
$$\frac{1}{r_2} \cdot \frac{1}{r_2} = \frac{1}{r_1 \cdot r_2} = -1$$

hence the equation is $x^2 - x - 1 = 0$.
Correct answer is $[C]$.

Problem (18). *The domain of the function* $f(x) = \sqrt{-x^2 + 5x - 4}$ *is*

- *[A]* $x < -4$
- *[B]* $x \leq -4$
- *[C]* $x \leq -4$ *or* $x \geq -1$
- *[D]* $-4 < x < -1$
- *[E]* $1 \leq x \leq 4$

Solution – Think for a while before you go on reading
The domain is all the $x \in \mathbb{R}$ such that

$$-x^2 + 5x - 4 \geq 0 \Leftrightarrow -(x-4)(x-1) \geq 0 \Leftrightarrow (x-4) \cdot (x-1) \leq 0 \Leftrightarrow 1 \leq x \leq 4.$$

$\cdot\, 1 \leq x \leq 4$

Correct answer is $[E]$.

Problem (19). *The range of* $f(x) = \dfrac{x^2 - x + 1}{x^2 + x + 1}$ *is*

- *[A]* $f(x) \leq \dfrac{1}{3}$
- *[B]* $f(x) \geq 3$
- *[C]* $f(x) \leq \dfrac{1}{3}$ *or* $f(x) > 3$
- *[D]* $\frac{1}{3} < f(x) < 3$
- *[E]* $\dfrac{1}{3} \leq f(x) \leq 3$

Solution – Think for a while before you go on reading
Let

$$y = \frac{x^2 - x + 1}{x^2 + x + 1}$$

to get a quadratic with parameter y:

$$(y-1)x^2 + (y+1)x + y - 1 = 0$$

This quadratic function has real solutions if and only if

$$\Delta(y) \geq 0 \Leftrightarrow (-y+3)(3y-1) \geq 0 \Leftrightarrow \frac{1}{3} \leq y \leq 3$$

$\circ\, \frac{1}{3} \leq y \leq 3$

Correct answer is $[E]$.

Problem (20). *If* $(\frac{1}{2})^{x^2+2x} > \frac{1}{8}$ *the values of* x *are*

- *[A]* $x < 3$
- *[B]* $-3 < x < 1$
- *[C]* $-3 < x \leq 1$
- *[D]* $-3 \leq x < 1$
- *[E]* $x > 1$

Solution – Think for a while before you go on reading
Write $d\frac{1}{8} = \frac{1}{2^3}$ to get:

$$(\frac{1}{2})^{x^2+2x} > \frac{1}{8} \Leftrightarrow (\frac{1}{2})^{x^2+2x} > (\frac{1}{2})^3 \overset{0<\frac{1}{2}<1}{\Leftrightarrow} x^2 + 2x < 3$$
$$\Leftrightarrow x^2 + 2x - 3 < 0 \Leftrightarrow (x+3)(x-1) < 0 \Leftrightarrow -3 < x < 1$$

Correct answer is $[B]$.

Problem (21). *The solution of the inequality*

$$|2x + 1| < |x - 2|$$

is

- *[A]* $x \leq -3$
- *[B]* $x < \frac{1}{3}$
- *[C]* $-3 < x < \frac{1}{3}$
- *[D]* $-3 < x$
- *[E]* *none of the above*

Solution – Think for a while before you go on reading
Since both sides are non negative we can square both members to get:

$$|2x+1| < |x-2| \Leftrightarrow (2x+1)^2 < (x-2)^2 \Leftrightarrow (x+3)(3x-1) < 0 \Leftrightarrow -3 < x < \frac{1}{3}$$

The roots of

$$(x+3)(3x-1) = 0$$

are $x_1 = -3$, $x_2 = \frac{1}{3}$ hence

$\circ -3 < x < \frac{1}{3}$

Correct answer is $[C]$.

5.4 Word Problems

Technique: **Think of the problem in short notes.**
Then use simple symbols.
Try to construct an equation or an inequality using the data.
Problem (1). *George sold a toy for 21$ and lost as much per cent as he gave to buy it. What is the greatest value he had?*

- *[A]* 30$
- *[B]* 40$
- *[C]* 50$
- *[D]* 70$
- *[E]* 80$

Solution:
Where is the subtle point?
Think for a while before you go on reading
Think of the problem in short notes, use simple symbols
George sold for 21$
Let the greatest value be x
Lost percentage x
He bought it x\$ and percentage $\frac{x}{100}$
Lost amount $x - 21$\$ hence

$$x - 21 = x \cdot \frac{x}{100} \Rightarrow x^2 - 100x + 2100 = 0 \Rightarrow x_1, x_2 \in \{30, 70\}$$

therefore the greatest value is 70$ and $[D]$ is the correct answer.

Problem (2). *The sum of two numbers is* 40 *and the sum of their reciprocal is* $\frac{5}{48}$. *The greater number is*

- *[A]* 16
- *[B]* 24
- *[C]* 30
- *[D]* 32
- *[E]* 36

Solution: *Think for a while before you go on reading*
Use short notes and simple symbols
Since their sum is 40 then the two numbers are $x \neq 0$ and $40 - x \neq 0$
From the sum of their reciprocals we get:

$$\frac{1}{x} + \frac{1}{40 - x} = \frac{5}{48} \Leftrightarrow x^2 - 40x + 384 = 0 \Leftrightarrow x_1, x_2 \in \{16, 24\}$$

and the correct answer is $[B]$.

Problem (3). *A train runs between two cities A and B, a distance of 480 km, at a constant speed. If the speed has been increased by 20 km it would have taken 0.80 hours less. Then, the speed of the train is*

- *[A]* 90
- *[B]* 100
- *[C]* 108
- *[D]* 112
- *[E]* 120

Solution: *Think for a while before you go on reading*
Use short notes and simple symbols
Let x and $x+20$ be the speed respectively,
the corresponding times are $\frac{480}{x}$ and $\frac{480}{x+20}$ hence

$$\frac{480}{x} - \frac{480}{x+20} = 0.80 \Rightarrow x = 100.$$

The correct answer is $[B]$.

Problem (4). *A tank can be filled with water, by two pipes working together in 20 minutes. By the smallest pipe alone the tank is filled in 10 minutes later than the larger one. The time taken by the larger pipe, in order to fill the tank, is*

- *[A]* $5\sqrt{17}$
- *[B]* $10 + 5 \cdot \sqrt{17}$
- *[C]* $15 + 5\sqrt{17}$
- *[D]* $35 - 5\sqrt{17}$
- *[E]* $35 + 5\sqrt{17}$

Solution: *Think for a while before you go on reading*
Use short notes and simple symbols

Clearly the larger pipe should work (alone) for $20 + x$ minutes to fill the tank and
For the smaller pipe it will take (alone) $20 + x + 10 = 30 + x$ minutes to fill the tank , so
In **one** minute the larger pipe fills $\frac{1}{20+x}$ of the tank and
In **one** minute the smaller pipe fills the $\frac{1}{30+x}$ of the tank and
since both pipes, working together, in **one** minute, fill the $\frac{1}{20}$ of the tank, we get:

$$\frac{1}{20+x} + \frac{1}{30+x} = \frac{1}{20} \Leftrightarrow x^2 + 40x - 400 = 0 \Leftrightarrow x_1, x_2 \in \{-5 - 5\sqrt{17}, -5 + 5\sqrt{17}\}$$

whith the solution $-5-5\sqrt{17}$ rejected.
So, **the time needed by the larger pipe** to fill the tank is

$$20+(-5+5\sqrt{17})=15+5\sqrt{17}$$

and the correct answer is $[C]$.

Problem (5). *Two numbers differ by* 5 *and the difference of their squares is* 105, *the smallest number is*

- *[A]* 8
- *[B]* 9
- *[C]* 10
- *[D]* 12
- *[E]* 13

Solution: *Think for a while before you go on reading*
Use short notes and simple symbols
Let x be the smallest number, then

$$(x+5)^2-x^2=105 \Leftrightarrow 10x+25=105 \Leftrightarrow x=8.$$

Correct answer is $[A]$.

Problem (6). *One number is three times another and the difference of their reciprocals is* $\frac{1}{3}$ *. Then, the sum of the numbers is*

- *[A]* 6
- *[B]* 8
- *[C]* 9
- *[D]* 10
- *[E]* 12

Solution: *Think for a while before you go on reading*
Use short notes and simple symbols
Let x be the smaller and $3x$ the larger, then:

$$\frac{1}{x}-\frac{1}{3x}=\frac{1}{3} \Leftrightarrow x=2,$$

hence their sum is $2+6=8$ and the correct answer is $[B]$.

Problem (7). *The daily production of* 40 *oil wells is* 16.000 *barrels. This production decreases by* 8 *barrels for every new well. If for* x *new wells we achieve the maximum production then* x *is*

- *[A]* 5

- *[B]* 10
- *[C]* 15
- *[D]* 20
- *[E]* 25

Solution: *Think for a while before you go on reading*
Use short notes and simple symbols
40 wells $\longrightarrow$ 16.000 barrels
1 well $\longrightarrow \dfrac{16.000}{40} = 400$ barrels
For every new x well $\longrightarrow 400 - 8x$ barrels
hence, total production $P(x)$

$$P(x) = (40 + x) \cdot (400 - 8x) = -8x^2 + 80x + 16000$$

which attains its maximum at $x = 5$.
The correct answer is $[A]$.
Remark. *Recall the formula:*

$$\textit{Maximum of } y = ax^2 + bx + c \textit{ is obtained at the point } x = \frac{-b}{2a} = \frac{-80}{-16} = 5.$$

5.5 Polynomial Functions

5.5.1 The remainder & factor theorems

Identification of polynomials: Equality of Degrees and equality of the coefficients of the corresponding terms.

For any polynomial $P(x)$, when divided by the monomial $x - r$, the remainder is $P(r)$.

If $r = \frac{p}{q}$ is a rational root of $P(x) = a_n x^n + \cdots\cdots + a_1 x + a_0$, then $p|a_0$ and $q|a_n$.

If $P(x)$ is a polynomial and r a root of P then $x - r$ is a factor of $P(x)$, i.e: $P(x) = (x - r) \cdot Q(x)$.
Example (1). *Find the roots of*

$$p(x) = 8x^3 + 4x + 3.$$

Step 1: *All rational roots are fractions with nominator the integer divisors of* 3 *and denominator the integers factors of* 8.
Step 2: *It is not hard to check that for* $\frac{-1}{2}$ *we have*

$$p(\frac{-1}{2}) = 0$$

hence $x + \frac{1}{2}$ is a factor of $p(x)$, i.e:.

$$p(x) = (x + \frac{1}{2})q(x)$$

***Step* 3:** *Divide*

$$p(x) : (x + \frac{1}{2}) = q(x)$$

to get

$$q(x) = 8x^2 - 4x + 6$$

and we observe that the discriminant is

$$\Delta_q = 16 - 192 = -176 < 0$$

and thus there are no real roots of $q(x)$.
***Step* 4:** *The complex conjugate roots are*

$$\frac{1 \pm i \cdot \sqrt{95}}{8}$$

Example (2). *If $P(x) = x^3 + x^2 - x + 1$ is divided by $x - 1$ the remainder is $P(1) = 1^3 + 1^2 - 1 + 1 = 2$.*

Example (3). *If $P(x) = x^3 + 3x^2 - 4$ then $P(1) = 0$, hence $x - 1$ divides $P(x)$ and*

$$x^3 + 3x^2 - 4 = (x - 1) \cdot (x^2 + px + q)$$

where $p = 4$ and $q = 4$.

5.5.2 Problems on the remainder theorem

Problem (1). *The remainder, when $p(x) = x^3 + x^2 + 2x - 1$ is divided by $x + 1$ is*

- *[A]* -3
- *[B]* 2
- *[C]* 3
- *[D]* 4
- *[E]* 5

Solution: By the remainder's theorem

$$p(x) = q(x)(x + 1) + r$$

so

$$r = p(-1) = -3$$

Correct answer is $[A]$.

Problem (2). *The remainder, when $p(x) = 8x^3 - 4x^2 + 2x - 1$ is divided by $2x + 1$ is*

- *[A]* -4

- *[B]* -3
- *[C]* 1
- *[D]* 3
- *[E]* 4

Solution: As in the previous problem

$$p(x) = (2x+1)q(x) + r$$

so

$$r = p(\frac{-1}{2}) = -4$$

Correct answer is $[A]$.

Problem (3). *If* $p(x) = x^3 + 2x^2 - ax + 4$ *has* $x+2$ *is a factor then the value of* a *is*

- *[A]* -2
- *[B]* 1
- *[C]* 2
- *[D]* 3
- *[E]* 4

Solution: Where is the subtle point?
$x + 2 | p(x)$ if and only if $p(-2) = 0$ hence

$$p(-2) = 0 \Leftrightarrow 2a + 4 = 0 \Rightarrow a = -2$$

Correct answer is $[A]$.

Problem (4). *If* $x^4 + ax^3 + bx^2 - 10x - 20$ *is divisible by* $x^2 - 1$ *then* $b =$

- *[A]* -19
- *[B]* -18
- *[C]* -10
- *[D]* 10
- *[E]* 19

Solution – Think for a while before you go on reading
Since $x^2 - 1 = (x-1) \cdot (x+1)$ it follows that $P(1) = 0$ and $P(-1) = 0$, hence

$$\left\{ \begin{matrix} P(1) & = 0 \\ P(-1) & = 0 \end{matrix} \right\} \Leftrightarrow \left\{ \begin{matrix} a+b & = 29 \\ b-a & = +9 \end{matrix} \right\} \Rightarrow b = 19.$$

Correct answer is $[E]$.

Problem (5). *If* $Q(x) = 64x^5 + 3m$ *is divisible by* $2x+1$ *then* $m =$

- *[A]* -3
- *[B]* -2
- *[C]* $-\frac{1}{3}$
- *[D]* $\frac{1}{3}$
- *[E]* $\frac{2}{3}$

Solution – Think for a while before you go on reading
Since $2x+1=0$ if $x=\frac{-1}{2}$ it follows that

$$Q(\frac{-1}{2})=0 \Leftrightarrow -2+3m=9 \Leftrightarrow m=\frac{2}{3}.$$

Correct answer is $[E]$.

Problem (6). *The remainder of the division* $Q(x)=50x^{2017}-40x^{2016}+80x^{2000}+90 \div (x+1)$ *is*

- *[A]* -50
- *[B]* -40
- *[C]* 40
- *[D]* 50
- *[E]* 80

Solution – Think for a while before you go on reading
Since $x+1=0 \Leftrightarrow x=-1$, using the remainder theorem it is enough to compute

$$Q(-1)=80$$

Correct answer is $[E]$.

Problem (7). *If* $P(x)=m^2x^4+5mx^2+4$ *is divisible by* $x-1$ *the value of* m *is*

- *[A]* -4
- *[B]* -1
- *[C]* -4 *or* -1
- *[D]* 1
- *[E]* 1 *or* 4

Solution – Think for a while before you go on reading
The monomial $x-1$ divides $P(x)$ and $x-1=0 \Leftrightarrow x=1$ hence

$$P(-1)=0 \Leftrightarrow m^2+5m+4=0 \Leftrightarrow (m+1)\cdot(m+4)=0 \Leftrightarrow m \in \{-1,-4\}$$

Correct answer is $[C]$.

Problem (8). *If $R(x) = x^3 + 3x^2 + 3x - 7$ is divisible by $x + m$ then m is*

- *[A]* -3
- *[B]* -2
- *[C]* -1
- *[D]* 1
- *[E]* 2

Solution – Think for a while before you go on reading
Since $x + m$ is a factor of $R(x)$ and $x + m = 0 \Leftrightarrow x = -m$ it follows that

$$R(-m) = 0 \Leftrightarrow -m^3 + 3m^2 - 3m - 7 = 0$$

Any **subtle point?** or any **implicit information?**
Yes, write

$$-m^3 + 3m^2 - 3m - 7 = 0 \Leftrightarrow -(m^3 - 3m^2 + 3m - 1) = 2^3 \Leftrightarrow$$
$$-(m-1)^3 = 2^3 \Leftrightarrow (m-1)^3 = (-2)^3 \Leftrightarrow m - 1 = -2 \Leftrightarrow m = -1$$

Correct answer is $[C]$.

Problem (9). *If $P(x) = 8x^3 + 4x + 3$ has a rational root r then r is*

- *[A]* $-\frac{1}{2}$
- *[B]* $-\frac{1}{4}$
- *[C]* $-\frac{1}{8}$
- *[D]* $\frac{1}{4}$
- *[E]* $\frac{1}{8}$

Solution: – Think for a while before you go on reading
Exact method If $\frac{l}{m}$ is a rational root then l is a divisor of 3 and m is a divisor of 8, thus

$$l \in \{\pm 1, \pm 3\}, \ m \in \{\pm, 1, \pm 2, \pm 4, \pm 8\}$$

and by testing the different combinations we derive that $r = \frac{-1}{2}$.
Correct answer is $[A]$.

Problem (10). *The remainder of $P(x) = x^3 + x^2 + 2x + m$ when is divided by $x + 1$ is 2. The vakue of m is*

- *[A]* -2

- *[B]* -1
- *[C]* 1
- *[D]* 2
- *[E]* 4

Solution – Think for a while before you go on reading
Use the formula

$$P(-1) = 2 \Rightarrow m = 4.$$

Correct answer is $[E]$.

Problem (11). *If $2x+1$ is a factor of $Q(x) = 8x^3 - 4x^2 + 2x - 3m$ then m is*

- *[A]* -3
- *[B]* -2
- *[C]* -1
- *[D]* 1
- *[E]* 2

Solution – Think for a while before you go on reading
Since $2x + 1 = 0 \Leftrightarrow x = \frac{-1}{2}$, use the formula for the remainder

$$Q(\frac{-1}{2}) = 0 \Leftrightarrow m = -1.$$

Correct answer is $[C]$.

Problem (12). *If $x^2 + x - 2$ is a factor of $P(x) = x^3 + 3x + ax + b$ then a and b are*

- *[A]* $-6, -2$
- *[B]* $-6, 2$
- *[C]* $6, -2$
- *[D]* $6, 2$
- *[E]* *none of the above*

Solution – Think for a while before you go on reading Formula for the remainder, $P(r) = 0$ where r belongs to the set of roots of $x^2 + x - 2 = (x+2)(x-1)$, so

$$P(-2) = 0 \text{ and } P(1) = 0$$

and thus

$$-2a + b = 14 \text{ and } a + b = -4$$

hence $a = -6$, $b = 2$. Correct answer is $[B]$.

Problem (13). *If* $0, 1, 2$ *are roots of a polynomial* $P(x)$ *then a factor of* $P(x)$ *is*

- *[A]* $-x^3 + 8x^2 + 2x$
- *[B]* $x^3 + 3x^2 - 2x$
- *[C]* $x^3 - 3x^2 + 2x$
- *[D]* $x^3 + 3x^2 + 2x$
- *[E]* $-x^3 + 3x^2 - 2x$

Solution – Think for a while before you go on reading
$P(x)$ is divisible by

$$(x - 0) \cdot (x - 1) \cdot (x - 2) = x^3 - 3x^2 + 2x.$$

hence the correct answer is $[C]$.

Problem (14). *The remainder of* $P(x) = (x - a)^3 + (x - b)^3 + (a + b)^3$ *when divided by* $x + a$ *is*

- *[A]* $-8a^3$
- *[B]* $-6a^3$
- *[C]* $-a^3$
- *[D]* a^3
- *[E]* $8a^3$

Solution – Think for a while before you go on reading
Apply formula for the remainder

$$P(-a) = -8a^3$$

Correct answer is $[A]$.

Problem (15). *The values of* a *and* b *for which* $f(x) = x^3 + ax^2 + 19x + b$ *is divisible by* $x^2 + 7x + 12$ *are*

- *[A]* $-8, -12$
- *[B]* $-8, 12$
- *[C]* $8, -12$
- *[D]* $8, 12$
- *[E] none of the above*

Solution – Think for a while before you go on reading
Note that $x^2 + 7x + 12 = (x + 3)(x + 4)$ so for $x = -4, x = -3$, applying the remainder theorem, we should have:

$$f(-3) = 0,$$
$$f(-4) = 0$$

hence

$$9a + b = 84$$
$$16a + b = 140$$

and thus $a = 8$, $b = 12$. Correct answer is $[D]$.

Problem (16). *If r_1, r_2 are the roots of $x^2 + x - 1$ then, an equation with roots $\frac{1}{r_1}$, $\frac{1}{r_2}$ is*

- *[A]* $x^2 + x + 1 = 0$
- *[B]* $x^2 - x + 1 = 0$
- *[C]* $-x^2 + x - 1 = 0$
- *[D]* $x^2 - x - 1 = 0$
- *[E]* $x^2 + x - 1 = 0$

Solution: It is

$$\frac{1}{r_1} + \frac{1}{r_2} = \frac{r_1 + r_2}{r_1 r_2},$$
$$\frac{1}{r_1} \cdot \frac{1}{r_2} = \frac{1}{r_1 \cdot r_2}$$

where

$$r_1 + r_2 = -1, \; r_1 \cdot r_2 = -1$$

hence

$$\frac{1}{r_1} + \frac{1}{r_2} = 1,$$
$$\frac{1}{r_1} \cdot \frac{1}{r_2} = -1,$$

and thus $\{\frac{1}{r_1}, \frac{1}{r_2}\}$ are roots of the polynomial

$$x^2 - x - 1 = 0.$$

Correct answer is $[D]$.

Problem (17). *If one root of $x^2 + ax + b = 0$ is $1 - i$, a, b are real numbers, then the value of a and b are*

- *[A]* $a = -2, \; b = -2$

- *[B]* $a = -2, \ , b = 2$
- *[C]* $a = 2, \ b = -2$
- *[D]* $a = 2, \ b = 2$
- *[E]* $a = 2, \ b = 1$

Solution: Recall the standard **fact:** i is the imaginary unit, $i^2 = -1$ (see: 7.2) if $z = x + yi \in \mathbb{C}$ a root of a polynomial then $\bar{z} = x - yi$ is also a root of the same polynomial. Since $\overline{1 - i} = 1 + i$:

$$(1-i)^2 + a(1-i) + b = 0$$
$$(1+i)^2 + a(1+i) + b = 0$$

hence

$$a + b - (2+a)i = 0$$
$$a + b + (2+a)i = 0$$

and thus $a = -2$, $b = 2$.
Way 2: Using the formulas for the sum and the product of the roots of polynomial of the second degree:

$$(1-i) + (1+i) = -a \Rightarrow a = -2,$$
$$(1-i)(1+i) = b \Rightarrow b = 2$$

Correct answer is $[B]$.

Problem (18). *If $1 - i$ is a zero of $x^2 - 2x + b = 0$ then the value of b is*

- *[A]* -2
- *[B]* -1
- *[C]* 0
- *[D]* 1
- *[E]* 2

Solution: Since $1 - i$ is a root then

$$(1-i)^2 - 2(1-i) + b = 0 \Rightarrow 1 - 2i - 1 - 2 + 2i + b = 0 \Rightarrow b = 2.$$

Correct answer is $[E]$.

Problem (19). *If $1 + i$ is a root of $x^2 - ax + b = 0$ and a, b are real numbers then the sum $a + b$ is*

- *[A]* -4
- *[B]* -2
- *[C]* 2

- *[D]* 3
- *[E]* 4

Solution: Since $1+i$ is a root then

$$(1+i)^2 - a(1+i) + b = 0 \Rightarrow 1 + 2i - 1 - a - ai + b = 0 \Rightarrow$$
$$b - a + (2-a)i = 0 \Rightarrow \left\{ \begin{array}{c} 2-a=0 \\ b-a=0 \end{array} \right\}$$

hence $a = b = 2$ so $a + b = 4$.
Correct answer is $[E]$.

Problem (20). *If $x+1$, $x-2$ are factors of $p(x) = x^3 + 3x^2 + ax + b$ then a, b are*

- *[A]* $-4, -8$
- *[B]* $-5, -8$
- *[C]* $-4, 4$
- *[D]* $4, 8$
- *[E]* *none of the above*

Solution: What is the subtle point?
$x + 1 = x - (-1)$ is a factor of $p(x)$, so $p(-1) = 0$
$x - 2$ is a factor of $p(x)$, so $p(2) = 0$ and thus

$$-1 + 3 - a + b = 0$$
$$8 + 12 + 2a + b = 0$$

hence $a = -6$, $b = -8$ and the correct answer is $[B]$.

Problem (21). *If the equation*

$$3x^2 - 4x + a = 0$$

has two equal real roots then, the value of a is

- *[A]* $\frac{-4}{3}$
- *[B]* $\frac{3}{4}$
- *[C]* $\frac{4}{3}$
- *[D]* $\frac{-3}{4}$
- *[E]* $\frac{4}{7}$

Solution: What is the subtle point?
Two equal roots means **discriminant** is null

$$r_1 = r_2 \Leftrightarrow \Delta = 0 \Leftrightarrow 16 - 12a = 0 \Leftrightarrow a = \frac{4}{3}.$$

Correct answer is $[C]$.

Problem (22). *If the equation $3x^2 + (a+1)x + 3 = 0$ has two different real roots, then the values of a are:*

- *[A]* $a < -7$
- *[B]* $a > 5$
- *[C]* $-7 < a < 5$
- *[D]* $a < 5$
- *[E]* $a < -7, a > 5$

Solution: What is the subtle point?
Two different real roots is equivalent to $\Delta > 0$, so

$$\Delta > 0 \Leftrightarrow (a+1)^2 - 36 > 0 \Leftrightarrow (a+7)(a-5) > 0 \Leftrightarrow a \in (-\infty, -7) \cup (5, \infty).$$

Correct answer is $[E]$.

Problem (23). *If the equation*

$$4x^2 + 8x - 4kx + 3k - 8 = 0$$

has no real roots, then the values of k are

- *[A]* $k < 3$
- *[B]* $k < 4$
- *[C]* $k < 3$ *or* $-k > 4$
- *[D]* $3 < k < 4$
- *[E]* $k > 4$

Solution: Where is the subtle point?
No real roots is equivalent to

$$\Delta < 0 \Leftrightarrow (8-4k)^2 - 4^2(3k-8) < 0 \Leftrightarrow k^2 - 7k + 12 < 0 \Leftrightarrow (k-3)(k-4) < 0$$

hence $3 < k < 4$ and the correct answer is $[D]$.

5.6 Rational Functions & Limits

5.6.1 Domains of functions

Given a function $f = f(x)$ we usually need to find its **domain**, that is the *largest* subset of $\mathbb{R}$ where f is defined, in other words, where the explicit expression $f(x)$ has a meaning.

How to find the domain of a function - Some useful instructions:

- If presence of **denominators** $\frac{A}{\phi(x)}$: Find the roots of $\phi(x) = 0$ and **exclude** them from the domain.

- In presence of **radicals** $\sqrt[2\nu]{\pi(x)}$: Solve the inequality $\pi(x) \geq 0$ and accept the values that verify the inequality.

- In presence of **logarithm** $\log \phi(x)$ or $\ln \phi(x)$: Accept the values x such that $\phi(x) > 0$.

- In presence of the trigonometric functions **tangent** $\tan \phi(x)$ and **cotangent** $\cot \phi(x)$: Accept the values x such that $\phi(x) \neq k\pi + \dfrac{\pi}{2}$ or $\phi(x) \neq k\pi,\ k \in \mathbb{Z}$ respectively.

Skip any of the following examples that seem difficult for you.

Example (1). *Find the domain $\mathcal{A}$ of the function $f(x) = \dfrac{1}{x}$.*

It is

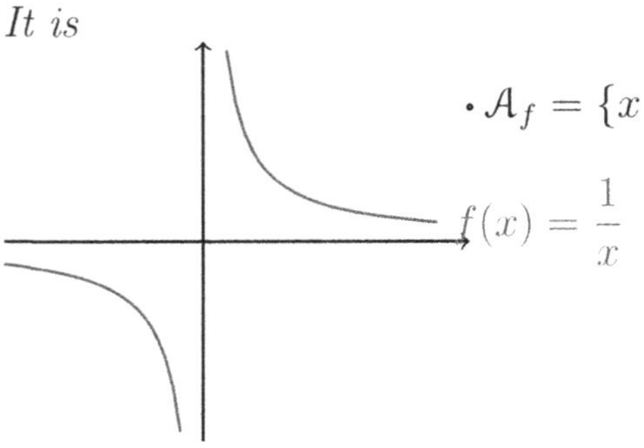

$\bullet\ \mathcal{A}_f = \{x : x \neq 0\}$

Example (2). *Find the domain of the function $g(x) = \dfrac{1}{x-1}$.*

It is

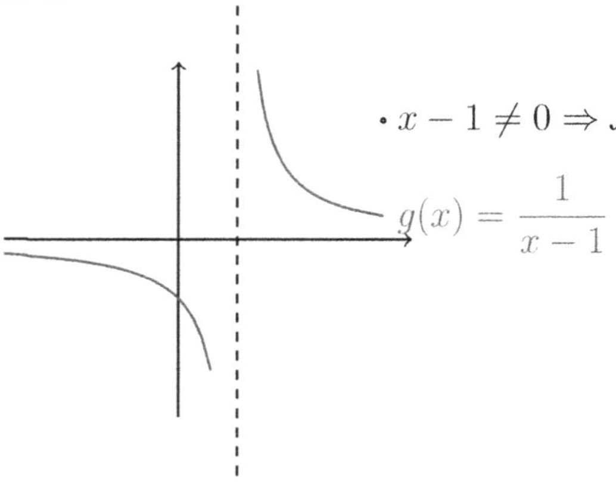

$\bullet\ x - 1 \neq 0 \Rightarrow \mathcal{A}_g = \{x : x \neq 1\}$

Example (3). *Find the domain of the function $h(x) = \sqrt{x-1}$.*

It is

$\bullet\ x - 1 \geq 0 \Rightarrow x \geq 1$

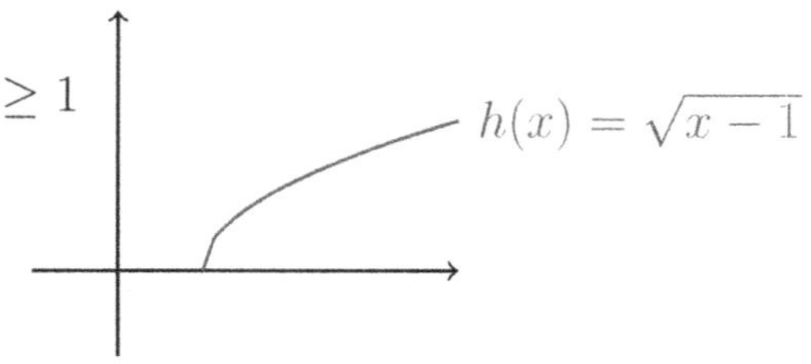

Example (4). *Find the domain $\mathcal{A}$ of the function $k(x) = \frac{1}{\sqrt{x-1}}$:*

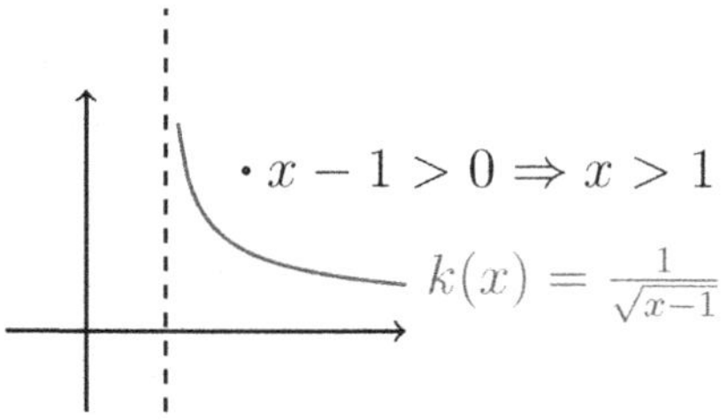

Few more examples:

Examples (5).

- *Find the domain $\mathcal{A}$ of the following function:*

$$l(x) = \sqrt{x^2 - x + 1},$$

$x^2 - x + 1 > 0 \Rightarrow \mathcal{A}_l = \mathbb{R}\cdot$

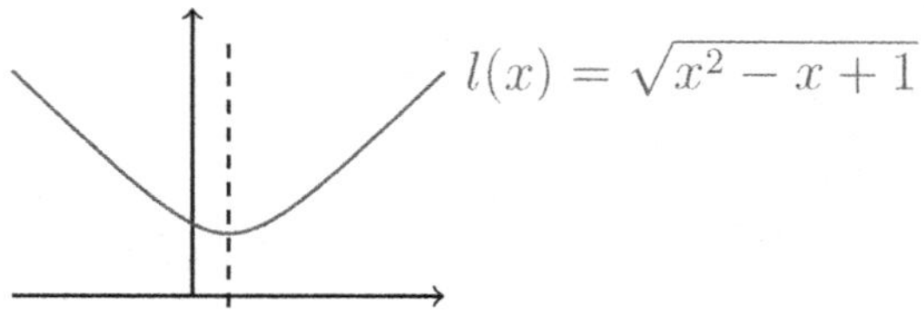

- *Find the domain of the function*

$$m(x) = \frac{1}{\sqrt{x^2 + x - 1}}$$

and solve the equation

$$m(x) = 1$$

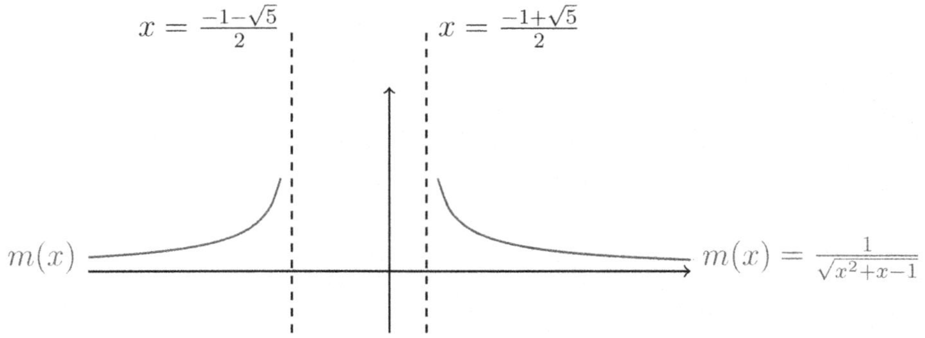

It is

$$x^2 + x - 1 > 0$$

where $\Delta = 5$ and the roots are

$$x_1 = \frac{-1-\sqrt{5}}{2}, \; x_2 = \frac{-1+\sqrt{5}}{2}$$

hence

$$x < \frac{-1-\sqrt{5}}{2} \text{ or } x > \frac{-1+\sqrt{5}}{2}$$

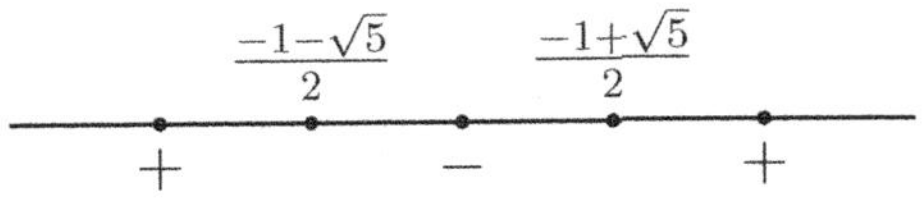

For the equation

$$m(x) = 1 \Rightarrow \frac{1}{\sqrt{x^2+x-1}} = 1 \Rightarrow x^2+x-1 = 1 \Rightarrow x^2+x-2 = 0$$

with roots

$$x_1 = -2 < \frac{-1-\sqrt{5}}{2}$$

***accepted** and*

$$x_2 = 1$$

with $0 < x_2 = 1 < \frac{-1+\sqrt{5}}{2}$ ***rejected** since it does not belong to* $\mathcal{A}_m$.

- *Find the domain of the function*

$$f(x) = \sqrt{1-\frac{1}{x}}$$

and solve the equation

$$f(x) = 1$$

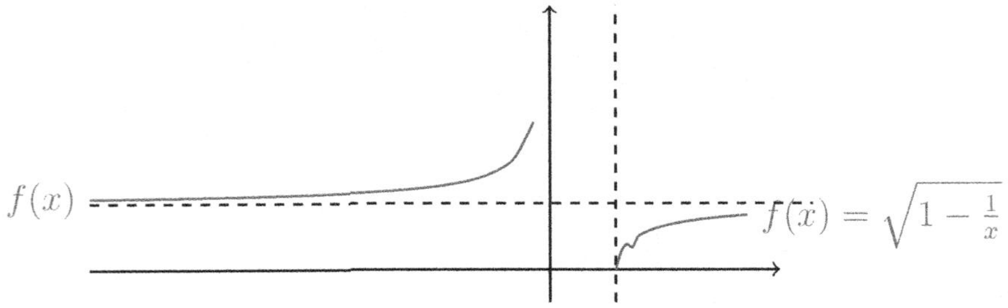

For the domain of f *:*

$$x: \ x \neq 0 \text{ and } \frac{x-1}{x} \geq 0 \Rightarrow \ x(x-1) \geq 0$$

hence $x < 0$ or $x \geq 1$.
To solve the equation

$$f(x) = 1 \Rightarrow \frac{x-1}{x} = 1 \Rightarrow x-1 = x \Rightarrow -1 = 0 \text{ impossible}$$

- *Find the domain of the function*

$$g(x) = \frac{2x+1}{x-1}$$

and solve the equation

$$g(x) = x$$

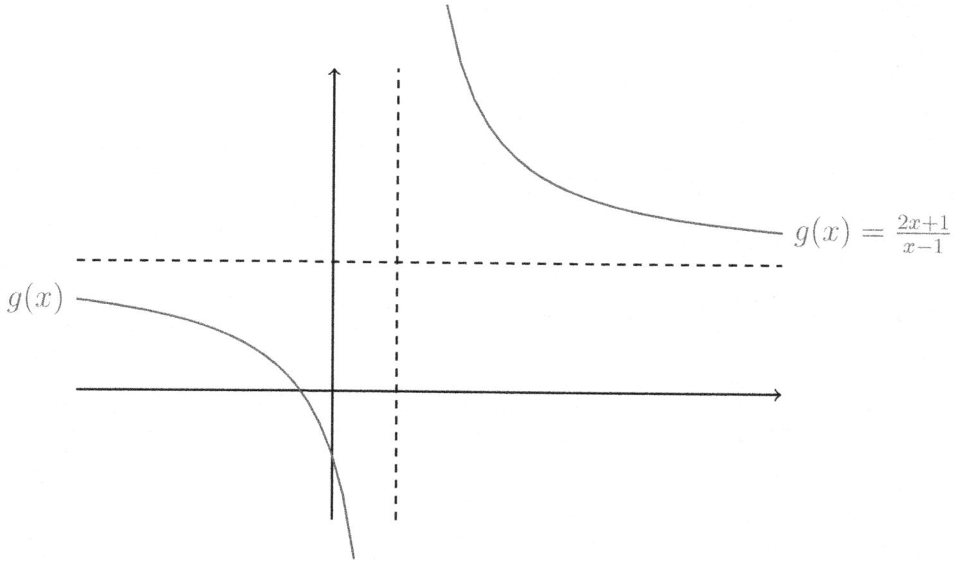

It is

$$x \neq 1$$

To solve the equation

$$g(x) = x \Rightarrow \frac{2x+1}{x-1} = x \Rightarrow x^2 - x = 2x + 1 \Rightarrow$$

$$x^2 - 3x - 1 = 0 \overset{\Delta = 3^2 - 4\cdot(-1) = 13}{\Rightarrow} x_{1,2} = \frac{3 \pm \sqrt{13}}{2}$$

- *Find the domain of the function*

$$h(x) = \ln\{f(x)\}$$

where

$$f(x) = \frac{x}{x+1}$$

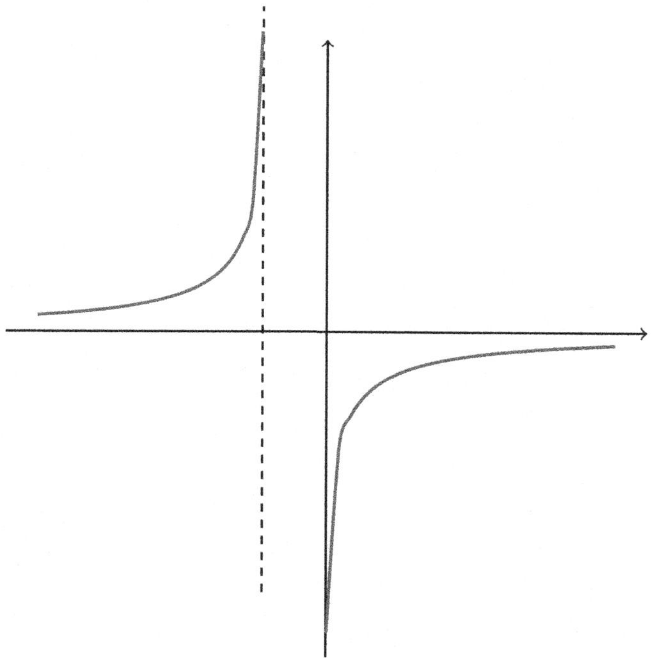

Domain of h :

$$x : \quad \frac{x}{x+1} > 0$$

hence

$$x < -1 \text{ or } x > 0$$

Next we deal with some problems

Problem (1). *The domain of the function*

$$f(x) = \frac{1}{(x-1)\sqrt{9-x^2}}$$

is

- *[A]* $-3 < x < 1$
- *[B]* $-3 < x < 3$ *and* $x \neq 1$
- *[C]* $x < -3$ *or* $x > 3$
- *[D]* $-3 < x < 3$
- *[E]* $1 < x < 3$

Solution: Domain $\mathcal{A}_f$ is determined by

$$x: \ x \neq 1 \text{ and } -3 < x < 3,$$

the correct answer is $[B]$.

Problem (2). *The domain of the function*

$$f(x) = \sqrt{1 - \frac{1}{x}}$$

is

- *[A]* $x < 0$ *or* $x \geq 1$
- *[B]* $x < 0$
- *[C]* $0 < x < 1$
- *[D]* $x > 1$
- *[E]* *none of the above*

Solution: It is

$$x: \ 1 - \frac{1}{x} \geq 0 \text{ and } x \neq 0 \Rightarrow \ x < 0 \text{ or } x \geq 1,$$

the correct answer is $[A]$.

Problem (3). *The domain of the function*

$$f(x) = \frac{1}{x - \sqrt{x^2}}$$

is

- *[A]* $\mathbb{R}$
- *[B]* $x \neq 0$
- *[C]* $-\infty < x < 0$
- *[D]* $x > 0$
- *[E]* *none of the above*

Solution: If

$$x \geq 0 \Rightarrow x - \sqrt{x^2} = x - |x| = 0, \text{ rejected}$$

hence $x < 0$ and the correct answer is $[C]$.

Problem (4). *The domain of the function*

$$f(x) = \sqrt{2 - \sqrt{x}}$$

is

- *[A]* $x \leq 4$
- *[B]* $x < 4$
- *[C]* $0 < x < 4$
- *[D]* $0 \leq x \leq 4$
- *[E]* *none of the above*

Solution: It is

$$2 - \sqrt{x} \geq 0 \text{ and } x \geq 0 \Rightarrow 4 \geq x \geq 0$$

and the correct answer is $[D]$.

Problem (5). *The domain of the function*

$$f(x) = \frac{2}{x} - \sqrt{4 - x^2}$$

is

- *[A]* $x \leq 2$
- *[B]* $x \geq -2$
- *[C]* $-2 \leq x \leq 2$

- *[D]* $-2 \leq x \leq 2$ *and* $x \neq 0$
- *[E]* *none of the above*

Solution: It is

$$x: \; x \neq 0 \text{ and } 4 \geq x^2 \Rightarrow \; x \neq 0 \text{ and } 2 \geq |x| \Rightarrow x \neq 0 \text{ and } -2 \leq x \leq 2$$

and the correct answer is $[D]$.

Problem (6). *The domain of the function*

$$f(x) = \frac{\sqrt{x+3}-2}{x^2-1}$$

is

- *[A]* $x \geq -3$
- *[B]* $x \neq \pm 1$
- *[C]* $x < -1$ *or* $x > 1$
- *[D]* $x \neq \pm 1$ *and* $x \geq -3$
- *[E]* *none of the above*

Solution: First, observe that

$$x^2 - 1 \neq 0 \Leftrightarrow x \neq \pm 1$$

Next, we should have

$$x + 3 \geq 0 \Leftrightarrow x \geq -3$$

hence

$$x \geq -3 \text{ and } x \neq \pm 1$$

and the correct answer is $[D]$.

Problem (7). *The domain of the function*

$$f(x) = \frac{x^2-5x+6}{x^2-6x+8}$$

is

- *[A]* $x \neq 2$
- *[B]* $x \neq 4$
- *[C]* $x \geq 4$
- *[D]* $x \leq 2$
- *[E]* $x \neq 2$ *and* $x \neq 4$

Solution: Observe that

$$x^2 - 6x + 8 \neq 0 \Leftrightarrow (x-2)(x-4) \neq 0 \Leftrightarrow x \neq 2,\ x \neq 4$$

hence, the correct answer is $[E]$.

Problem (8). *The domain of the function*

$$f(x) = \ln \frac{x}{\sqrt{x^2 - 1}}$$

is

- *[A]* $x \neq \pm 1$
- *[B]* $x > 1$
- *[C]* $x < -1$
- *[D]* $x > 0$ *and* $x \neq 1$
- *[E]* *none of the above*

Solution: It is

$$x^2 - 1 > 0 \Leftrightarrow \sqrt{x^2} > 1 \Leftrightarrow |x| > 1 \Leftrightarrow x < -1 \text{ and } x > 1 \quad \text{(Denominator)}$$

and

$$\frac{x}{\sqrt{x^2 - 1}} > 0 \Leftrightarrow x > 0 \quad \text{(Logarithm)}$$

since $\sqrt{x^2 - 1} \geq 0$ whenever it is defined, hence

$$x : \ x^2 - 1 > 0 \text{ and } x > 0 \Rightarrow \ |x| > 1 \text{ and } x > 0 \Rightarrow x : \ x > 1$$

Correct answer is $[B]$.

Problem (9). *The domain of the function*

$$f(x) = \sqrt{1 - \ln\left(\frac{x-1}{x+1}\right)}$$

is

- *[A]* $x > \dfrac{1-e}{1+e}$
- *[B]* $x > \dfrac{1+e}{1-e}$
- *[C]* $x < \dfrac{1+e}{1-e}$
- *[D]* $x < \dfrac{1+e}{1-e}$ *or* $x > 1$
- *[E]* *none of the above*

Solution: Starting with

$$\frac{x-1}{x+1} > 0 \Leftrightarrow x < -1 \text{ or } x > 1 \text{ (Logarithm)}, \text{ (1)}$$

and

$$x+1 \neq 0 \Leftrightarrow x \neq -1 \quad \text{(Denominator)},$$

it remains to check

$$1 - \ln\left(\frac{x-1}{x+1}\right) \geq 0 \Leftrightarrow 1 \geq \ln\left(\frac{x-1}{x+1}\right) \Leftrightarrow e \geq \frac{x-1}{x+1} \Leftrightarrow$$

$$\frac{x(e-1)+1+e}{x+1} \geq 0 \Rightarrow x \leq \frac{1+e}{1-e} \text{ or } x \geq -1 \qquad (2)$$

From (1) and (2) we conclude that

$$x < \frac{1+e}{1-e} \text{ or } x > 1,$$

hence the correct answer is $[D]$.

5.6.2 How to find limits - 2 mnemonic rules

- **The easy case:** To calculate a limit of the form

$$\lim_{x \to x_{x_0}} f(x),$$

if the formula of $f(x)$ contains known functions as polynomial, exponential, logarithmic, trigonometric etc., since they all are continuous, we substitute in place of the variable x the value x_0 to obtain

$$\lim_{x \to x_{x_0}} f(x) = f(x_0).$$

- **Rule** 1: For expressions of the form

$$\frac{0}{0}, \frac{\alpha}{0}$$

Factorize & Simplify

Example (1). ***(i)***

$$\lim_{x \to 3} \frac{x^2-9}{x-3} = \frac{3^2-9}{3-3} = \frac{0}{0} \overset{??}{=} \text{undefined}$$

Now, ***factorize*** *and* ***simplify***

$$\lim_{x \to 3} \frac{x^2-9}{x-3} = \lim_{x \to 3} \frac{(x-3)(x+3)}{x-3} = \lim_{x \to 3} x+3 = 3+3 = 6.$$

(ii)

$$\lim_{x \to 2} \frac{x^2-5x+6}{x^2-4} = \frac{0}{0} = \overset{\text{factorize}}{\lim} {}_{x \to 2} \frac{(x-3)(x-2)}{(x-2)(x+2)} \overset{\text{simplify}}{=} \lim_{x \to 2} \frac{x-3}{x+2} = \frac{-1}{4}$$

(iii)

$$\lim_{x\to 1}\frac{x^2-1}{(x-1)^2}=\frac{0}{0}\overset{\text{undefined}}{=}\lim_{x\to 1}\frac{(x-1)(x+1)}{(x-1)^2}\overset{simplify}{=}\lim_{x\to 1}\frac{x+1}{x-1}=\frac{2}{0}$$

Next, we examine two very simple cases:
Case 1: *If* $x\to 1$ *and* $x<1$ *then* $x-1<0$ *(i.e: x approaches 1 from the left) we get:*

$$\lim_{x\to 1^-}\frac{x^2-1}{(x-1)^2}=\lim_{x\to 1^-}\frac{2}{0^-}=-\infty$$

Case 2: *If* $x\to 1$ *and* $x>1$ *then* $x-1>0$ *(i.e: x approaches 1 from the right) we get:*

$$\lim_{x\to 1^+}\frac{x^2-1}{(x-1)^2}=\lim_{x\to 1^+}\frac{2}{0^+}=+\infty$$

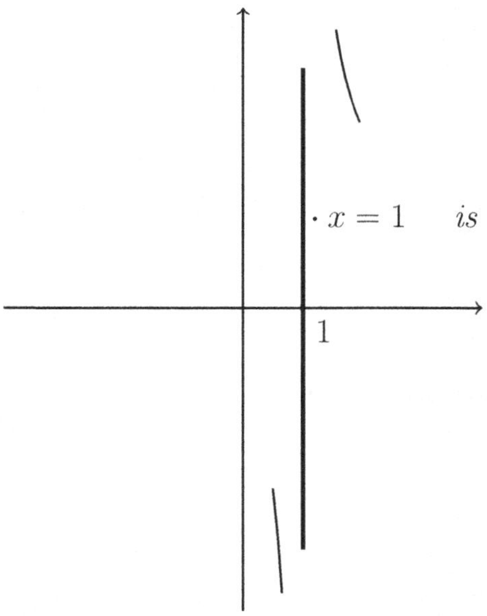

$\cdot\, x=1$ *is called a **vertical asymptote***

(iv) *Find quickly the limit and the asymptote*

$$\lim_{x\to 3}\frac{x+2}{x-3}=\frac{5}{0},$$

We examine the following cases
Case 1: *If* $x<3$, $x-3<0$ *then*

$$\lim_{x\to 3^-}\frac{x+2}{x-3}=\frac{5}{0^-}=-\infty$$

Case 2: *If* $x>3$, $x-3>0$ *then*

$$\lim_{x\to 3^+}\frac{x+2}{x-3}=\frac{5}{0^+}=+\infty$$

So, $x=3$ is a vertical asymptote

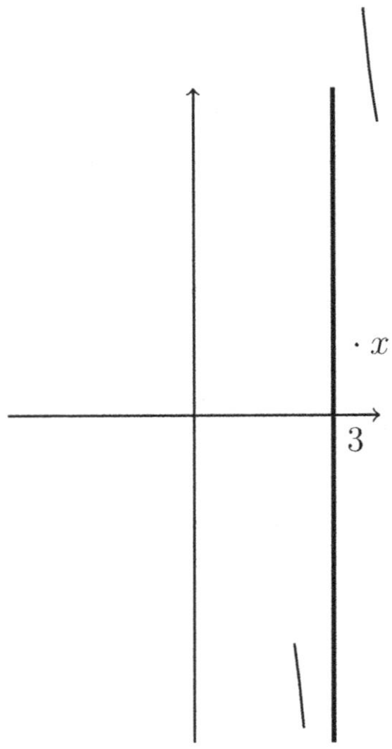

$\cdot\, x = 3$ *is called a* ***vertical asymptote***

- **Rule** 2: For expressions of the form

$$\frac{\infty}{\infty}$$

we provide the following

Example (**2**). ***(i)*** *Find the limit*

$$\lim_{x\to+\infty}\frac{2x^2+x-1}{3x^2-x-1}=\lim_{x\to+\infty}\frac{2x^2}{3x^2}=\frac{2}{3},\ y=\frac{2}{3}\ \text{horizontal aymptote}$$

(ii) *Find* $\lim_{x\to+\infty}\frac{3x^5+x-1}{2x^3-x+4}$. *It is:*

$$\lim_{x\to+\infty}\frac{3x^5+x-1}{2x^3-x+4}=\lim_{x\to+\infty}\frac{3x^5}{2x^3}=\lim_{x\to+\infty}\frac{2}{3}x^2=+\infty.$$

(iii) *Find* $\lim_{x\to+\infty}\frac{x^2+x+5}{4x^4-x+1}$. *It is:*

$$\lim_{x\to+\infty}\frac{x^2+x+5}{4x^4-x+1}=\lim_{x\to+\infty}\frac{x^2}{4x^4}=\lim_{x\to+\infty}\frac{1}{4x^2}=0,\ y=0\ \text{horizontal asymptote.}$$

A simple Rule: *If* $\alpha\neq 0$*:*

$$\frac{\alpha}{\pm\infty}=0$$

and since

$$\frac{1}{\pm\infty}=0$$

then, obviously, the

$$\frac{0}{\pm\infty}=0$$

holds true.

Remark. *Summarizing: if*

$$f(x) = \frac{P(x)}{Q(x)}$$

where $P(x) = a_n x^n + a_{n-1}x^{n-1} + \cdots + a_1 x + a_0$, $Q(x) = b_m x^m + b_{m-1}x^{m-1} + \cdots b_1 x + b_0$ *are polynomial functions then:*

$$\lim_{x\to\pm\infty} f(x) = \lim_{x\to\pm\infty} \frac{P(x)}{Q(x)} = \left\{ \begin{array}{lll} \frac{a_m}{b_m}, & \text{if} & n = m \\ \pm\infty, & \text{if} & n > m \\ 0, & \text{if} & n < m \end{array} \right\}$$

Example (3).

$$\lim_{x\to+\infty} \frac{5x^5+\cdots}{6x^5+\cdots} = \frac{5}{6},$$
$$\lim_{x\to+\infty} \frac{5x^5+\cdots}{6x^4+\cdots} = \lim_{x\to+\infty} \frac{5x}{6} = +\infty,$$
$$\lim_{x\to+\infty} \frac{-5x^5+\cdots}{6x^4+\cdots} = \lim_{x\to+\infty} \frac{-5}{6}x = -\infty, \lim_{x\to-\infty} \frac{4x^5+\cdots}{7x^6+\cdots} = \lim_{x\to-\infty} \frac{4}{7x} = 0,$$
$$\lim_{x\to-\infty} \frac{10x^4+\cdots}{20x^3+\cdots} = \lim_{x\to-\infty} \frac{x}{2} = -\infty$$

Problem (1). *The limits*

$$\lim_{x\to+\infty} \frac{x^2-5x+6}{x^2-6x+8} \quad \textit{(a)}$$

and

$$\lim_{x\to 2} \frac{x^2-5x+6}{x^2-6x+8} \quad \textit{(b)}$$

are

- *[A]* 1 *and* 1
- *[B]* $+\infty$ *and* $\frac{1}{2}$
- *[C]* 1 *and* $-\frac{1}{2}$
- *[D]* 1 *and* $\frac{1}{2}$
- *[E]* 1 *and* -1

respectively

Solution: First, it is

$$\lim_{x\to+\infty} \frac{x^2-5x+6}{x^2-6x+8} = \lim_{x\to+\infty} \frac{x^2}{x^2} = 1$$

Secondly, it is

$$\lim_{x\to 2}\frac{x^2-5x+6}{x^2-6x+8}=\frac{0}{0}\overset{\text{Factorize}}{\Rightarrow}$$
$$\lim_{x\to 2}\frac{x^2-5x+6}{x^2-6x+8}=\lim_{x\to 2}\frac{(x-2)(x-3)}{(x-2)(x-4)}$$
$$=\lim_{x\to 2}\frac{x-3}{x-4}=\frac{1}{2}$$

Correct answer is $[D]$.

Problem (2). *The limit*

$$\lim_{x\to 1}\frac{\sqrt{x+3}-2}{x^2-1}$$

is

- *[A]* $\frac{-1}{2}$
- *[B]* $\frac{-1}{4}$
- *[C]* $\frac{1}{8}$
- *[D]* $\frac{1}{4}$
- *[E]* *none of the above*

Solution: It is

$$\lim_{x\to 1}\frac{\sqrt{x+3}-2}{x^2-1}=\frac{\sqrt{1+3}-2}{1-1}=\frac{0}{0}$$

Apply special technique:
Multiply both, numerator and denominator by the conjugate $\sqrt{x+3}+2$ to get:

$$\lim_{x\to 1}\frac{\sqrt{x+3}-2}{x^2-1}=\lim_{x\to 1}\frac{\left(\sqrt{x+3}-2\right)\left(\sqrt{x+3}+2\right)}{(x^2-1)\left(\sqrt{x+3}+2\right)}$$
$$=\lim_{x\to 1}\frac{(\sqrt{x+3})^2-2^2}{(x-1)(x+1)(\sqrt{x+3}+2)}$$
$$=\lim_{x\to 1}\frac{(x+3)-4}{(x-1)(x+1)(\sqrt{x+3}+2)}$$
$$=\lim_{x\to 1}\frac{x-1}{(x-1)(x+1)(\sqrt{x+3}+2)}$$

and simplifying $x-1$ on denominator and numerator we get

$$\lim_{x\to 1}\frac{\sqrt{x+3}-2}{x^2-1}=\lim_{x\to 1}\frac{1}{(x+1)(\sqrt{x+3}+2)}=\frac{1}{8}$$

Correct answer is $[C]$.

Problem (3). *The limit*

$$\lim_{x\to 1}\frac{\sqrt{x}-1}{x-1}$$

is

- *[A]* -1
- *[B]* $\frac{-1}{4}$
- *[C]* $-\frac{1}{2}$
- *[D]* $\frac{-1}{8}$
- *[E]* $\frac{1}{2}$

Solution: It is, by substitution:

$$\lim_{x\to 1}\frac{\sqrt{x}-1}{x-1}=\frac{0}{0}\ \text{Indefinite form}$$

Use factorization:

$$\begin{aligned}\lim_{x\to 1}\frac{\sqrt{x}-1}{x-1} &= \lim_{x\to 1}\frac{\sqrt{x}-1}{(\sqrt{x}-1)(\sqrt{x}+1)}\\ &= \lim_{x\to 1}\frac{1}{\sqrt{x}+1}=\frac{1}{2}\end{aligned}$$

Correct answer is $[E]$.

Problem (4). *If*

$$\lim_{x\to x_0} f(x)=\alpha,\ \text{, and}\ \lim_{x\to x_0}=\beta,$$

then:

$$\begin{aligned}\lim_{x\to x_0} 3f(x) &= \ldots\ldots\\ \lim_{x\to x_0}\{3f(x)+2g(x)\} &= \ldots\ldots\\ \lim_{x\to x_0}\frac{3f(x)+2g(x)}{f(x)+g(x)} &= \ldots\ldots\ \text{if}\ \alpha+\beta\neq 0\\ \lim_{x\to x_0}\frac{(f(x))^2+64}{f(x)+8} &= \ldots\ldots\ \text{if}\ \alpha+8\neq 0\end{aligned}$$

Solution: *It is*

$$\begin{aligned}\lim_{x\to x_0} 3f(x) &= 3\lim_{x\to x_0} f(x)=3\alpha,\\ \lim_{x\to x_0}\{3f(x)+2g(x)\} &= \lim_{x\to x_0}\{3f(x)\}+\lim_{x\to x_0}\{2g(x)\}\\ &= 3\lim_{x\to x_0} f(x)+2\lim_{x\to x_0} g(x)\\ &= 3\alpha+2\beta\end{aligned}$$

and similarly

$$\begin{aligned}\lim_{x\to x_0}\frac{3f(x)+2g(x)}{f(x)+g(x)} &= \frac{\lim_{x\to x_0}\{3f(x)+2g(x)\}}{\lim_{x\to x_0}\{f(x)+g(x)\}}\\ &= \frac{3\alpha+2\beta}{\alpha+\beta},\ \text{for } \alpha+\beta\neq 0\\ \lim_{x\to x_0}\frac{(f(x))^2+64}{f(x)+8} &= \frac{\lim_{x\to x_0}\{(f(x))^2+64\}}{\lim_{x\to x_0}\{f(x)+8\}}\\ &= \frac{\lim_{x\to x_0}(f(x))^2+64}{\lim_{x\to x_0}f(x)+8}\\ &= \frac{(\lim_{x\to x_0}f(x))^2+64}{\lim_{x\to x_0}f(x)+8} = \frac{\alpha^2+64}{\alpha+8},\quad \text{for } \alpha+8\neq 0\end{aligned}$$

Problem (5). *If*

$$f(x) = \left\{\begin{array}{ll}\dfrac{x^2-1}{x+1}, & x\neq -1\\ a-2, & x=-1\end{array}\right\}$$

and f is continuous then $a\in\mathbb{R}$ is equal to

- *[A]* -2
- *[B]* 0
- *[C]* 2
- *[D]* 1
- *[E]* 4

Solution: **Step 1:** By definition, f is continuous at a point x_0 in its domain if and only if

$$\lim_{x\to x_0} f(x) = f(x_0)$$

Step 2: For $x_0\neq -1$ it is

$$\lim_{x\to x_0} f(x) = \frac{x_0^2-1}{x_0+1} = x_0-1 = f(x_0)$$

hence f is continuous at every point $x_0\neq -1$.
Step 3: It remains to check under which restrictions, on the value of the parameter a, f is continuous at $x_0=-1$. It is:

$$\begin{gathered}\lim_{x\to -1} f(x) = f(-1) \Leftrightarrow \lim_{x\to -1}\frac{x^2-1}{x+1} = a-2 \overset{\text{factorize}}{\Longrightarrow}\\ \lim_{x\to -1}\frac{(x-1)(x+1)}{x+1} = a-2 \Rightarrow \lim_{x\to -1}(x-1) = a-2 \Rightarrow\\ -2 = a-2 \Rightarrow a=0\end{gathered}$$

and the correct answer is $[B]$.

Problem (6). *Find the values of a, b for which the function*

$$f(x) = \left\{ \begin{array}{ll} 2ax - b, & x \neq 2 \\ b + 2, & x = 2 \end{array} \right\}$$

is continuous and $f(3) = 3$.

Solution:

Step 1 : From $f(3) = 3$ we get

$$6a - b = 3 \ \textbf{(i)}$$

Step 2: Continuity at the point $x_0 = 2$:

$$\lim_{x \to 2} f(x) = f(2) \Leftrightarrow 4a - 2b = 2 \ \textbf{(ii)}$$

we thus reduce to the system

$$\left\{ \begin{array}{rr} 6a - b = & 3 \\ 4a - 2b = & 2 \end{array} \right\}$$

hence

$$a = \frac{1}{2}, \ b = 0.$$

Problem (7). *If*

$$\lim_{x \to x_0} f(x) = 5 \ \text{and} \ \lim_{x \to x_0} \{2f(x) + 3g(x)\} = 19$$

then

$$\lim_{x \to x_0} g(x)$$

is equal to

- *[A]* -3
- *[B]* 1
- *[C]* 2
- *[D]* 3
- *[E]* 4

Solution: It is

$$\lim_{x \to x_0} \{2f(x) + 3g(x)\} = 19 \Rightarrow$$

$$\lim_{x \to x_0} 2f(x) + \lim_{x \to x_0} 3g(x) = 19$$

$$3 \lim_{x \to x_0} f(x) + 2 \lim_{x \to x_0} g(x) = 19 \overset{\lim_{x \to x_0} f(x) = 5}{\Rightarrow}$$

$$2 \cdot 5 + 3 \lim_{x \to x_0} g(x) = 19 \Rightarrow 3 \lim_{x \to x_0} g(x) = 19 - 10 = 9$$

hence

$$\lim_{x \to x_0} g(x) = 3$$

and the correct answer is $[D]$.

Problem (8). *Let*

$$f(x) = \frac{4x^2 - 100}{x - 5},$$

then $\lim_{x \to 5} f(x)$ *is*

- *[A]* 10
- *[B]* 20
- *[C]* 30
- *[D]* 40
- *[E]* 50

Solution: It is

$$\lim_{x \to 5} \frac{4x^2 - 100}{x - 5} \overset{\frac{0}{0} - \text{factorize}}{=} \lim_{x \to 5} \frac{4(x^2 - 5^2)}{x - 5} = 4 \lim_{x \to 5} \frac{(x-5)(x+5)}{x - 5}$$

so

$$\lim_{x \to 5} \frac{4x^2 - 100}{x - 5} = 4 \lim_{x \to 5} (x + 5) = 40$$

and the correct answer is $[D]$.

5.7 Exponential & Logarithmic function

Example (1). *If* $f(x) = \ln x$ *then*

$$y = \ln x \Rightarrow x = e^y$$

hence

$$y = e^x$$

is the inverse of the ln *function.*
Do not forget the mnemonic rule:

$$\log_a b = c \Leftrightarrow a^c = b,$$
$$\ln_e b = c \Leftrightarrow e^c = b$$

5.7.1 Summary

$$a^x a^y = a^{x+y}, \;\; a^0 = 1$$
$$\frac{a^x}{a^y} = a^{x-y}, \;\; a^{-x} = \frac{1}{a^x}$$
$$(a^x)^y = a^{xy}, \;\; a^x b^x = (ab)^x$$

for any $a, b > 0$.

Mnemonic Rule Example: Find x if $x^{0.3} = 4$ then: use calculator to compute $x = 4^{\frac{1}{0.3}}$.

5.7.2 Logarithms

Mnemonic rule for $\boxed{\log}$: If $\log_b a = c$ then $b^c = a$.

Example (2).

- $\log_a a = 1$ *because* $a^1 = a$
- $\log_3 a = 4 \to 3^4 = a \to a = 81$
- $\log_x 8 = 3 \to x^3 = 8 \to x = 8^{\frac{1}{3}} = 2$

Note: *If no base is indicated then base* $= 10$.

Properties:

- [1] $\log_c(ab) = \log_c a + \log_c b$
- **Example:** Given that $\log_{10} 5 = 0.7$ find $\log_{10} 50$.
- **Solution:**

$$\begin{aligned}\log_{10} 50 = \log_{10}(5 \times 10) &= \log_{10} 5 + \log_{10} 10\\ &= 0,7 + 1 = 1,7.\end{aligned}$$

- [2] $\log_c(\frac{a}{b}) = \log_c a - \log_c b$
- **Example:** Given that $\log_{10} 3 = 0.5$ find $\log_{10} 0.03$.
- **Solution:**

$$\begin{aligned}\log_{10} 0.03 = \log_{10} \frac{3}{100} &= \log_{10} 3 - \log_{10} 100\\ &= 0.5 - \log_{10} 10^2 = 0.5 - 2\log_{10} 10 = 0.5 - 2 = -1.5.\end{aligned}$$

- [3] $\log_c(a^x) = x \log_c a$
- **Example:**

$$\log_{10} \sqrt[3]{5} = \frac{1}{3} \log_{10} 5 = \frac{1}{3} 0.7$$

Do not forget:

$$\begin{aligned}\log_c 1 &= 0\\ \log_c c &= 1\\ \log_c c^p &= p\\ c^{\log_c p} &= p\end{aligned}$$

- $\log_{10} 1000 = \log_{10} 10^3 = 3$
- $\log_3 3^5 = 5$
- $10^{\log_{10} 5} = 5$

Change of base

$$\log_b a = \frac{\log_c a}{\log_c b}$$

- **Example** : $\log_3 5 = ?$ It is

$$\log_3 5 = \frac{\log_{10} 5}{\log_{10} 3}$$

5.7.3 Problem Solving now

Problem (1). *If $f(x) = \ln ex^2$ and $g(x) = e^x$ then $f(g(x))$ is*

- *[A]* $2x + 1$
- *[B]* $2x + e$
- *[C]* $2x^2 + x$
- *[D]* $x^2 + 2x + 1$
- *[E]* *none of the above*

Solution: It is

$$f(g(x)) = f(e^x) = \ln e(e^x)^2 = \ln ee^{2x} = \ln e^{2x+1} = 2x + 1.$$

Correct answer is [*A*].

Problem (2). *If $f(x) = e^{2x}$ and $g(x) = 2\ln x$ then $f(g(x))$ is*

- *[A]* $2x$
- *[B]* $4x$
- *[C]* x^2
- *[D]* x^3
- *[E]* x^4

Solution:

$$f(g(x)) = f(2\ln x) = e^{2\cdot 2\ln x} = e^{4\cdot \ln x} = (e^{\ln x})^4 = x^4$$

Correct answer is [*E*].
Remark. *Remember the formulae*

$$e^{\ln f} = f, \ \ln e^f = f$$
$$\log 10^x = x, \ 10^{\log x} = x$$
$$\ln x = a \Rightarrow e^a = x$$
$$\log x = a \Rightarrow 10^a = x$$

Problem (3). *If* $f(x) = \log^2(2x+6)$ *and* $g(x) = e^{x^2} + 1$ *then* $f(g(0))$ *is*

- *[A]* 1
- *[B]* 2
- *[C]* 10
- *[D]* 20
- *[E]* *none of the above*

Solution: Since $\log x = \log_{10} x$ we get:

$$g(0) = 2 \Rightarrow f(g(0)) = f(2) = \log^2(2 \times 2 + 6) = \log^2 10 = 1$$

Correct answer is $[A]$.

Problem (4). *If* $f(x) = 10^x + 9$, $x > 0$ *then* $f^{-1}(109)$ *is*

- *[A]* -2
- *[B]* -1
- *[C]* 1
- *[D]* 2
- *[E]* 9

Solution: Solving $y = f(x)$ in x :

$$y = 10^x + 9 \Rightarrow 10^x = y - 9 \Rightarrow x = \log(y-9)$$

hence

$$f^{-1}(x) = \log(x-9) \Rightarrow f^{-1}(109) = \log 100 = 2$$

Correct answer is $[D]$.

Problem (5). *If* $f(x) = \log(3x+7)$ *then* $f^{-1}(2)$ *is*

- *[A]* 27
- *[B]* 30
- *[C]* 31
- *[D]* 32
- *[E]* 33

Solution:

- **Way** 1: Exact method - Standard technique: Interchange y and x:

$$y = \log(3x+7) \Rightarrow x = \log(3y+7) \Rightarrow 10^x = 3y+7 \Rightarrow y = \frac{10^x - 7}{3}$$

so $f^{-1}(x) = \frac{10^x - 7}{3}$ and consequently

$$f^{-1}(2) = \frac{10^2 - 7}{3} = \frac{93}{3} = 31.$$

- **Way** 2: Alternatively, we have

$$2 = \log(3x+7) \Rightarrow 10^2 = 3x+7 \Rightarrow 3x = 93 \Rightarrow x = 31.$$

Correct answer is $[C]$.

Problem (6). *If* $25^x = 16^{x+y}$ *then* $\frac{x}{y}$ *is equal to*

- *[A]* $\dfrac{\log 16}{\log 16 - \log 25}$
- *[B]* $\dfrac{\log 16 - \log 25}{\log 16}$
- *[C]* $\dfrac{\log 25 - \log 16}{\log 16}$
- *[D]* $\log 16 - \log 25$
- *[E]* $\dfrac{\log 16}{\log 25 - \log 16}$

Think for a while before reading on

Solution

Exponential expressions

Reminds us of logarithms

hence

$$\log 25^x = \log 16^{x+y} \Rightarrow x \log 25 = (x+y)\log 16 \Rightarrow$$
$$\log 25 = (1 + \frac{y}{x})\log 16 \Rightarrow \frac{y}{x} = \frac{\log 25 - \log 16}{\log 16}$$

Correct answer is $[C]$.

Problem (7). *For* $a > 0$ *the value of* $\log_{a^2} a$ *is equal to*

- *[A]* $\dfrac{1}{2}$
- *[B]* 1
- *[C]* 2
- *[D]* 3

- *[E]* *none of the above*

Think for a while before reading on
Solution: It is

$$\log_{a^2} a = x \Rightarrow (a^2)^x = a \Rightarrow a^{2x} = a^1 \Rightarrow 2x = 1 \Rightarrow x = \frac{1}{2}.$$

Correct answer is $[A]$.

Problem (8). *For* $a > 0$ *the value of* $\log_{a^3} \sqrt[3]{a^2}$ *is equal to*

- *[A]* -1
- *[B]* $\frac{-1}{9}$
- *[C]* $\frac{-2}{9}$
- *[D]* $\frac{2}{9}$
- *[E]* $\frac{1}{3}$

Think for a while before reading on
Solution: It is

$$\log_{a^3} \sqrt[3]{a^2} = x \Rightarrow a^{3x} = a^{\frac{2}{3}} \Rightarrow 3x = \frac{2}{3} \Rightarrow x = \frac{2}{9}$$

Correct answer is $[D]$.

Problem (9). *If* $\log_{10} a = 3$ *then* $\log_{10} 100 \cdot a^3$ *is*

- *[A]* 6
- *[B]* 8
- *[C]* 9
- *[D]* 10
- *[E]* 11

Solution-Procedure

$$\log_{10} a = 3 \Rightarrow a = 10^3$$

hence if we take into account the formula

$$\log_c c^p = p$$

we get

$$\log_{10} 100 \cdot a^3 = \log_{10}\{10^2 \cdot (10^3)^3\}$$
$$= \log_{10} 10^2 + \log_{10} 10^9 = 2 + 9 = 11$$

and the correct answer is $[E]$.

Problem (10). *If* $\log_5 a = 7$ *then* $\log_5 125a^3$ *is*

- *[A]* 6
- *[B]* 8
- *[C]* 12
- *[D]* 16
- *[E]* 24

Solution-Procedure

$$\log_5 a = 7 \Rightarrow a = 5^7$$

hence, similarly to the previous problem, we get:

$$\begin{aligned} \log_5 125a^3 &= \log_5\{5^3 \cdot (5^7)^3\} \\ &= \log_5 5^3 + \log_5 5^{21} \\ &= 3 + 21 = 24 \end{aligned}$$

and the correct answer is $[E]$.

A second way

Starting from the goal we have

$$\begin{aligned} \log_5 125a^3 &= \log_5 125 + \log_5 a^3 = \log_5 5^3 + 3 \cdot \log_5 a \\ &= 3 + 3 \cdot 7 = 3 + 21 = 24. \end{aligned}$$

Note:

It is often easier to give more emphasis to the goal, starting actually from the goal.

Problem (11). *If* $\log_b a = 3$ *then* $\log_b b^2 a^4$ *is*

- *[A]* 5
- *[B]* 6
- *[C]* 8
- *[D]* 10
- *[E]* 14

Solution-Procedure

Let us start from the goal and try to simplify it by using formulas

$$\begin{aligned} \log_b b^2 a^4 &= \log_b b^2 + \log_b a^4 \\ &= 2 + 4 \cdot \log_b a \\ &= 2 + 4 \cdot 3 = 14 \end{aligned}$$

and the correct answer is $[E]$.

Problem (12). *The value of*

$$\sum_{r=2}^{100} \log 5^r$$

is

- *[A]* $5000 \log 5$
- *[B]* $5049 \log 5$
- *[C]* $5050 \log 5$
- *[D]* $5060 \log 5$
- *[E]* $5100 \log 5$

Solution-Procedure

Let us try to simplify the goal:

$$\begin{aligned}\sum_{r=2}^{100} \log 5^r &= \log 5^2 + \log 5^3 + \log 5^4 + \cdots\cdots + \log 5^{100} \\ &= 2\log 5 + 3\log 5 + 4\log 5 + \cdots\cdots + 100\log 5 \\ &= [2 + 3 + 4 + \cdots\cdots + 100]\log 5\end{aligned}$$

where $2, 3, 4, \ldots, 100$ is an arithmetic sequence
with $a_1 = 2$ and $d = 1$,
therefore by using formula

$$S_n = \frac{n}{2}\{a_1 + a_n\}$$

we get

$$\sum_{r=2}^{100} r = \frac{99}{2}\{2 + 100\} = 99 \cdot 51 = 5049$$

hence

$$\sum_{r=2}^{100} \log 5^r = 5049 \cdot \log 5$$

and the correct answer is $[B]$.

Problem (13). *If* $\log_8 x^3 = \log_4 64$ *then* x *is*

- *[A]* 2
- *[B]* 4
- *[C]* 6
- *[D]* 8

- *[E]* 16

Solution-Procedure

What is the subtle point?
We see different bases here, 8 and 4
We also note that $8, 4, 64$ are all powers of 2:

$$8 = 2^3, \; 4 = 2^2, \; 64 = 2^6$$

and from another point of view

$$64 = 4^3$$

Let us try some of these:

$$\log_4 64 = \log_4 4^3 = 3$$

so

$$\begin{aligned} \log_8 x^3 &= \log_4 64 \Rightarrow \\ \log_8 x^3 &= 3 \Rightarrow \\ x^3 &= 8^3 \Rightarrow \\ x &= 8 \end{aligned}$$

and the correct answer is D.

Another approach

Let us try to *express both terms in the same base*, say with respect to base 2:

$$\log_8 x^3 = \log_4 64 \Leftrightarrow \frac{\log_2 x^3}{\log_2 8} = \frac{\log_2 64}{\log_2 4} \Leftrightarrow$$
$$\frac{\log_2 x^3}{\log_2 2^3} = \frac{\log_2 2^6}{\log_2 2^2} \Leftrightarrow \frac{3\log_2 x}{3\log_2 2} = \frac{6\log_2 2}{2\log_2 2} \Leftrightarrow$$
$$\log_2 x = 3 \Leftrightarrow x = 2^3 = 8.$$

Problem (14). *If*

$$\ln(x+1) + \ln(x+2) = \ln 2(x+1)^2$$

then x is equal to

- [A] -2 or -1
- [B] -1 or 0
- [C] 0 or 1
- [D] 1 or 2
- [E] 2 or *none of the above*

Solution-Procedure

- Recall that $\ln a + \ln b = \ln(a \cdot b)$ and $\ln x^k = k \ln x$, so

- from $\ln(x+1) + \ln(x+2) = \ln 2(x+1)^2$ we get

$$\ln\left[(x+1)\cdot(x+2)\right] = \ln 2(x+1)^2 \Rightarrow (x+1)\cdot(x+2) = 2(x+1)^2$$

hence

- $x^2 + 3x + 2 = 2x^2 + 4x + 2 \Rightarrow x^2 + x = 0 \Rightarrow x(x+1) = 0$
- therefore $x = 0$ or $x = -1$.
- **Attention!** $\ln(x+1)$ is defined if and only if

$$x + 1 > 0 \Leftrightarrow x > -1$$

so the solution $x = -1$ is rejected and we thus obtain that $x = 0$ is the only acceptable solution.

Correct answer is $[E]$.

Problem (15). *The sum*

$$\log\frac{1}{2} + \log\frac{2}{3} + \log\frac{3}{4} + \cdots\cdots + \log\frac{99}{100}$$

is equal to

- *[A]* -2
- *[B]* 1
- *[C]* 0
- *[D]* $\frac{1}{9}$
- *[E]* 3

Solution-procedure:

Use property

$$\log x + \log y = \log(x\cdot y),\ \ x, y > 0$$

to derive

$$\begin{aligned}\log\frac{1}{2} + \log\frac{2}{3} + \log\frac{3}{4} + \cdots\cdots + \log\frac{99}{100} &= \log\left(\frac{1\times 2\times 3\times\cdots\times 99}{1\times 2\times 3\times\cdots\times 99\times 100}\right)\\ &= \log\frac{1}{100} = \log 10^{-2}\\ &= -2\end{aligned}$$

because

$$\log x^k = k\log x,\ \ x > 0$$

hence, we deduce that the correct answer is $[A]$

Problem (16). *If* $\log_8 3 = k$ *then* $\log_8 18$ *in terms of* k *is given by:*

- *[A]* $2k - \frac{1}{9}$
- *[B]* $k + \frac{1}{9}$
- *[C]* $2k + \frac{1}{3}$
- *[D]* $2k + \frac{1}{9}$
- *[E]* $-2k - \frac{1}{9}$

Solution-procedure:

It is

$$\log_8 3 = k \Leftrightarrow 8^k = 3$$

and $18 = 3 \times 3 \times 2$ hence

$$\begin{aligned}\log_8 18 &= \log_8(3^2 \times 2)\\ &= \log_8 3^2 + \log_8 2\\ &= 2\log_8 3 + \log_8 2\\ &= 2k + \log_8 2\\ &= 2k + \log_8 8^{\frac{1}{3}}\\ &= 2k + \frac{1}{3}\end{aligned}$$

and the correct answer is $[C]$.

Problem (17). *The points $A(x_1, y_1)$, $B(x_2, y_2)$ lie on the graph of the function*

$$y = \log_a x$$

Through the midpoint of AB a horizontal line is drawn to cut the curve $y = \log_a x$ at the point $C(x_3, y_3)$. Then the x_i's are related by

- *[A]* $x_1^2 = x_2 \cdot x_3$
- *[B]* $x_1 = x_2 = x_3^2$
- *[C]* $x_1 \cdot x_2 = x_3^3$
- *[D]* $x_2^2 = x_1 \cdot x_3$
- *[E]* $x_3^2 = x_1 \cdot x_2$.

Solution-procedure:

The midpoint $M(x_\mu, y_\mu)$ of AB is

$$(\frac{x_1+x_2}{2}, \frac{y_1+y_2}{2})$$

an since the line ϵ_M drawn through M is horizontal its equation is

$$\epsilon_M : \; y = \frac{y_1+y_2}{2}$$

hence its intersection with the curve $y = \log_a x$ is given by

$$\begin{aligned}\log_a x &= \frac{y_1+y_2}{2} = \frac{1}{2} log_a x_1 + \frac{1}{2}\log_a x_2 \\ &= \frac{1}{2}\log_a(x_1 \cdot x_2) \\ &= \log_a (x_1 \cdot x_2)^{\frac{1}{2}}\end{aligned}$$

and thus

$$x = x_3 = (x_1 \cdot x_2)^{\frac{1}{2}} \Leftrightarrow x_3^2 = x_1 \cdot x_2$$

and the correct answer is $[E]$.

5.8 Parametric Equations

Example (1). *If $y^2 = 3x+1$ and $y = 2t+3$ find $x(t)$.*

Solution: It is:

$$\begin{aligned}y^2 = 3x+1 &\overset{y(t)=2t+3}{\Rightarrow} (2t+3)^2 = 3x+1 \Rightarrow \\ x(t) &= \frac{1}{3}\left\{4t^2+12t+8\right\}\end{aligned}$$

Example (2). *Eliminate the parameter t from the equations $x = t^3+2t+3, \quad y = \sqrt[3]{2t+1}$*

Solution: It is:

$$y = \sqrt[3]{2t+1} \Rightarrow y^3 = 2t+1 \Rightarrow t = \frac{y^3-1}{2}$$

hence

$$x = \left(\frac{y^3-1}{2}\right)^3 + y^3 - 1 + 3 = \left(\frac{y^3-1}{2}\right)^3 + y^3 + 2$$

Example (3). *If $x = 6\cos\theta\sin\theta$ and $y = \cos^2\theta$ find the relation between x, y by eliminating θ.*

Solution: We shall use the formulas

$$\frac{1+\cos 2\theta}{2} = \cos^2\theta, \; \sin 2\theta = 2\cos\theta\sin\theta,$$

then:

$$x = 6\cos\theta\sin\theta \overset{\sin 2\theta = 2\cos\theta\sin\theta}{\Longrightarrow} \frac{x}{3} = \sin 2\theta$$

and

$$y = \cos^2\theta \overset{\frac{1+\cos 2\theta}{2} = \cos^2\theta}{\Longrightarrow} y = \frac{1+\cos 2\theta}{2} \Rightarrow \cos 2\theta = 2y - 1$$

hence

$$(2y-1)^2 + (\frac{x}{3})^2 = 1$$

Problem (1). *If we eliminate t in the parametric equations*

$$\left.\begin{aligned} x &= t^2 + t \\ y &= t^2 - t \end{aligned}\right\}, \quad \textit{then}$$

- *[A]* $x^2 + y^2 + 2xy - 2x - 2y = 0$
- *[B]* $x^2 + y^2 - 2xy - 2x - 2y = 0$
- *[C]* $x^2 - y^2 + 2xy - 2x - 2y = 0$
- *[D]* $x^2 + y^2 + 2xy + 2x + 2y = 0$
- *[E]* $x^2 - y^2 - 2xy - 2x - 2y = 0$

Solution: It is :

$$\left\{ \begin{array}{c} x = t^2 + t \\ y = t^2 - t \end{array} \right\} \Rightarrow \left\{ \begin{array}{c} x + y = 2t^2 \\ x - y = 2t \end{array} \right\} \Rightarrow \left\{ \begin{array}{c} \frac{x+y}{2} = t^2 \\ \frac{x-y}{2} = t \end{array} \right\} \Rightarrow \left(\frac{x-y}{2}\right)^2 = \frac{x+y}{2}$$

hence

$$x^2 + y^2 - 2xy - 2x - 2y = 0 \Leftrightarrow (x-y)^2 = 2(x+y)$$

Correct answer is $[B]$.

Problem (2). *If we eliminate the parameter λ from the equations*

$$x = 2\lambda - 1,$$
$$y = 3\lambda + 2$$

then

- *[A]* $y = \frac{3}{2}x + 3$
- *[B]* $y = \frac{3}{2}x + \frac{7}{2}$
- *[C]* $y = -\frac{3}{2}x + \frac{7}{2}$

- *[D]* $y = -\frac{3}{2}x - 3$
- *[E]* $y = -\frac{3}{2}x + 7$

Solution:
Think for a while before reading on
Solve, for example, the first equation in λ and substitute to the second:

$$\lambda = \frac{x+1}{2} \overset{y=3\lambda+2}{\Longrightarrow}$$
$$y = 3\frac{x+1}{2} + 2 \Rightarrow y = \frac{3}{2}x + \frac{7}{2}$$

Correct answer is $[B]$.

Problem (3). *If we eliminate λ from the equations*

$$x = 3\lambda + 1,$$
$$y = \lambda^2 - 2\lambda + 4$$

then

- *[A]* $y = \frac{x^2 + 8x + 46}{3}$
- *[B]* $y = \frac{x^2 - 8x + 40}{3}$
- *[C]* $y = \frac{x^2 - 8x + 43}{3}$
- *[D]* $y = \frac{x^2 - 8x + 46}{9}$
- *[E]* $y = \frac{x^2 - 8x + 43}{9}$

Solution:
Think for a while before reading on
Step 1: Solve the first equation in λ:

$$x = 3\lambda + 1 \Rightarrow \lambda = \frac{x-1}{3}$$

Step 2: Substitute in the second equation to get:

$$y = \lambda^2 - 2\lambda + 4 \overset{\lambda=\frac{x-1}{3}}{\Rightarrow} y = \left(\frac{x-1}{3}\right)^2 - 2\left(\frac{x-1}{3}\right) + 4$$

and the correct answer is $[E]$.

Problem (4). *If we eliminate λ from the equations*

$$\left.\begin{aligned} x &= \lambda^2 + 2\lambda + 2, \\ y &= 2\lambda^2 + \lambda - 1 \end{aligned}\right\} \quad \textit{then}$$

- *[A]* $-3x+6y+12=\dfrac{(2x+y-5)^2}{3}$
- *[B]* $-9x+18y+36=\dfrac{(2x-y+5)^2}{1}$
- *[C]* $0x-18y+36=\dfrac{(2x-y-5)^2}{1}$
- *[D]* $\dfrac{-x+2y+4}{3}=(\dfrac{2x-y-5}{3})^2$
- *[E]* *none of the above*

Solution:

Think for a while before reading on

Step 1: Multiply the first equation by 2 and subtract the second

$$\left\{\begin{array}{rl} x= & \lambda^2+2\lambda+2 \\ y= & 2\lambda^2+\lambda-1 \end{array}\right\} \Rightarrow \left\{\begin{array}{rl} 2x= & 2\lambda^2+4\lambda+4 \\ y= & 2\lambda^2+\lambda-1 \end{array}\right\} \Rightarrow 2x-y=3\lambda+5 \Rightarrow$$
$$\lambda=\frac{2x-y-5}{3}$$

Step 2: Substitute $\lambda=\dfrac{2x-y-5}{3}$ in one of the two equations to get

$$\frac{-x+2y+4}{3}=(\frac{2x-y-5}{3})^2$$

and the correct answer is $[D]$.

Problem (5). *Eliminating* λ *from the equations*

$$x=\frac{4\lambda}{1+\lambda^2},$$
$$y=3+2\frac{1-\lambda^2}{1+\lambda^2}$$

then

- *[A]* $\left(\dfrac{x}{4}\right)^2+\left(\dfrac{y-3}{2}\right)^2=1$
- *[B]* $\left(\dfrac{x}{2}\right)^2+\left(\dfrac{y-3}{2}\right)^2=1$
- *[C]* $\left(\dfrac{x}{2}\right)^2+\left(\dfrac{y+3}{2}\right)^2=1$
- *[D]* $x^2+y^2 6y+9=2$
- *[E]* *none of the above*

Solution:
Think for a while before reading on
Step 1: Use the second equation to get

$$\frac{y-3}{2} = \frac{1-\lambda^2}{1+\lambda^2}$$

Step 2: It is easy to check, taking the squares of the above expression and of the first equation, that

$$\left(\frac{x}{2}\right)^2 + \left(\frac{y-3}{2}\right)^2 = 1$$

and the correct answer is $[B]$.

Problem (6). *Eliminating θ from the equations*

$$x = 2\sin\theta,$$
$$y = 2\cos\theta$$

then

- *[A]* $x^2 + y^2 - 1 = 0$
- *[B]* $x^2 + y^2 - 2 = 0$
- *[C]* $x^2 + y^2 = 8$
- *[D]* $\frac{x^2 + y^2}{4} = 1$
- *[E]* *none of the above*

Solution: Use relative formulas from trigonometry
Think for a while before reading on
Step 1: Solve both the equations in $\cos\theta$, $\sin\theta$:

$$\frac{x}{2} = \sin\theta, \quad \frac{y}{2} = \cos\theta$$

and taking the squares we obtain

$$\frac{x^2 + y^2}{4} = 1,$$

the correct answer is $[D]$.

Problem (7). *If we eliminate a, b from the equations*

$$x = \frac{2ab}{a^2 + b^2},$$
$$y = \frac{a^2 - b^2}{a^2 + b^2}$$

- *[A]* $x^2 + y^2 = 2$
- *[B]* $x^2 + y^2 = 4$

- *[C]* $x^2 - y^2 = 1$
- *[D]* $(x + y)^2 = 1$
- *[E]* $x^2 + y^2 = 1$

Solution: Use relative formulas from trigonometry

Think for a while before reading on

Step 1: Take the squares of both equations

$$x^2 = \frac{4a^2b^2}{(a^2 + b^2)^2},$$
$$y^2 = \frac{a^4 - 2a^2b^2 + b^4}{(a^2 + b^2)^2}$$

Step 2: Add the above equations to get:

$$x^2 + y^2 = \frac{a^4 + 2a^2b^2 + b^4}{(a^2 + b^2)^2} = \frac{(a^2 + b^2)^2}{(a^2 + b^2)^2} = 1$$

hence the correct answer is $[E]$.

Problem (8). *Eliminating θ from the equations*

$$x^2 = 3\tan^2\theta,$$
$$y^2 = 4 - \frac{3}{\cos^2\theta}$$

we get

- *[A]* $x^2 + y^2 = 1$
- *[B]* $x^2 + y^2 = 2$
- *[C]* $x^2 + y^2 = 4$
- *[D]* $x^2 - y^2 = 1$
- *[E]* *none of the above*

Solution: Use relative formulas from trigonometry

Think for a while before reading on

Step 1: Solve the equations in $\cos^2\theta$ and $\tan^2\theta$:

$$\frac{x^2}{3} = \tan^2\theta = \frac{\sin^2\theta}{\cos^2\theta},$$
$$\frac{y^2 - 4}{3} = \frac{-1}{\cos^2\theta}$$

and thus

$$\frac{x^2}{3} + \frac{y^2 - 4}{3} = \frac{\sin^2\theta - 1}{\cos^2\theta} \overset{\cos^2\theta + \sin^2\theta = 1}{\Longrightarrow}$$
$$\frac{x^2 + y^2 - 4}{3} = -\frac{\cos^2\theta}{\cos^2\theta} = -1 \Rightarrow x^2 + y^2 - 4 = -3$$

and the correct answer is $[A]$.

Problem (9). *Eliminating a $(a \neq 0)$ from the equations*

$$x = a - \frac{2}{a},$$
$$y = a + \frac{2}{a}$$

we get

- *[A]* $x^2 - y^2 = 8$
- *[B]* $(x + y)(x - y) = -8$
- *[C]* $x^2 + y^2 = 8$
- *[D]* $x^2 - y^2 = -8$
- *[E]* $x^2 - y^2 = 4$

Solution:
Think for a while before reading on
Step 1: Add the equations to get

$$x + y = 2a$$

and subtract the second from the first

$$x - y = -\frac{4}{a}$$

Step 2: Multiply $x + y$, $x - y$ to finally obtain

$$(x + y) \cdot (x - y) = 2a \cdot (\frac{-4}{a}) = -8$$

and the correct answer is $[B]$.

Problem (10). *If*

$$y^2 = 3x + 1,$$
$$y = 2t + 3$$

then $x(t)$ is

- *[A]* $x(t) = \frac{1}{3}\{4t^2 + 6t + 8\}$
- *[B]* $x(t) = \frac{1}{3}\{4t^2 + 12t + 8\}$
- *[C]* $x(t) = \frac{1}{3}\{4t^2 - 6t + 8\}$
- *[D]* $x(t) = \frac{1}{3}\{2t^2 + 6t + 9\}$
- *[E]* *none of the above*

Solution:
Think for a while before reading on
It is

$$3x + 1 = y^2 = (2t + 3)^2$$

hence

$$x(t) = \frac{1}{3}\{(2t + 3)^2 - 1\} = \frac{1}{3}\{4t^2 + 12t + 8\}$$

and the correct answer is $[B]$.

Chapter 6

Topic 3

6.1 Trigonometry

6.1.1 Basic formulae for trigonometry

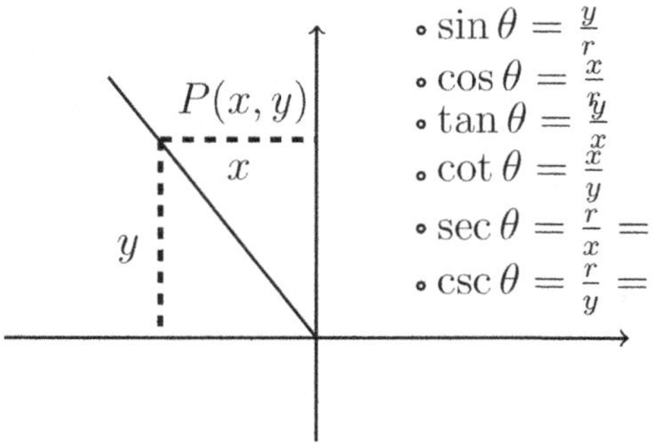

- $r = \sqrt{x^2 + y^2}$
- $\sin\theta = \frac{y}{r}$
- $\cos\theta = \frac{x}{r}$
- $\tan\theta = \frac{y}{x}$
- $\cot\theta = \frac{x}{y}$
- $\sec\theta = \frac{r}{x} = \frac{1}{\cos\theta}$
- $\csc\theta = \frac{r}{y} = \frac{1}{\sin\theta}$

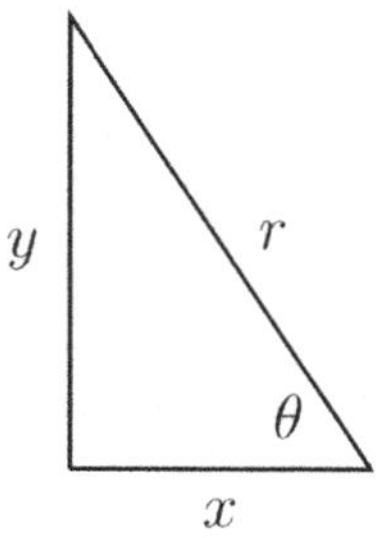

<u>Sign of trigonometric functions</u>

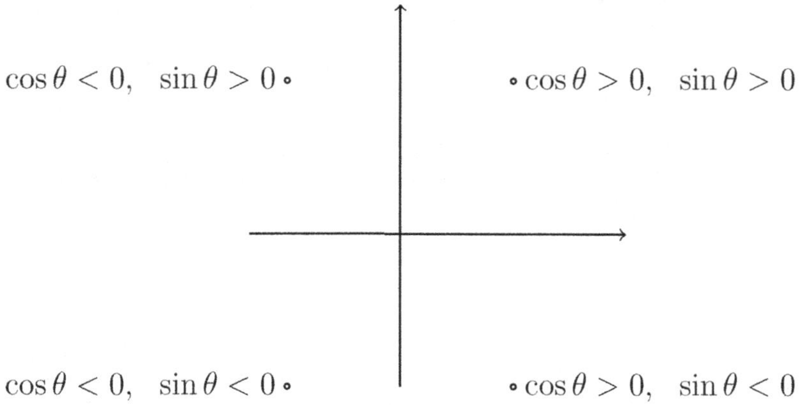

- **Mnemonic Rule**
- 1st quadrant, all positive
- 2nd quadrant, sin is positive
- 3rd quadrant, all negative
- 4^{rth} quadrant, cos is positive

- **Transformation** of degrees θ^o to radius x:

$$\theta^o \to \theta \frac{\pi}{180}$$

or, equivalently

$$\frac{\pi}{x} = \frac{180}{\theta}$$

If θ is given then the following relations hold for the angle:

- $\frac{\pi}{2} - \theta$:

$$\sin(\frac{\pi}{2} - \theta) = \cos\theta, \ \cos(\frac{\pi}{2} - \theta) = \sin\theta, \ \tan(\frac{\pi}{2} - \theta) = \cot\theta, \ \cot(\frac{\pi}{2} - \theta) = \tan\theta$$

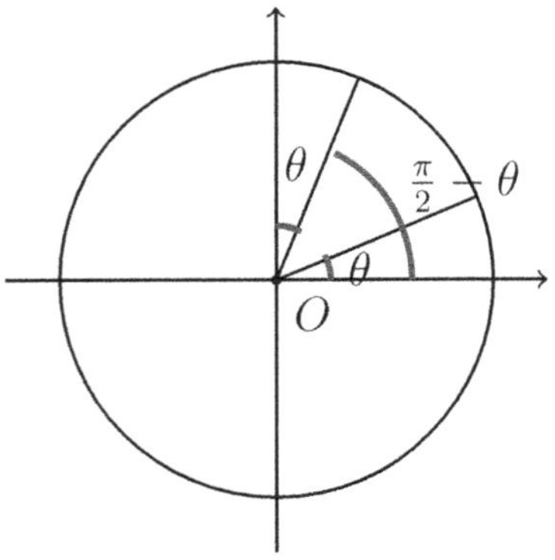

- **Mnemonic Rule**
- $-\theta$: $\sin(-\theta) = -\sin\theta, \cos(-\theta) = \cos\theta$ **only** cos **is positive**
- $\pi - \theta$: $\sin(\pi - \theta) = \sin\theta, \cos(\pi - \theta) = -\cos\theta$ **only** sin **is positive**
- $\pi + \theta$: $\sin(\pi + \theta) = -\sin\theta, \cos(\pi + \theta) = -\cos\theta$ **only** tan **or** cot **is positive**
- $\pi + \theta$: $\tan(\pi + \theta) = \tan\theta, \ \cot(\pi + \theta) = \cot\theta$ **only** tan **or** cot **is positive**
- $2\pi - \theta$: $\sin(2\pi - \theta) = -\sin\theta, \cos(2\pi - \theta) = \cos\theta$ **only** cos **is positive**

· Arc Lenght $= \theta \cdot r$, θ in radian
· Sector Area $= \frac{1}{2}\theta \cdot r^2 = \frac{1}{2}(\theta \cdot r) \cdot r$, like in triangle - **mnemonic rule**
$s = \theta \cdot r$

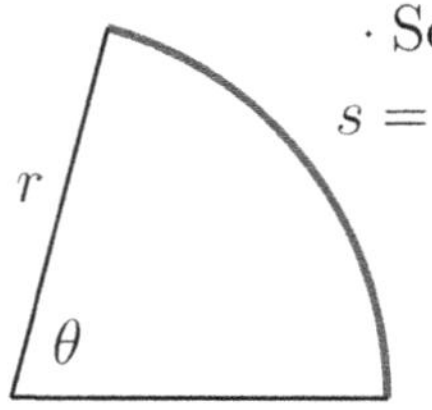

Radians	0	$\frac{\pi}{2}$	π	2π
Degrees	0°	90°	180°	360°
sin	0	1	0	0
cos	1	0	-1	1
tan	0	$\pm\infty$	0	0

Also

Radians	$\frac{\pi}{6}$	$\frac{\pi}{3}$	$\frac{\pi}{4}$
Degrees	30°	60°	45°
sin	$\frac{1}{2}$	$\frac{\sqrt{3}}{2}$	$\frac{\sqrt{2}}{2}$
cos	$\frac{\sqrt{3}}{2}$	$\dfrac{1}{2}$	$\dfrac{\sqrt{2}}{2}$
tan	$\dfrac{\sqrt{3}}{3}$	$\sqrt{3}$	1

- If

$$y = A\sin(Bx + c)$$

then $|A|$ is its **amplitude** and $\dfrac{2\pi}{B}$ its **period**.

- **Pythagorean** type rules:

$$\sin^2 + \cos^2 x = 1, \ \sec^2 x = 1 + \tan^2 x, \ \csc^2 x = 1 + \cot^2 x$$

- **Double angle:**

$$\sin 2x = 2\sin x\cos x, \ \cos 2x = \cos^2 x - \sin^2 x = 2\cos^2 x - 1 = 1 - 2\sin^2 x$$

- **Addition - Subtraction** rules:

$$\sin(A \pm B) = \sin A\cos B \pm \cos A\sin B,$$
$$\cos(A \pm B) = cosA\cos B \mp \sin A\sin B,$$
$$\tan(A \pm B) = \frac{\tan A \pm \tan B}{1 \mp \tan A \cdot \tan B}$$

- **Inverse trigonometric functions:** $\sin x : \left[-\frac{\pi}{2}, \frac{\pi}{2}\right] \to [-1, 1]$ and its inverse

$$\sin^{-1} x : [-1, 1] \to \left[-\frac{\pi}{2}, \frac{\pi}{2}\right]$$

For example:

$$\sin x = a \Rightarrow \sin^{-1} a = x,$$
$$\sin^{-1}(0.5) = x \Rightarrow \sin x = 0.5 \Rightarrow x = \frac{\pi}{6}$$

and $\cos x : [-\pi, 0] \to [-1, 1]$ has it inverse

$$\cos^{-1} x : [-1, 1] \to [-\pi, 0]$$

$$\cos\theta = x \Rightarrow \cos^{-1} x = \theta$$

- **Triangle trigonometry:**

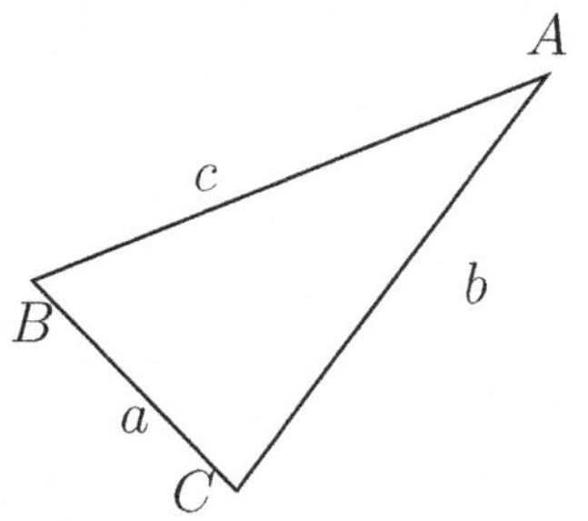

- $\dfrac{\sin A}{a} = \dfrac{\sin B}{b} = \dfrac{\sin C}{c}$
- $a^2 = b^2 + c^2 - 2bc\cos A$
- $b^2 = a^2 + c^2 - 2ac\cos B$
- $c^2 = a^2 + b^2 - 2ab\cos C$
- Area $= \frac{1}{2}bc\sin A = \frac{1}{2}ac\sin B = \frac{1}{2}ab\sin C$

- **An ambiguous case:**

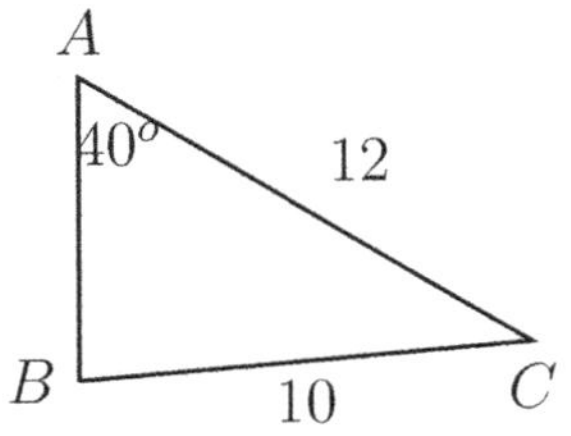

$$\frac{\sin B}{12} = \frac{\sin 40^o}{10} \Rightarrow \sin B \sim 0.77 \Rightarrow$$
$$B = \sin^{-1} 0.77 = 50^o \text{ or } B = 180^o - 50^o = 130^o.$$

6.1.2 Examples & problems for trigonometry

Remark. *If* $f(x) = \sin x$ *then* $x = \sin^{-1}(f(x))$

Example (1).

- *It is* $\sin\frac{\pi}{6} = \frac{1}{2}$ *and thus* $\frac{\pi}{6} = \sin^{-1}(\frac{1}{2})$

- *It is* $\sin 20^o \sim 0.34$ *hence* $\sin^{-1}(0.34) = 20^o$

Example (2). *If* $\sin\theta = 0.3$ *then* $\tan\theta = ?$

Solution: **Way** 1: Since $\sin\theta = 0.3$ we get

$$\theta = \sin^{-1}(0.3) \sim 17.5^o$$

hence

$$\tan\theta = \tan(\sin^{-1}(0.3)) \sim 0.315$$

Way 2: It is

$$\tan\theta = \frac{\sin\theta}{\cos\theta} = \frac{\sin\theta}{\sqrt{1-\sin^2\theta}} = \frac{0.3}{\sqrt{0.91}} = 0.315$$

Example (3). *If*

$$f(x) = \frac{1}{\csc x}$$

then check if

$$f(-x) = -f(x)$$

Solution: It is $\csc x = \frac{1}{\sin x}$ so $f(x) = \sin x$ hence

$$f(-x) = \sin(-x) = -\sin x$$

Example (4). *In the figure bellow they are given*

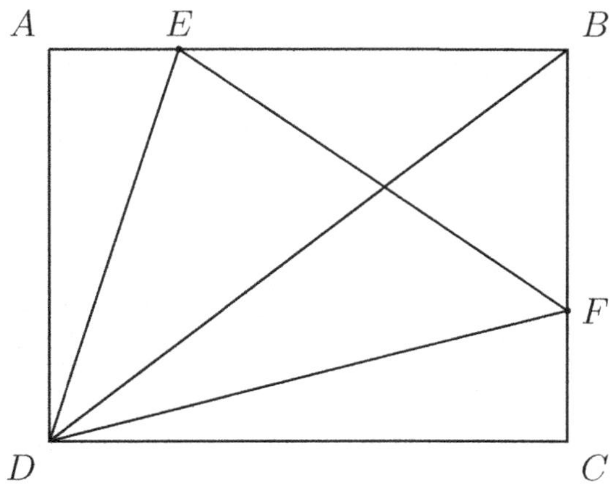

- $AE = 2$
- $AB = 8$
- $BD = 10$
- $FC = 2$

compute the angle

$$\theta = E\hat{D}F$$

Solution: **Use the formula**

$$\cos\theta = \frac{a^2 + b^2 - c^2}{2ab}$$

What is missing? AD, DE, DF and EF
Use Pythagorean:

$$AD^2 = BD^2 - AB^2 \Rightarrow AB = \sqrt{10^2 - 8^2} = \sqrt{36} = 6$$

also

$$DE^2 = AE^2 + AD^2 = 2^2 + 6^2 = 40 \Rightarrow DE = \sqrt{40} = 2\sqrt{10}$$

and

$$DF^2 = FC^2 + DC^2 = 4 + 64 = 68 \Rightarrow DF = \sqrt{68}$$

and

$$EF^2 = EB^2 + BF^2 = 6^2 + 4^2 \Rightarrow EF = \sqrt{52}$$

hence

$$\cos\theta \sim 0.53$$

and thus

$$\theta = \cos^{-1}(0.53) \sim 58^o$$

Problem (1). *If* $x = \cos^{-1}\dfrac{2}{3}$ *and* $5x + y = 320^o$ *then* y *is equal to*

- *[A]* 79°
- *[B]* 81°
- *[C]* 70°
- *[D]* 90°
- *[E]* 270^0

Solution: It is

$$x = \cos^{-1}\frac{2}{3} \Rightarrow \cos x = \frac{2}{3} \Rightarrow x \sim 48.2^o$$

so

$$y = 320^o - 5x \overset{x\sim 48.2^o}{\Rightarrow} y = 320^o - 241^o = 79^o$$

and the correct answer is $[A]$.

Problem (2). *Given the graph*

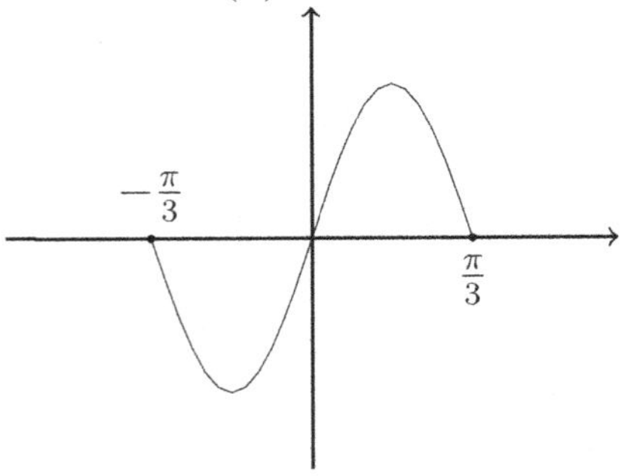

which of the following could be the equation of one cycle of the graph

- *[A]* $y = \cos 3x$
- *[B]* $y = \sin 3x$
- *[C]* $y = \cos\left(\frac{\pi}{2} + 3x\right)$
- *[D]* $y = -\sin 3x$
- *[E]* *none of the above*

Solution: **Period of** $\sin ax$ is equal to $\frac{2\pi}{a}$ hence

$$d\frac{2\pi}{3} = \frac{2\pi}{a} \Rightarrow a = 3$$

and since the function vanishes at $\pm\frac{\pi}{3}$ is follows that the solution is

$$y = \sin 3x = \cos\left(\frac{\pi}{2} - 3x\right), \text{ Corrext answer is [B]}$$

Problem (3). *The smallest possible value of the angle* $\hat{B}$ *is*

- *[A]* 59°
- *[B]* 62°
- *[C]* 80°
- *[D]* 88°
- *[E]* 89°

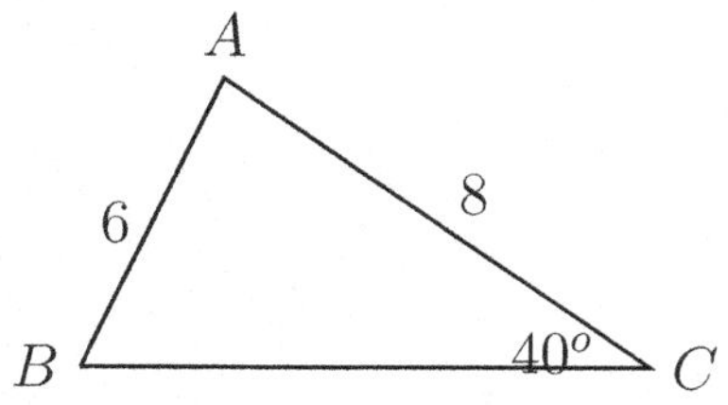

Solution: By the **sin** law

$$\frac{\sin 40^o}{6} = \frac{\sin \hat{B}}{8} \Rightarrow \sin \hat{B} = \frac{8}{6} \sin 40^o = 0.86$$

hence $\hat{B} = 59^o$ or $\hat{B} = 121^o$.
Correct answer is $[A]$.

Problem (4). *The greatest possible value of the area of the above triangle is*

- *[A]* 20
- *[B]* 22
- *[C]* 23.7
- *[D]* 23.9
- *[E]* 24

Solution: The area A is given by the formula

$$A = \frac{1}{2} AB \cdot AC \cdot \sin \hat{A},$$

where $\hat{A} = 180^o - (40^o + 59^o) = 81^o$, hence

$$A = \frac{1}{2} AB \cdot AC \cdot \sin \hat{A} = \frac{1}{2} 6 \cdot 8 \cdot \sin 81^o \sim 23.7.$$

Correct answer is $[C]$.

Problem (5). *The area of the shaded figure is*

- *[A]* $\frac{4\pi}{3} - \sqrt{2}$
- *[B]* $\frac{4\pi}{3} - \sqrt{3}$
- *[C]* $\frac{4\pi}{3} + \sqrt{2}$
- *[D]* $\frac{4\pi}{3} + 3$
- *[E]* $\frac{4\pi}{3} + 1$

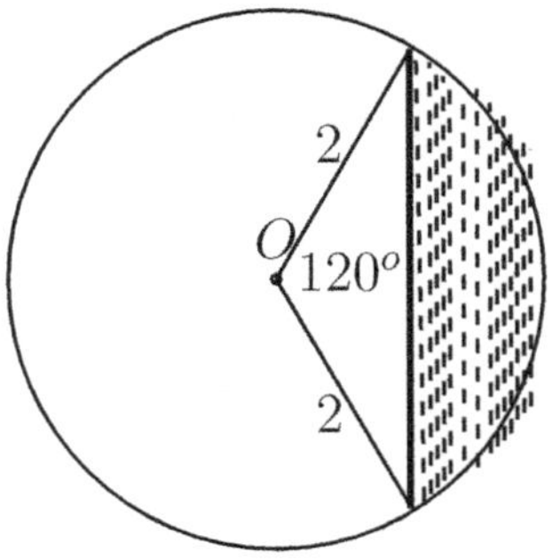

Solution: The area of the whole circle is $A_{2\pi} = \pi r^2$ so, by symmetry, the area corresponding to an angle θ is

$$A_\theta = \frac{1}{2}\theta r^2$$

Also, the area corresponding to the isosceles triangle of central angle 120° is

$$A_{triangle} = \frac{1}{2}r^2 \sin 120^o$$

hence the shaded area is

$$A_{\text{shaded}} = \frac{1}{2}\theta r^2 - \frac{1}{2}r^2 \sin 120^o = \frac{1}{2}2^2\frac{2\pi}{3} - \frac{1}{2}2^2 \sin\frac{2\pi}{3} = \frac{4\pi}{3} - \sqrt{3}.$$

Correct answer is $[B]$.

Problem (6). *If* $\sin x = -\frac{1}{2}$ *and* $\frac{3\pi}{2} < x < 2\pi$ *then* x *is*

- *[A]* $\frac{9\pi}{6}$
- *[B]* $\frac{5\pi}{6}$
- *[C]* $\frac{11\pi}{6}$
- *[D]* $\frac{11\pi}{5}$
- *[E]* $\frac{11\pi}{3}$

Solution: Since $\sin\frac{\pi}{6} = \frac{1}{2}$

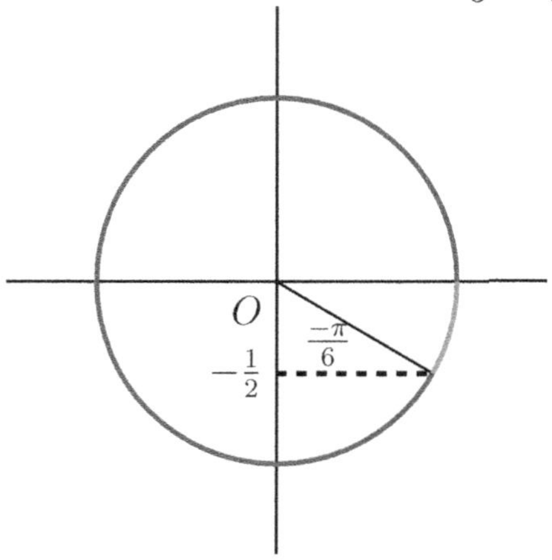

it follows that $\sin\frac{-\pi}{6} = -\frac{1}{2}$ hence

$$x = 2\pi - \frac{\pi}{6} = \frac{11\pi}{6}.$$

Correct answer is $[C]$.

Problem (7). *If* $\sin(2x + \frac{\pi}{4}) = \frac{1}{2}$ *and* $0 < x < \frac{\pi}{2}$ *then* x *is*

- *[A]* $\frac{5\pi}{24}$
- *[B]* $\frac{6\pi}{24}$
- *[C]* $\frac{7\pi}{24}$
- *[D]* $\frac{4\pi}{24}$
- *[E]* $\frac{11\pi}{24}$

Solution: It is $\frac{1}{2} = \sin\frac{\pi}{6} = \sin\frac{5\pi}{6}$ hence $\frac{5\pi}{6} = 2x + \frac{\pi}{4}$ and thus

$$2x = \frac{7\pi}{12} \Rightarrow x = \frac{7\pi}{24},$$

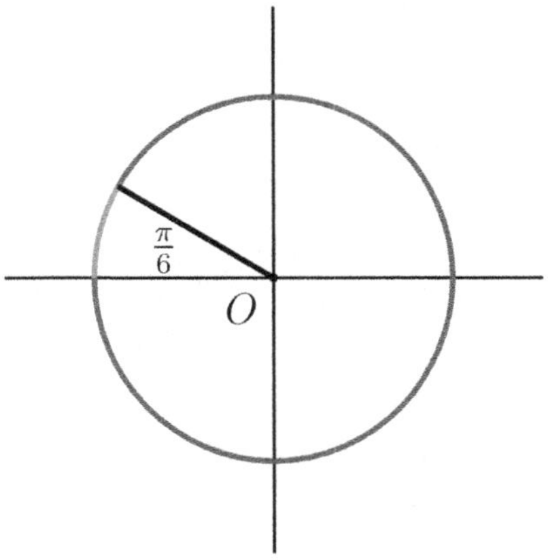

Correct answer is $[C]$.

Problem (8). *If* $\cos 2x = -\frac{\sqrt{3}}{2}$ *and* $0 < x < \frac{\pi}{2}$ *then* x *is*

- *[A]* $\frac{\pi}{6}$
- *[B]* $\frac{5\pi}{12}$
- *[C]* $\frac{2\pi}{3}$
- *[D]* $\frac{3\pi}{4}$

- *[E]* $\frac{5\pi}{6}$

Solution: It is

$$\cos\frac{5\pi}{6} = \cos(\pi - \frac{\pi}{6}) = -\frac{\sqrt{3}}{2} \Rightarrow 2x = \frac{5\pi}{6} \Rightarrow x = \frac{5\pi}{12}.$$

Correct answer is $[B]$.

Problem (9). *The equation of the graph*

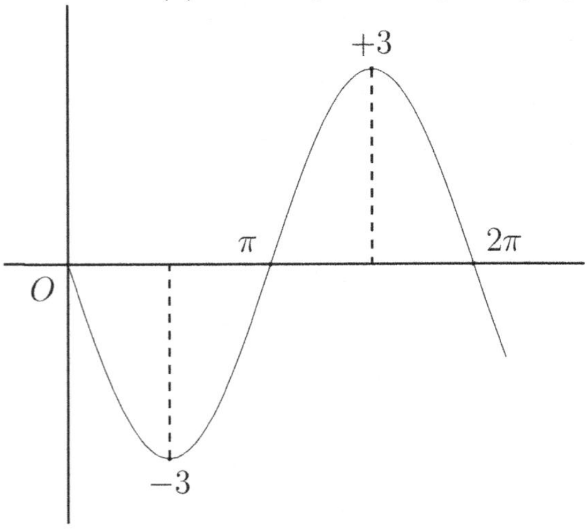

is

- *[A]* $f(x) = -3\sin x$
- *[B]* $f(x) = -\frac{3}{2}\sin x$
- *[C]* $f(x) = 3\sin x$
- *[D]* $f(x) = 3\cos x$
- *[E]* $f(x) = \cos 3x$

Solution: The function $f(x)$ should be of the form:
either $f(x) = \alpha\sin\beta x$ or $f(x) = \alpha\cos\beta x$.
Use the data:
$f(0) = 0$, $f(\frac{\pi}{2}) = -3$, $f(\pi) = f(2\pi) = 0$, $f(\frac{3\pi}{2}) = 3$, then: **Obviously, they all verify only** $[A]$.
Now a complete explanation is:

$$f(0) = 0 \Rightarrow \left\{\begin{array}{c} \alpha\sin 0 = 0 \\ \text{or} \\ \alpha\cos 0 = 0 \end{array}\right\} \Rightarrow \left\{\begin{array}{c} 0 = 0 \quad \text{accepted} \\ \text{or} \\ \alpha = 0 \quad f \text{ identically vanishing} \end{array}\right\}$$

and thus $f(x) = \alpha\cos\beta x$ is **rejected**.
For

$$f(x) = \alpha\sin(\beta x)$$

we **determine** the parameters α, β:

$$\left\{\begin{array}{l} f(\frac{\pi}{2}) = -3 \\ f(\frac{3\pi}{2}) = +3 \end{array}\right\} \Rightarrow \left\{\begin{array}{l} \alpha \sin\beta\frac{\pi}{2} = -3 \\ \alpha \sin\beta\frac{3\pi}{2} = +3 \end{array}\right\} \Rightarrow \sin\frac{\beta\pi}{2} = -\sin\frac{3\beta\pi}{2} \Rightarrow$$

$$\left\{\begin{array}{c} \frac{3\beta\pi}{2} = \frac{\beta\pi}{2} + \pi \\ \text{or} \\ \frac{\beta\pi}{2} = -\frac{3\beta\pi}{2} \end{array}\right\} \Rightarrow \left\{\begin{array}{l} \frac{2\beta\pi}{2} = \pi \\ \beta\pi = 0, \quad \text{rejected} \end{array}\right\} \Rightarrow \beta\pi = \pi$$

hence $\beta = 1$ and consequently from $f(x) = \alpha \sin x$ and $f(\frac{\pi}{2}) = -3$ we get $\alpha = -3$ and thus

$$f(x) = -3\sin x$$

Further trigonometric formulas & remarks

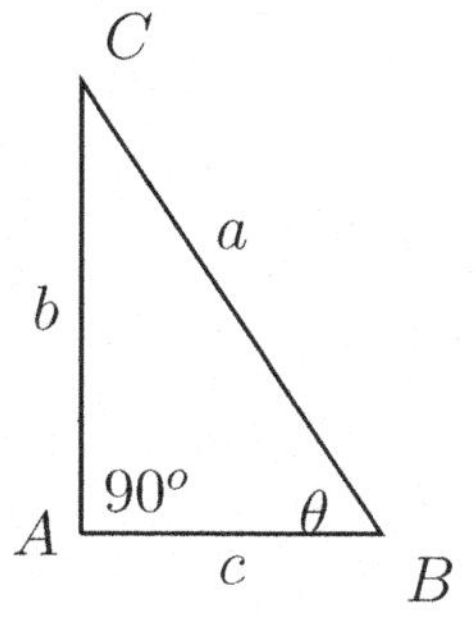

$$\sin\theta = \tfrac{b}{a}$$
$$\cos\theta = \tfrac{c}{a}$$
$$\tan\theta = \tfrac{b}{c}$$
$$\cot\theta = \tfrac{c}{b} = \tfrac{1}{\tan\theta}$$
$$\sec\theta = \tfrac{a}{c} = \tfrac{1}{\cos\theta}$$
$$\csc\theta = \tfrac{a}{b} = \tfrac{1}{\sin\theta}$$

Remember: Using graphs the mind works fifteen (15) times **faster** and you **do not forget!**

For example let us take a look at the graph of

$$y = \sin x$$

$$\sin x \text{ increases in } (-\frac{\pi}{2}, \frac{\pi}{2})$$
$$\sin x \text{ decreases in } (\frac{\pi}{2}, \frac{3\pi}{2})$$

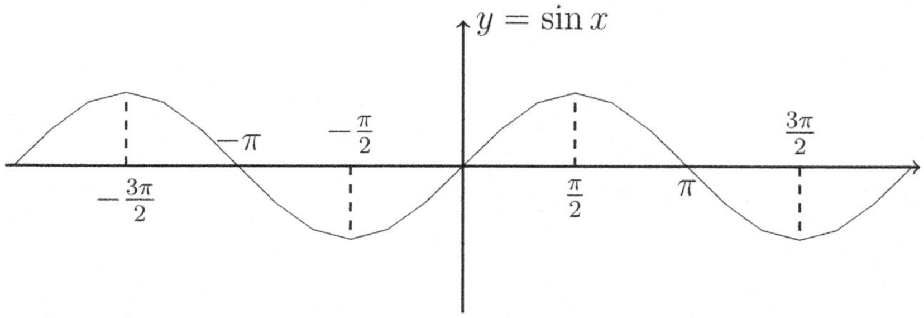

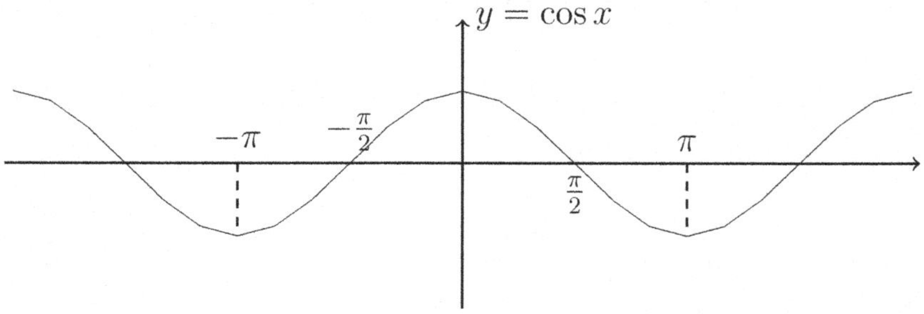

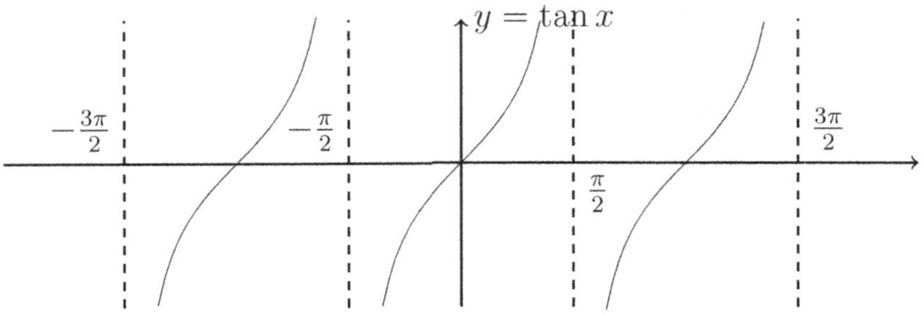

Remember the graph and use it in solving problem, whenever it is necessary. Also, **do not forget** that

$$\tan x = \frac{\sin x}{\cos x}, \ \cot x = \frac{\cos x}{\sin x}$$

Problem (1). *The exact value of* $\tan(-120^o)$ *is*

- *[A]* $-\sqrt{3}$
- *[B]* -1
- *[C]* $-\frac{\sqrt{3}}{3}$
- *[D]* 1
- *[E]* $\sqrt{3}$

Solution: It is

$$\tan(-120^o) = -\tan(120^o) = -\tan(180^o - 60^o) = \tan 60^o = \sqrt{3}$$

Correct answer is $[E]$.

Problem (2). *The amplitude of the function*

$$f(x) = \sin x + \sqrt{3}\cos x$$

is

- *[A]* 1
- *[B]* $\sqrt{2}$
- *[C]* $\sqrt{3}$
- *[D]* 2
- *[E]* 3

Solution: Use the fact that $\tan\frac{\pi}{3} = \sqrt{3}$ and write

$$\sin x + \sqrt{3}\cos x = \sin x + \tan\frac{\pi}{3}\cos x = \sin x + \frac{\sin\frac{\pi}{3}}{\cos\frac{\pi}{3}}\cos x$$

$$= \frac{\sin(x + \frac{\pi}{3})}{\frac{1}{2}} = 2\sin(x + \frac{\pi}{3})$$

hence the amplitude is 2 and the correct answer is $[D]$.

Problem (3). *If $f(x) = 4\sin mx$ and the period is 3π then m is equal to*

- *[A]* 1/3
- *[B]* 2/3
- *[C]* 1
- *[D]* 2
- *[E]* 3

Solution: Use the formula for the period

$$3\pi = \frac{2\pi}{m} \Rightarrow m = \frac{2}{3}$$

Correct answer is $[B]$.

Problem (4). *If $\sin x = -0.6$ and $\pi < x < \frac{3\pi}{2}$ then $\cos x = \ldots$*

- *[A]* -0.9
- *[B]* -0.8
- *[C]* -0.7
- *[D]* 0.7
- *[E]* 0.8

Solution: It is

$$\cos x = -\sqrt{1 - \sin^2 x} = -0.8$$

Correct answer is $[B]$.

Problem (5). *Which of the following is true*

$$\begin{aligned} \tan^{-1}(1) + \tan^{-1}(-1) = 0, \ (I) \\ \sin^{-1}(1) - \sin^{-1}(-1) = 0, \ (II) \\ \sin^{-1}(\frac{1}{2}) - \cos^{-1}(\frac{1}{2})0, \ (III) \end{aligned}$$

Solution: Think for a while before go on reading
Step 1: It is

$$\tan^{-1}(1) = \phi \Leftrightarrow \tan\phi = 1 \Leftrightarrow \phi = \frac{\pi}{4},$$
$$\tan^{-1}(-1) = \phi \Leftrightarrow \tan\phi = -1 \Leftrightarrow \phi = -\frac{\pi}{4}$$

hence

$$\tan^{-1}(1) + \tan^{-1}(-1) = 0$$

and I is true.
Similarly working we see that II and III **do not hold true**.

Problem (6). *If the amplitude of the function*

$$f(x) = 2a\sin(a-1)x, \ a > 0$$

is 3 *then the period is*

- *[A]* $\frac{\pi}{2}$
- *[B]* π
- *[C]* 2π
- *[D]* 4π
- *[E]* 6π

Solution: Think for a while before reading on
What is the subtle point?
Since the amplitude is 3 is follows that

$$2a = 3 \Rightarrow a = \frac{3}{2}$$

hence

$$f(x) = 3\sin\left(\frac{1}{2}x\right)$$

and the period T is given by the standard formula

$$T = \frac{2\pi}{\frac{1}{2}} = 4\pi,$$

correct answer is $[D]$.

Problem (7). *If the amplitude of the function*

$$f(x) = a^3\cos\frac{3a}{2}x, \ a > 0$$

is 8 *then the period is*

- *[A]* $\frac{\pi}{2}$
- *[B]* $\frac{2\pi}{3}$
- *[C]* 2π
- *[D]* 3π
- *[E]* 6π

Solution: What is the subtle point?
Amplitude is 8 so

$$a^3 = 8 \Rightarrow a = 2$$

hence

$$f(x) = 8\cos\frac{3\times 2}{2}x = 8\cos 3x$$

and thus the period is given by

$$T = \frac{2\pi}{3}$$

Correct answer is $[B]$.

Problem (8). *Let*

$$f(x) = \sin(ax),$$

with graph

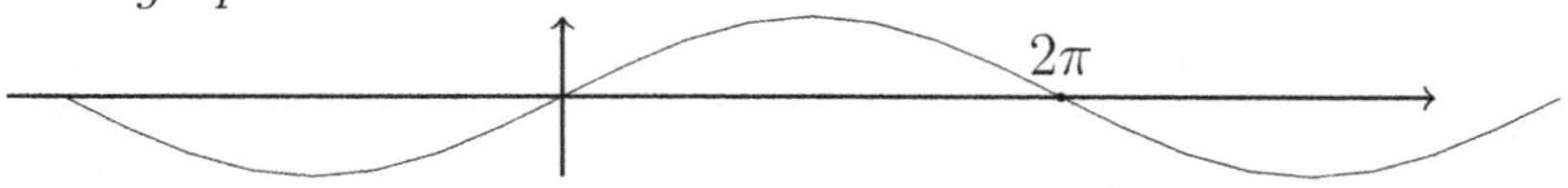

then $a =$

- *[A]* $\frac{1}{3}$
- *[B]* $\frac{1}{2}$
- *[C]* 1
- *[D]* 2
- *[E]* 4

Solution: What is the subtle point?
By the graph we see that the period T is

$$T = 4\pi$$

hence

$$4\pi = \frac{2\pi}{a} \Rightarrow a = \frac{1}{2}$$

Correct answer is $[B]$.

Problem (9). *The minimum value of*

$$f(x) = 2\cos^2 x - 1$$

is

- *[A]* -1
- *[B]* $-\frac{1}{2}$
- *[C]* 0
- *[D]* 1

- *[E]* 2

Solution: Think for a while before reading on
Use formulas from trigonometry - $\cos 2x$: It is

$$\cos 2x = \cos^2 x - \sin^2 x = 2\cos^2 x - 1$$

hence

$$-1 \leq \cos 2x \leq +1$$

and the minimum value is -1. Correct answer is $[A]$.

Problem (10). *The maximum value of*

$$f(x) = \cos^2 x - \sin^2 x$$

is

- *[A]* $\frac{1}{2}$
- *[B]* 1
- *[C]* $\frac{3}{2}$
- *[D]* 2
- *[E]* 3

Solution: Think for a while before reading on
Use formulas from trigonometry - $\cos 2x$: It is

$$\cos^2 x - \sin^2 x = \cos^2 x - (1 - \cos^2 x) = 2\cos^2 x - 1 = \cos 2x$$

hence,

$$-1 \leq \cos 2x \leq +1,$$

so the maximum value is $+1$ and the correct answer is $[B]$.

Problem (11). *The maximum value of*

$$f(x) = 2\tan x - 1,\ 0 \leq x \leq \frac{\pi}{4}$$

is

- *[A]* $\frac{1}{2}$
- *[B]* 1
- *[C]* $\sqrt{3}$
- *[D]* 3
- *[E]* 4

Solution: Think for a while before reading on
Use formulas from trigonometry:
What is the subtle point?
$\tan x$ is increasing in $[0, \frac{\pi}{4}]$ therefore

$$\tan 0^o \leq \tan x \leq \tan \frac{\pi}{4}$$

so

$$0 \leq \tan x \leq 1 \Rightarrow$$
$$0 \leq 2\tan x \leq 2, \text{ for } 0 \leq x \leq \frac{\pi}{4}$$

hence

$$-1 \leq 2\tan x - 1 \leq 1, \text{ for } 0 \leq x \leq \frac{\pi}{4}$$

and thus, the correct answer is $[B]$.

Problem (12). *The minimum value of*

$$f(x) = 5\tan x - 4, \ \frac{3\pi}{4} \leq x \leq \pi$$

is

- *[A]* -9
- *[B]* -5
- *[C]* -4
- *[D]* -1
- *[E]* 0

Solution: Think for a while before reading on
Use formulas from trigonometry:
What is the subtle point?
$\tan x$ is increasing in $[\frac{3\pi}{4}, \pi]$ therefore

$$\tan \frac{3\pi}{4} \leq \tan x \leq \tan \pi$$

so

$$-1 \leq \tan x \leq 0, \text{ for } \frac{3\pi}{4} \leq x \leq \pi \Rightarrow$$
$$-5 \leq 5\tan x \leq 0, \text{ for } \frac{3\pi}{4} \leq x \leq \pi \Rightarrow$$
$$-9 \leq 5\tan x - 4 \leq -4,$$

for $\frac{3\pi}{4} \leq x \leq \pi$ and the correct answer is $[A]$.

6.2 Analytic & Coordinate Geometry

6.2.1 Basic equations - Analytic Geometry

Standard equation of the circle

$$(x - x_0)^2 + (y - y_0)^2 = r^2$$

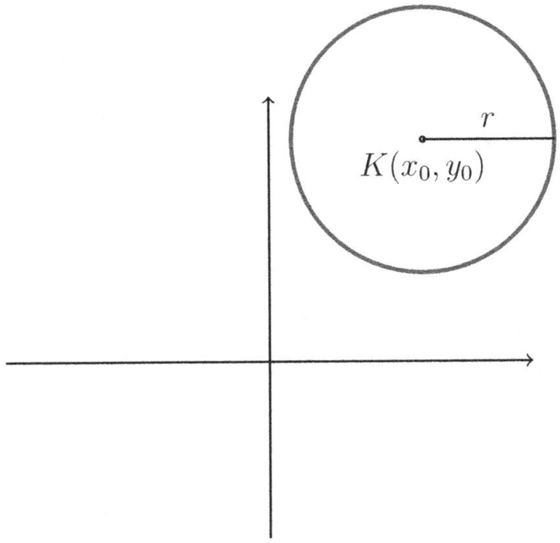

is the equation of the circle with center $K(x_0, y_0)$ and radius r.

Standard equation of the ellipse

$$\frac{(x - x_0)^2}{a^2} + \frac{(y - y_0)^2}{b^2} = 1$$

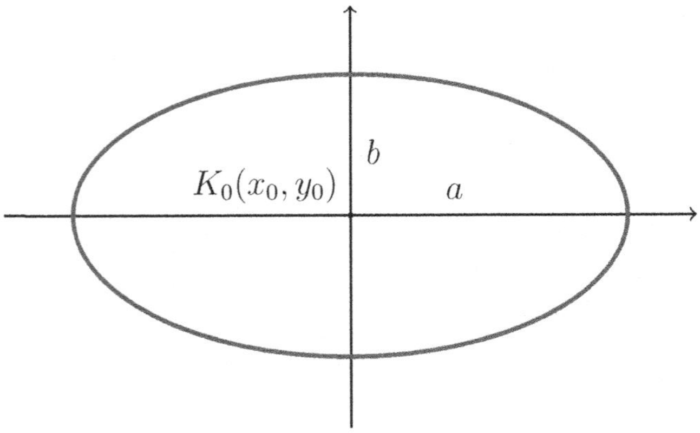

$c = \pm\sqrt{a^2 - b^2}, \quad a \geq b > 0$

$\epsilon = \frac{c}{a}$ eccentricity

$(x_0 \pm a, y_0)$ vertices

$F_1(x_0 - c, y_0)$, $F_2(x_0 + c, y_0)$ foci

$K_0(x_0, y_0)$ center

is the equation of an ellipse with a horizontal major axis and

$$a^2 = b^2 + c^2$$

and

$$\frac{(x - x_0)^2}{b^2} + \frac{(y - y_0)^2}{a^2} = 1,$$

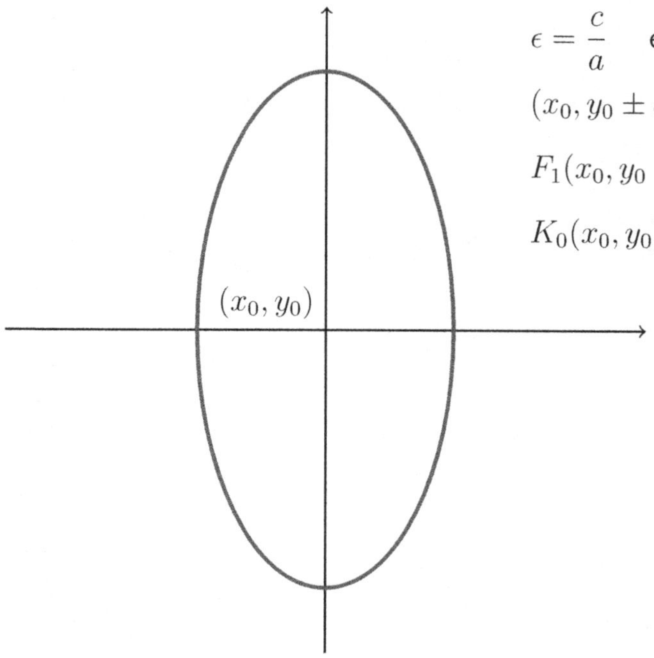

$c = \pm\sqrt{a^2 - b^2}, \quad a \geq b > 0$

$\epsilon = \dfrac{c}{a}$ eccentricity

$(x_0, y_0 \pm a)$ vertices

$F_1(x_0, y_0 - c), \ F_2(x_0, y_0 + c)$ foci

$K_0(x_0, y_0)$ center

is the equation of an ellipse with a vertical major axis.

Standard equation of the hyperbola

$$\frac{(x - x_0)^2}{a^2} - \frac{(y - y_0)^2}{b^2} = 1$$

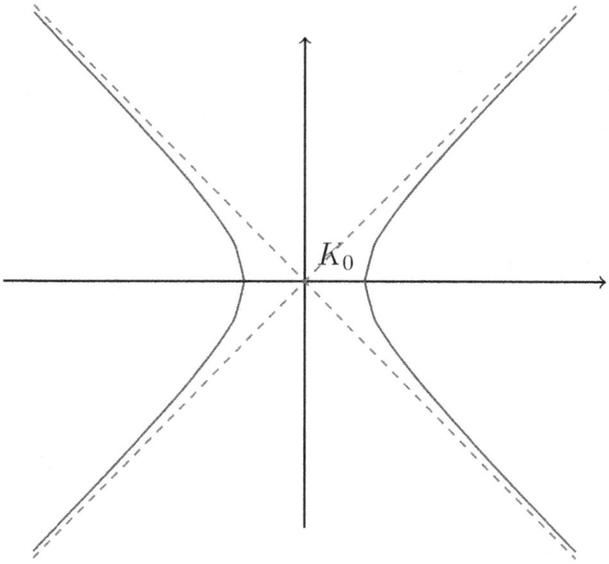

$K_0(x_0, y_0)$, center

$c^2 = a^2 + b^2$

$(x_0 - c, y_0)$, $(x_0 + c, y_0)$ foci

$(x_0 - a, y_0)$, $(x_0 + a, y_0)$ vertices

$y - y_0 = \pm\dfrac{b}{a} \cdot (x - x_0)$ asymptotes

$\epsilon = \sqrt{1 + \dfrac{b^2}{a^2}}$ eccentricity

$$\frac{(y - y_0)^2}{a^2} - \frac{(x - x_0)^2}{b^2} = 1$$

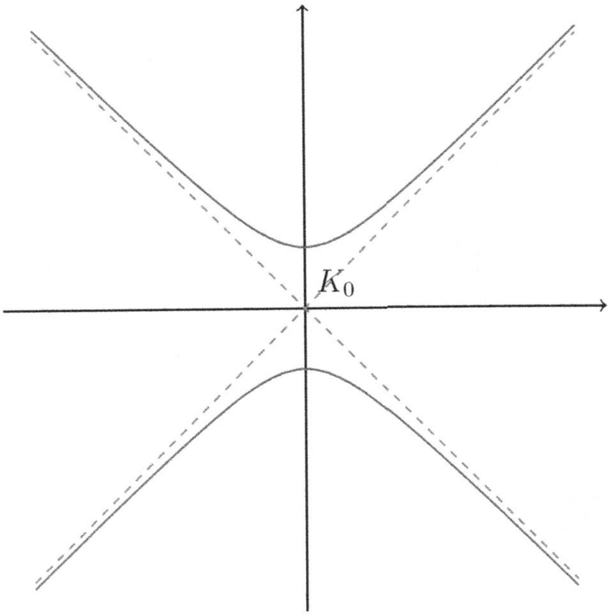

$K_0(x_0, y_0)$, center

$c^2 = a^2 + b^2$

$(x_0, y_0 - c)$, $(x_0, y_0 + c)$ foci

$(x_0, y_0 - a)$, $(x_0, y_0 + a)$ vertices

$y - y_0 = \pm\frac{a}{b} \cdot (x - x_0)$ asymptotes

Standard equation of the parabola

$$x^2 = 4py, \; p \in \mathbb{R}$$

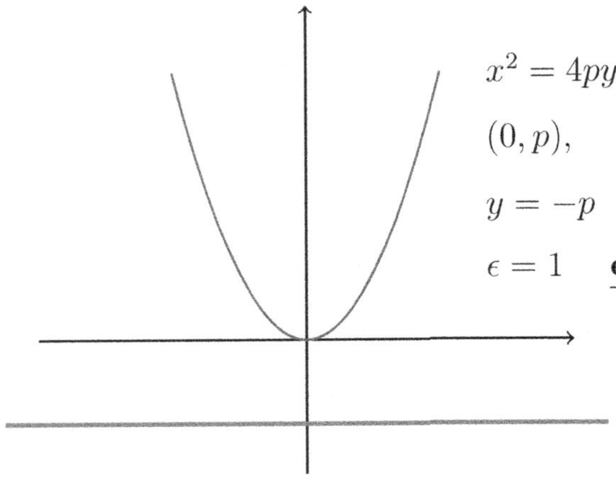

$x^2 = 4py, \quad p > 0$

$(0, p)$, focus

$y = -p$ directrix

$\epsilon = 1$ **eccentricity is equal to 1, by definition**

6.2.2 Problems on 2 dimensional coordinate plane

If any problem seems difficult for you, skip it ...
Problem (1). *Equation*

$$x^2 + y^2 - 2x + 4y + 1 = 0$$

is the equation of a circle with

- *[A] center $K(1, 2)$ and radius $\rho = 2$*
- *[B] center $K(-1, 2)$ and radius $\rho = 1$*
- *[C] center $K(1, 2)$ and radius $\rho = 4$*
- *[D] center $K(1, -2)$ and radius $\rho = 2$*
- *[E] none of the above*

Solution: Think for a while before reading on.
It is

$$\begin{aligned} x^2+y^2-2x+4y+1=0=0 &\Leftrightarrow \\ (x^2-2x+1)-1+(y^2+4y+4)-4+1=0 &\Leftrightarrow \\ (x-1)^2-1+(y+2)^2-4+1=0 & \end{aligned}$$

hence

$$(x-1)^2+(y+2)^2=4$$

and the corerct answer is $[D]$.

Problem (2). *Which of the following conditions should hold true so that the equation*

$$x^2+y^2+2\alpha x+2\beta y+2\gamma=0$$

represents a circle?

- *[A]* $\alpha^2+\beta^2=4\gamma^2$
- *[B]* $\alpha^2+\beta^2=2\gamma$
- *[C]* $\alpha^2+\beta^2+2\gamma>0$
- *[D]* $\alpha^2+\beta^2>2\gamma$
- *[E] none of the above*

Solution: Think for a while before reading on.
What is the subtle point?
Express the given equation as a sum of perfect squares plus extra terms:

$$\begin{aligned} x^2+y^2+2\alpha x+2\beta y+2\gamma=0 &\Leftrightarrow \\ (x+\alpha)^2-\alpha^2+(y+\beta)^2-\beta^2+2\gamma=0 &\Leftrightarrow \\ (x+\alpha)^2+(y+\beta)^2=\alpha^2+\beta^2-2\gamma & \end{aligned}$$

hence the condition

$$\alpha^2+\beta^2-2\gamma>0$$

should hold true and the correct answer is $[D]$.

Problem (3). *The radius of the circle*

$$x^2+y^2=8$$

is

- *[A]* 2
- *[B]* $2\sqrt{2}$
- *[C]* $4\sqrt{2}$
- *[D]* 4

- *[E]* 8

Solution: Think for a while before reading on.
It is

$$x^2 + y^2 = 8 \Leftrightarrow x^2 + y^2 = (2\sqrt{2})^2$$

and the correct answer is B.

Problem (4). *The center of the circle*

$$x^2 + y^2 - 6x + 4y + 10 = 0$$

is the point

- *[A]* $(3, -1)$
- *[B]* $(3, -2)$
- *[C]* $(2, 3)$
- *[D]* $(-2, 3)$
- *[E]* $(3, 2)$

Solution: Think for a while before reading on. What is the subtle point? Try to express the given equation as a sum of perfect squares plus extra constant terms:

$$\begin{aligned} x^2 + y^2 - 6x + 4y + 10 = 0 \Leftrightarrow \\ (x-3)^2 - 9 + (y+2)^2 - 4 + 10 = 0 \Leftrightarrow \\ (x-3)^2 + (y+2)^2 = 3 \end{aligned}$$

and the correct answer is $[B]$.

Problem (5). *A circle passing through the point* $(3, 9)$ *and of radius* 9*, has an equation of the form*

- *[A]* $x^2 + y^2 = 81$
- *[B]* $x^2 + y^2 = 3$
- *[C]* $x^2 + y^2 = 9$
- *[D]* $(x-3)^2 + y^2 = 81$
- *[E]* $(x-3)^2 + y^2 = 49$

Solution: Think for a while before reading on.
What is the subtle point?
Just check that for $x = 3$, $y = 9$ the equation in $[D]$ is satisfied.

Problem (6). *Given the equation*

$$9x^2 + 16y^2 - 18x - 64y - 71 = 0$$

the center, vertices and the focuses of the ellipse are

- *[A]* *center* $K(1,2)$, *vertices* $(-3,2)$, $(5,2)$ *and focuses* $(-\sqrt{7}+1,2)$, $(\sqrt{7}+1,2)$
- *[B]* *center* $K(1,2)$, *vertices* $(3,2)$, $(5,2)$ *and focuses* $(-\sqrt{7}+1,2)$, $(\sqrt{7}+1,2)$
- *[C]* *center* $K(-1,2)$, *vertices* $(-3,2)$, $(5,2)$ *and focuses* $(-\sqrt{7}+1,2)$, $(\sqrt{7}+1,2)$
- *[D]* *center* $K(1,-2)$, *vertices* $(-3,2)$, $(5,2)$ *and focuses* $(-\sqrt{7}-1,2)$, $(\sqrt{7}+1,2)$
- *[E]* *none of the above*

Solution: Think for a while before reading on.
The given equation is written:

$$9x^2+16y^2-18x-64y-71=0 \Leftrightarrow 9(x-1)^2-9+16(y-2)^2-64=71 \Leftrightarrow$$
$$9(x-1)^2+16(y-2)^2=144 \Leftrightarrow 3^2(x-1)^2+4^2(y-2)^2=12^2$$

hence

$$\frac{(x-1)^2}{\left(\frac{12}{3}\right)^2}+\frac{(y-2)^2}{\left(\frac{12}{4}\right)^2}=1 \Leftrightarrow \frac{(x-1)^2}{4^2}+\frac{(y-2)^2}{3^2}=1$$

and thus the center of the ellipse is $K(1,2)$, $a=4>3=b$, $c^2=a^2-b^2=4^2-3^2=7$ so $c=\pm\sqrt{7}$
Since the vertices of the ellipse are in a distance of $a=4$ from its center we conclude that

$$(-4+1,2),\ (4+1,2)$$

are vertices of the ellipse.
To find the foci of the ellipse we need $c=\pm\sqrt{7}$ and thus the foci are given by the points

$$(-\sqrt{7}+1,2),\ (\sqrt{7}+1,2)$$

Correct answer is $[A]$.

Problem (7). *If the circle*

$$x^2+y^2-6x-8ky+k^2-2k+1=0$$

passes through the origin of the coordinates axis, then the value of k *is*

- *[A]* 0
- *[B]* 1
- *[C]* 2
- *[D]* 3
- *[E]* 4

Solution: Think for a while before reading on.
What is the subtle point?
The origin $O(0,0)$ belongs to the circle and by substitution of $x = 0, y = 0$ into the equation, we get:

$$k^2 - k + 1 = 0 \Rightarrow (k-1)^2 = 0$$

hence $k = 1$ and the correct answer is $[B]$.

Problem (8). *Given the following ellipses find the ellipses with foci on the yy' and symmetry center the origin that has a focal distance of* 6

- *[A]* $\dfrac{x^2}{25} + \dfrac{y^2}{16} = 1$
- *[B]* $\dfrac{x^2}{3} + \dfrac{y^2}{5} = 1$
- *[C]* $\dfrac{y^2}{5} + \dfrac{x^2}{3} = 1$
- *[D]* $\dfrac{y^2}{6} + \dfrac{x^2}{8} = 1$
- *[E]* $\dfrac{x^2}{16} + \dfrac{y^2}{25} = 1$

Solution: Think for a while before reading on.
First step: Note that all the given ellipses have the origin $(0,0)$ as center.
Second step: The foci of the ellipse lie on the yy' axis if the equation is of the form

$$\frac{x^2}{b^2} + \frac{y^2}{a^2}, \quad a \geq b > 0$$

so the remaining candidates are $[B]$, $[D]$ and $[E]$.
Third step: Compute the $c's$ of the candidate ellipses

$$c_B = \sqrt{5-3} = \sqrt{2},$$
$$c_D = \sqrt{8-6} = \sqrt{2}$$
$$c_E = \sqrt{25-16} = \sqrt{9} = 3$$

therefore the corresponding foci are

$$F_1(0, -\sqrt{2}), \;\; F_2(0, \sqrt{2})$$

for both the ellipses $[B]$ and $[D]$ and with foci distance

$$d(F_1, F_2) = 2\sqrt{2}$$

and thus these two ellipses do not satisfy the required properties.
Fourth step: For the ellipse $[E]$ we see that their foci are $F_1(0,-3)$, $F_2(0,3)$ and the distance is

$$d(F_1, F_2) = 3 - (-3) = 6,$$

hence the correct answer is $[E]$.

Problem (9). *The ellipse that has the same eccentricity with the ellipse*

$$C : \frac{x^2}{3^2} + \frac{y^2}{5^2} = 1$$

is

- *[A]* $\frac{x^2}{4} + \frac{y^2}{3} = 1$
- *[B]* $\frac{x^2}{4^2} + \frac{y^2}{3^2} = 1$
- *[C]* $\frac{y^2}{4} + \frac{x^2}{3} = 1$
- *[D]* $\frac{y^2}{4^2} + \frac{x^2}{3^2} = 1$
- *[E]* $\frac{4x^2}{3^2} + \frac{4y^2}{5^2} = 1$

Solution: Think for a while before reading on.

Step 1: Compute c's of the ellipses $[A]$-$[E]$ and the corresponding eccentricities:

$$c_A = \sqrt{4-3} = 1 \Rightarrow \epsilon_A = \frac{1}{5},$$
$$c_B = \sqrt{16-9} = \sqrt{7} \Rightarrow \epsilon_B = \frac{\sqrt{7}}{4},$$
$$c_C = \sqrt{4-3} = 1 \Rightarrow \epsilon_C = \frac{1}{4},$$
$$c_D = \sqrt{16-9} = \sqrt{7} \Rightarrow \epsilon_D = \frac{\sqrt{7}}{4},$$
$$c_E = \sqrt{\frac{25}{4} - \frac{9}{4}} = 2 \Rightarrow \epsilon_E = \frac{4}{5}$$

and in the meanwhile, for the ellipse $\mathcal{C} : \frac{x^2}{3^2} + \frac{y^2}{5^2} = 1$ is

$$c_{\mathcal{C}} = \sqrt{25-9} = \sqrt{16} = 4$$

and

$$\epsilon_{\mathcal{C}} = \frac{4}{5},$$

hence, the correct answer is $[E]$.

Problem (10). *A focus of the ellipse* $\mathcal{C} : \frac{x^2}{2} + \frac{y^2}{3} = 1$ *is*

- *[A]* $(2, 3)$
- *[B]* $(0, \sqrt{2})$
- *[C]* $(\sqrt{3}, 0)$

- *[D]* $(-1, 0)$
- *[E]* $(0, -1)$

Solution: Think for a while before reading on.
Compute c**:** $c = \sqrt{3-2} = 1$
Observe: The origin $(0, 0)$ is the center of the given ellipse.
Foci of $\mathcal{C}$**:** $(0, -1)$ and $(0, 1)$
Conclusion: the correct answer is $[E]$.

Problem (11). *For the ellipse* $\mathcal{C} : \dfrac{x^2}{a^2} + \dfrac{y^2}{b^2} = 1$ *and foci distance* $2c$ *and major axis* $2a$ *there holds*

- *[A]* $a > b > c$
- *[B]* $b^2 = c^2 - a^2$
- *[C]* $0 < a < b$
- *[D]* $c > a$
- *[E]* $c < a$

Solution: Think for a while before reading on.
What is the subtle point? We are given that $2a$ is the major axis, it follows that

$$c = \sqrt{a^2 - b^2} \Rightarrow c < \sqrt{a^2} \Rightarrow c < a$$

and thus, the correct answer is $[E]$.

Problem (12). *The ellipses*

$$\mathcal{C}_1 : \frac{x^2}{a^2} + \frac{y^2}{b^2} = 1 \text{ and } \mathcal{C}_2 : \frac{x^2}{b^2} + \frac{y^2}{a^2} = 1$$

have

- *[A]* *only two common points*
- *[B]* *four common points*
- *[C]* *only one common point*
- *[D]* *no common points*
- *[E]* *an infinity of common points*

Solution: Think for a while before reading on.
Observe that: the two ellipses have the origin as their common center and the one is *generated* by the other by a rotation of $\frac{\pi}{2}$ around their common center of symmetry.
Conclusion: the correct answer is $[B]$.

Problem (13). *If* $x = 2\cos\theta - 2$ *and* $y = 3\sin\theta - 1$*, then, the ellipse defined by these two parametric equations has:*

- *[A]* *center* $(2,-1)$ *and foci distance* $2\sqrt{5}$
- *[B]* *center* $(-2,1)$ *and foci distance* $2\sqrt{5}$
- *[C]* *center* $(-2,-1)$ *and foci distance* $\sqrt{5}$
- *[D]* *center* $(-2,-1)$ *and foci distance* $2\sqrt{5}$
- *[E]* *none of the above*

Solution: **Think for a while before reading on.**
Using the parametric equations:

$$\left.\begin{aligned}\cos\theta &= \frac{x+2}{2}\\ \sin\theta &= \frac{y+1}{3}\end{aligned}\right\} \Rightarrow \frac{(x+2)^2}{2^2}+\frac{(y+1)^2}{3^2}=1$$

hence the given ellipse has center $(-2,-1)$, $c=\sqrt{9-4}=\sqrt{5}$ with focal distance

$$d(F_1,F_2)=2\sqrt{5}$$

and thus the correct answer is $[D]$.

Problem (14). *The equation of the ellipse with center* $(1,-1)$*, a focus on* $(3,-1)$ *and eccentricity* $\epsilon=\dfrac{1}{2}$ *is:*

- *[A]* $\dfrac{(x-1)^2}{16}+\dfrac{(y-1)^2}{12}=1$
- *[B]* $\dfrac{(x-1)^2}{12}+\dfrac{(y+1)^2}{16}=1$
- *[C]* $\dfrac{(x-1)^2}{16}+\dfrac{(y+1)^2}{12}=1$
- *[D]* $\dfrac{(x+1)^2}{16}+\dfrac{(y+1)^2}{12}=1$
- *[E]* *none of the above*

Solution: **Think for a while before reading on.**
First step: Its center is $K(1,-1)$ so, the equation of the ellipse is of the form

$$\frac{(x-1)^2}{a^2}+\frac{(y+1)^2}{b^2}=1,$$

hence only $[B]$ and $[C]$ do have the required form.
Use information for the focal points: Since one of the foci is

$$(3,-1)=(1+c,-1)$$

we conclude that the remaining candidate is $[C]$.
Compute: It is

$$c_{\mathfrak{e}}=\sqrt{16-12}=\sqrt{4}=2$$

hence the foci are

$$(1-2,-1),\ (1+2,-1)$$

and eccentricity is

$$\epsilon = \frac{2}{4} = \frac{1}{2}$$

So, $[C]$ is indeed the correct answer.

Problem (15). *The mid-point of the chord of the hyperbola*

$$\mathcal{C} : 4x^2 - 9y^2 = 36,$$

which has equation

$$\zeta :\ 2y = x + 1,$$

is

- *[A]* $(\frac{9}{7}, \frac{7}{8})$
- *[B]* $(\frac{7}{9}, \frac{8}{7})$
- *[C]* $(\frac{8}{7}, \frac{9}{7})$
- *[D]* $(\frac{9}{7}, \frac{8}{7})$
- *[E]* *none of the above*

Solution: Think for a while before reading on.
Intersection $\mathcal{C} \cap \zeta$:

$$\left.\begin{array}{l} 4x^2 - 9y^2 = 36 \\ 2y = x + 1 \end{array}\right\} \Rightarrow 7y^2 - 16y - 32 = 0$$

so, if y_1, y_2 are the roots of this equations then $y_1 + y_2 = \frac{16}{7}$ therefore, for the midpoint $Q(x_q, y_q)$ of the chord:

$$y_q = \frac{8}{7}$$

Using relation $2y = x + 1$ we also get

$$x_q = \frac{9}{7}$$

and the correct answer is $[D]$.

Problem (16). *The points* $\left(\dfrac{a}{\cos\phi}, b\tan\phi\right)$, *when ϕ varies, are moving on the hyperbola*

- *[A]* $ax^2 - by^2 = a^2b^2$
- *[B]* $ax^2 - by^2 = 1$
- *[C]* $bx^2 - ay^2 = 1$

- *[D]* $\mathcal{C} : b^2x^2 - a^2y^2 = a^2b^2$
- *[E]* *none of the above*

Solution: Think for a while before reading on.

Let: $x = \dfrac{a}{\cos\phi}$, $y = b\tan\phi$, then:

$$\frac{x^2}{a^2} = \frac{1}{\cos^2\phi} \quad \text{and} \quad \frac{y^2}{b^2} = \frac{\sin^2\phi}{\cos^2\phi}$$

hence

$$\frac{x^2}{a^2} - \frac{y^2}{b^2} = \frac{1}{\cos^2\phi} - \frac{\sin^2\phi}{\cos^2\phi} = \frac{1-\sin^2\phi}{\cos^2\phi} = 1$$

and the correct answer is $[D]$.

Problem (17). *The points* $\left(\frac{a}{2}(t+\frac{1}{t}), \frac{b}{2}(t-\frac{1}{t})\right)$, *as* t *varies, are moving along the hyperbola*

- *[A]* $\dfrac{x^2}{b^2} - \dfrac{y^2}{a^2} = 1$
- *[B]* $a^2x^2 - b^2y^2 = 1$
- *[C]* $\mathcal{C} : \dfrac{x^2}{a^2} - \dfrac{y^2}{b^2} = 1$
- *[D]* $x^2b^2 - y^2a^2 = 1$
- *[E]* *none of the above*

Solution: Think for a while before reading on.

Work as in the previous case:

$$x = \frac{a}{2}(t+\frac{1}{t}) \Rightarrow 2\frac{x}{a} = t + \frac{1}{t},$$
$$y = \frac{b}{2}(t-\frac{1}{t}) \Rightarrow 2\frac{y}{b} = t - \frac{1}{t}$$

hence

$$4\frac{x^2}{a^2} = \left(t+\frac{1}{t}\right)^2 = t^2 + \frac{1}{t^2} + 2$$
$$4\frac{y^2}{b^2} = \left(t-\frac{1}{t}\right)^2 = t^2 + \frac{1}{t^2} - 2$$

Subtract the second equation form the first to get

$$4\left(\frac{x^2}{a^2} - \frac{y^2}{b^2}\right) = 4$$

hence

$$\frac{x^2}{a^2} - \frac{y^2}{b^2} = 1$$

and the correct answer is $[C]$.

Problem (18). *The isosceles hyperbola $\mathcal{C}: x^2 - y^2 = a^2$ has eccentricity*

- *[A]* $\epsilon = 1$
- *[B]* $\epsilon = \sqrt{2}$
- *[C]* $\epsilon = \frac{3}{2}$
- *[D]* $\epsilon = 2$
- *[E]* *none of the above*

Solution: **Think for a while before reading on.**

The eccentricity ϵ of the hyperbola is given by $\epsilon = \sqrt{1 + \frac{b^2}{a^2}}$, where:

$$c^2 = a^2 + b^2 \Rightarrow c = \sqrt{a^2 + b^2} \overset{a=b}{=} \sqrt{2}a \Rightarrow \epsilon = \frac{\sqrt{2}a}{a} = \sqrt{1 + \frac{a^2}{b^2}} = \sqrt{2}$$

and the correct answer is $[B]$.

Problem (19). *The hyperbola*

$$\mathcal{C}: \frac{y^2}{b^2} - \frac{x^2}{a^2} = 1$$

has as asymptotic lines the

- *[A]* $y = \frac{b}{a}x,\ y = \frac{a}{b}x$
- *[B]* $y = \frac{b}{a}x,\ y = \frac{-a}{b}x$
- *[C]* $y = \frac{b}{a}x,\ y = \frac{a}{b}x$
- *[D]* $ay = bx,\ by = ax$
- *[E]* $y = \frac{b}{a}x,\ y = -\frac{b}{a}x$

Solution: **Think for a while before reading on.**
Indeed the asymptotes of the given hyperbola are

$$\frac{y}{b} - \frac{x}{a} = 0 \text{ and } \frac{y}{b} + \frac{x}{a} = 0$$

hence, the correct answer is $[E]$.

Problem (20). *The equation*

$$x^2 - 9y = 0$$

represents a

- *[A]* *circle*
- *[B]* *ellipses*

- *[C]* *parabola*
- *[D]* *straight line*
- *[E]* *none of the above*

Solution: Think for a while before reading on.
It is enough to write

$$y = \frac{1}{9}x^2$$

and the correct answer is $[C]$.

Problem (21). *The equation*

$$kx^2 - ly^2 = 0$$

represents, for all values of $k, l \in \mathbb{R}$*:*

- *[A]* *an ellipse*
- *[B]* *a hyperbola*
- *[C]* *a parabola*
- *[D]* *a circle*
- *[E]* *none of the above*

Solution: Think for a while before reading on.
We observe that for $k = -l$ the equation is written

$$x^2 + y^2 = 0$$

and thus, the graph degenerates to a single point . So the correct answer is $[E]$.

Problem (22). *The straight line* $\epsilon : x - 2y = 0$ *is an asymptote of the hyperbola*

- *[A]* $4x^2 - y^2 = 4$
- *[B]* $x^2 - 4y^2 = 4$
- *[C]* $4x^2 - 4y^2 = 1$
- *[D]* $4x^2 - y^2 = -4$
- *[E]* $x^2 - y^2 = 1$

Solution: Think for a while before reading on.
Since the asymptotes of the hyperbola

$$\frac{x^2}{a^2} - \frac{y^2}{b^2} = 1$$

are

$$y = \pm\frac{b}{a}x$$

then, for the given straight line is

$$y = \frac{1}{2}x$$

hence $a = 2$, $b = 1$ and the correct answer is $[B]$, i.e:

$$\frac{x^2}{2^2} - \frac{y^2}{1^2} = 1.$$

Problem (23). *The asymptotes of the hyperbola*

$$\mathcal{C} : \frac{y^2}{4} - \frac{x^2}{9} = 1$$

are

- *[A]* $y = \pm\frac{2}{3}x$
- *[B]* $y = \pm\frac{3}{2}x$
- *[C]* $y = \pm\frac{9}{4}x$
- *[D]* $y = \pm\frac{4}{9}x$
- *[E]* *none of the above*

Solution: Think for a while before reading on.
The given hyperbola is written

$$\frac{y^2}{2^2} - \frac{x^2}{3^2} = 1$$

therefore the asymptotes are given by the equations

$$\left(\frac{y}{2} - \frac{x}{3}\right) \cdot \left(\frac{y}{2} + \frac{x}{3}\right) = 0$$

hence

$$y = \pm\frac{2}{3} \cdot x$$

and the correct answer is $[A]$.

Problem (24). *The hyperbolas with equations*

$$\mathcal{C}_1 : \frac{x^2}{4} - \frac{y^2}{9} = 1 \text{ and } \mathcal{C}_2 : \frac{y^2}{9} - \frac{x^2}{4} = 1$$

have

- *[A]* *the same asymptotes and eccentricity*
- *[B]* *different asymptotes and different eccentricities*
- *[C]* *the same asymptotes and different eccentricities*

- *[D]* *different asymptotes and the same eccentricity*
- *[E]* *none of the above*

Solution: **Think for a while before reading on.**
The asymptotes of $\mathcal{C}_1$ are given by

$$\left(\frac{x}{2}-\frac{y}{3}\right)\cdot\left(\frac{x}{2}+\frac{y}{3}\right)=0 \Leftrightarrow y=\pm\frac{3}{2}x$$

and the asymptotes of $\mathcal{C}_2$ are given also by

$$y=\pm\frac{3}{2}x$$

so $\mathcal{C}_1$, $\mathcal{C}_2$ have the same asymptotes.
For the eccentricities, $\epsilon_2=\sqrt{1+\frac{4}{9}}$ and $\epsilon_1=\sqrt{1+\frac{9}{4}}$ so, they are different.
Correct answer is $[C]$.

Problem (25). *The number* $e=2.7$ *is the eccentricity of*

- *[A]* *an ellipse*
- *[B]* *a hyperbola*
- *[C]* *a circle*
- *[D]* *an ellipse or a hyperbola*
- *[E]* *none of the above*

Solution: **Think for a while before reading on.**
The correct answer is $[B]$ since $e>1$.

6.2.3 Polar Coordinates in plane geometry

Example (1). *Find the equation of the straight line*

$$y=3x+2$$

in polar coordinates.
Solution:
It is

$$\left.\begin{aligned} x &= r\cos\theta,\\ y &= r\sin\theta \end{aligned}\right\} \Rightarrow r\sin\theta=3r\cos\theta+2$$

and solving in r we get

$$r=\frac{2}{\sin\theta-3\cos\theta}$$

Example (2). *Find the equation of the circle*

$$\left(x-\frac{1}{2}\right)^2+y^2=1$$

in polar coordinates.

Solution: It is

$$\left.\begin{aligned} x &= r\cos\theta, \\ y &= r\sin\theta \end{aligned}\right\} \Rightarrow \left(r\cos\theta-\frac{1}{2}\right)^2+r^2\sin^2\theta=1\Rightarrow$$

$$r^2\cos^2\theta-r\cos\theta+\frac{1}{4}+r^2\sin^2\theta=1\Rightarrow$$

$$r^2-r\cos\theta=\frac{3}{4}$$

Example (3). *Find a cartesian representation of the graph of the polar equation*

$$r=\theta$$

Solution:

The given equation is a representation of the curve named as the Spiral of Archimedes[1]. It is

$$x=r\cos r$$
$$y=r\sin r$$

so

$$x^2+y^2=r^2$$
$$\frac{y}{x}=\tan r$$

therefore

$$r=\sqrt{x^2+y^2}$$
$$r=\tan^{-1}\left(\frac{y}{x}\right)$$

and thus we obtain the implicit cartesian representation

$$\sqrt{x^2+y^2}=\tan^{-1}\left(\frac{y}{x}\right)$$

Example (4). *Draw the graph of the curve given by*

$$r=2$$

[1]Archimedes used calculus about 2000 years before it was formally established by Newton and Leibnitz, hence he is considered as the greatest mathematician of the world

Solution:
It is

$$x = r\cos\theta = 2\cos\theta$$
$$y = r\sin\theta = 2\sin\theta$$

hence

$$x^2 + y^2 = 4$$

and the graph is a circle with center to the origin and radius 2.

Example (5). *The polar equation*

$$r = a\cos\theta$$

is a circle with diameter a with its left- most edge at the pole.

Solution:
Write

$$x = r\cos\theta$$
$$y = r\sin\theta$$
$$\cos\theta = \frac{x}{r}$$

so

$$x^2 + y^2 = r^2 \overset{r=a\cos\theta}{\Rightarrow} \sqrt{x^2+y^2} = \frac{a \cdot x}{\sqrt{x^2+y^2}} \Rightarrow$$
$$x^2 + y^2 = ax \Rightarrow (x - \frac{a}{2})^2 + y^2 = \frac{a^2}{4}$$

Example (6). *The polar equation*

$$r = a\sin\theta$$

is a circle with diameter a with its bottom - most edge at the pole.

Solution:
Working similarly as in preceding example we get:

$$\sqrt{x^2+y^2} = \frac{a \cdot y}{\sqrt{x^2+y^2}} \Rightarrow x^2 + y^2 = ay \Rightarrow$$
$$x^2 + \left(y - \frac{a}{2}\right)^2 = \frac{a^2}{4}$$

6.2.4 3 - dimensional geometry

Let $P(x_1, y_1, z_1)$, $Q(x_2, y_2, z_2)$ be two points in 3 - dimensional Euclidean space, then:

1. The **distance** between the points P, Q is

$$d(P,Q) = \sqrt{(x_1 - x_2)^2 + (y_1 - y_2)^2 + (z_1 - z_2)^2}$$

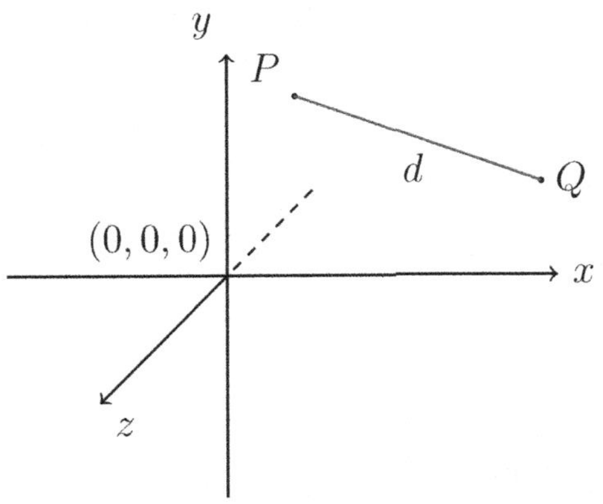

2. Any **plane** in the space has an equation of the form

$$Ax + By + Cz + D = 0$$

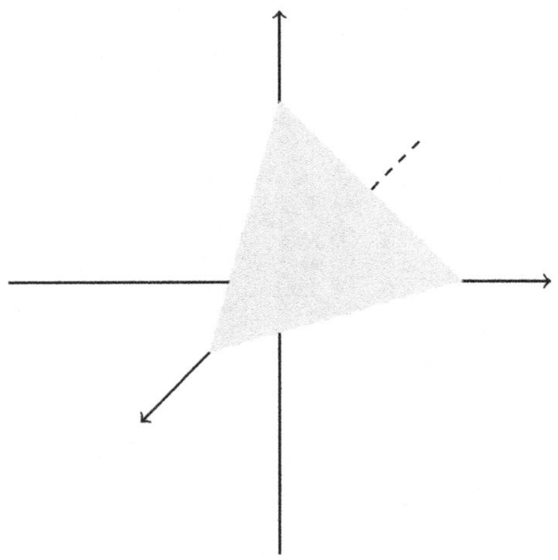

3. Any plane π **parallel** to the Oxz coordinate plane has an equation of the form

$$y = p$$

   and $(0, p, 0)$ is the point where π intersects the Oy - axis.

4. If $A = 0$, the

$$By + Cz + D = 0$$

   is the equation of a plane parallel to the Ox - axis.

5. The **distance** between the point $Q(x_1, y_1, z_1)$ and the plane

$$Ax + By + Cz + D = 0$$

   is given by

$$d(Q, \pi) = \frac{|Ax_1 + By_1 + Cz_1 + D|}{\sqrt{A^2 + B^2 + C^2}}$$

Problem (1). *The equation $x = 3$ represents a plane in 3 - dimensional space, parallel to the*

- *[A]* y *- axis*

- *[B]* z *- axis*
- *[C]* Oxy *plane*
- *[D]* Oxz *plane*
- *[E]* Oyz *plane*

Solution: The correct answer is $[E]$.

Problem (2). *The equation*

$$2x + y = 4$$

represents a plane

- *[A]* *parallel to the* Oxy *plane*
- *[B]* *parallel to the* Oyz *plane*
- *[C]* *parallel to the* Oxz *plane*
- *[D]* *parallel to the* Oz *- axis*
- *[E]* *none of the above*

Solution: The correct answer is $[D]$.

Problem (3). *The equation*

$$y = 4$$

represents a plane

- *[A]* *perpendicular to the* Oz *- axis*
- *[B]* *perpendicular to the* Oxy *plane*
- *[C]* *parallel to the* Ox *axis*
- *[D]* *parallel to the* Oxz *plane*
- *[E]* *none of the above*

Solution: The correct answer is $[D]$

Chapter 7

Topic 4 - Sequences - Complex numbers - Vectors - Matrices

7.1 Sequences

7.1.1 Summary

1. $a, a + d, a + 2d, \ldots$
   it is an **arithmetic sequence** (or **progression**)
   the nth term is given by
$$a_n = a + (n-1) \cdot d$$
2. The **sum** of the n first terms of an arithmetic sequence is given by
$$S_n = \frac{n}{2}(a + a_n) = \frac{n}{2}[2a + (n-1)d]$$
3. $a, ar, ar^2, \ldots$
   it is a **geometric sequence**
   the nth term is given by
$$a_n = a \cdot r^{n-1}$$
4. The **sum** of the n first terms of a geometric sequence is given by
$$S_n = \frac{a \cdot (r^n - 1)}{r - 1} = \frac{a \cdot (1 - r^n)}{1 - r}, \; r \neq 1$$
5. In a geometric sequence, if $0 < |r| < 1$ then the **sum of all terms** is given by
$$S = \frac{a}{1 - r}$$

7.1.2 Arithmetic sequences

Example. *[1] Let*

$$3, 5, 7, 9, \ldots,$$

be given.
The ***difference between two successive terms*** *is always the* ***same.***
In this example

$$d = 2$$

Example. *[2] Let* $-5, -2, 1, 4, 7, \ldots\ldots$ *The difference of any pair of successive terms is*

$$d = -2 - (-5) = 1 - (-2) = 3$$

More generally we write

$$a, a + d, a + 2d, \ldots\ldots$$

What is the nth term? It is

$$a_n = a + (n-1)d$$

For instance, in example 7.1.2 the 100th term is

$$a_{100} = 3 + 2 \cdot (100 - 1) = 3 + 99 \cdot 2 = 201$$

What is the sum of the first 100 terms ?
The formula is

$$S_{100} = \frac{100}{2}[a + a_{100}] = \frac{100}{2}[3 + 201] = 50 * 204 = 10.200$$

Generally

$$S_n = \frac{n}{2}[a + a_n] = \frac{n}{2}[2a + (n-1)d]$$

Multiple Choice Problems

Problem (1). *The* 7th *term of an arithmetic sequence is* 15 *and the* 15th *is* 31. *The sum of the first* 20 *terms is*

- *[A]* 380
- *[B]* 400
- *[C]* 410
- *[D]* 440
- *[E]* 460

Solution - procedure:

We have to compute

$$S_{20} = \frac{20}{2}[2a + (20-1) \cdot d]$$

What is missing?
a and d are missing

$$a_7 = 15 \rightarrow 15 = a + (7-1) \cdot d$$

$$a_{15} = 31 \rightarrow 31 = a + (15-1) \cdot d$$

Look at everything we have written. System:

$$15 = a + 6d$$

$$31 = a + 14d$$

subtract
$16 = 8 \cdot d$ to get
$d = 2$ and $a = 3$ hence

$$S_{20} = 10 \cdot (2 \cdot 3 + 19 \cdot 2) = 440$$

the correct answer is $[D]$.

Problem (2). *Given that a, b and c are successive terms in an arithmetic sequence and*

$$(b-c)x^2 + (a-c)x + (a-b) = 0$$

then x is equal to:

- *[A]* -1
- *[B]* 0
- *[C]* 1
- *[D]* $\frac{1}{2}$
- *[E]* 2

Solution-procedure:

Use that a, b, c are successive terms
of an arithmetic sequence
Let $d \neq 0$ be the *difference* of the sequence
Write

$$b = a + d, \quad c = a + 2d$$

hence

$$-dx^2 - 2dx - d = 0$$

so

$$-d\left(x^2 + 2x + 1\right) = 0 \Leftrightarrow$$
$$-d(x+1)^2 = 0 \Leftrightarrow$$
$$x + 1 = 0 \Leftrightarrow x = -1$$

and the correct answer is $[A]$.

Problem (3). *If f is a function such that $f(1) = 2$ and*

$$f(n+1) = \frac{3f(n)+1}{3}$$

then $f(100)$ is

- *[A]* -35

- *[B]* 15
- *[C]* 35
- *[D]* 40
- *[E]* 45

Solution-procedure:

Let $f(n) = a_n$ and observe

$$f(n+1) = \frac{3f(n)+1}{3} \Leftrightarrow$$
$$3(a_{n+1} - a_n) = 1 \Leftrightarrow$$
$$d = a_{n+1} - a_n = \frac{1}{3}$$

therefore (a_n) is an arithmetic sequence with $d = \frac{1}{3}$ and thus

$$a_{100} = f(100) = a_1 + (100-1) \times \frac{1}{3} = 2 + \frac{99}{3} = 35$$

and the correct answer is $[C]$.

Problem (4). *If the interior angles of a convex pentagon form an arithmetic sequence and the smallest interior angle is* 90° *the possible values of the largest angle in the pentagon are*

- *[A]* 90°
- *[B]* 116°
- *[C]* 126°
- *[D]* 136°
- *[E]* 180°

Solution-procedure:

Recall first that for an n-polygon
the sum of its interior angles is

$$\Sigma_n = n \times 180^o - 360^o = 180^o \times (n-2)$$

so for a pentagon

$$\Sigma_5 = 3 \times 180^o = 540^o$$

Since $5 \times 90^o = 450 < 540^o$ we describe the set of the angles by

$$\{90^o, 90^o + d, 90^o + 2d, 90^o + 3d, 90^o + 4d\}$$

hence

$$90^o + (90^o + d) + (90^o + 2d) + (90^o + 3d) + (90^o + 4d) = 540^o \Rightarrow$$
$$450^o + 10d = 540^o \Rightarrow$$
$$10d = 90^o$$
$$\Rightarrow d = 9^o$$

therefore the largest angle shall be

$$90^o + 4 \times 9^o = 90^o + 36^o = 126^o$$

and the correct answer if $[C]$.

Problem (5). *The sum*

$$\sum_{n-1}^{100} (5 + 2n)$$

is equal to

- *[A]* 10105
- *[B]* 10500
- *[C]* 10550
- *[D]* 10600
- *[E]* 10800

Solution:

Where is the subtle point?
Remember

$$\sum_{n=1}^{100} 5 = 100 \times 5 = 500$$

so it

$$\sum_{n-1}^{100} (5 + 2n) = \sum_{n=1}^{100} + 2\sum_{n=1}^{100} n = 500 + 2\sum_{n=1}^{100} n$$
$$= 500 + 2\frac{1 \times (100 + 1)}{2} \times 100 = 500 + 101 \times 100 = 10600$$

and the correct answer is $[D]$.

Problem (6). *The sum of the first* 50 *terms of the arithmetic sequence*

$$\frac{1}{36}, \frac{1}{18}, \frac{1}{12}, \ldots$$

is

- *[A]* 30

- *[B]* 31.4
- *[C]* 32.5
- *[D]* 33.5
- *[E]* 35.5

Solution: We use the formula

$$S_n = \frac{n}{2}\left[2a_1 + (n-1)d\right]$$

where $a_1 = \frac{1}{36}$, $n = 50$ and $d = \frac{1}{18} - \frac{1}{36} = \frac{1}{36}$, hence

$$S_{50} = \frac{50}{2}\left[\frac{2}{36} + 49 \times \frac{1}{36}\right] = 35.5$$

Correct answer is $[E]$.

Problem (7). *The sum of the first 20 positive integers, divisible by 7, is*

- *[A]* 1400
- *[B]* 1470
- *[C]* 1500
- *[D]* 1570
- *[E]* 1840

Solution: Where is the subtle point?
Make it explicit

$$\text{Sum} = 7 + 14 + 21 + \cdots$$

Formula

$$S = \frac{n}{2}\left[2a_1 + (n-1)d\right], \text{ where } a_1 = 7,\ d = 7,\ n = 20$$

so

$$S_{20} = \frac{20}{2}\left[2 \times 7 + (20-1) \times 7\right] = 1470$$

Problem (8). *The sum of the first 20 terms in an arithmetic sequence whose 7th and 10th terms are 21 and 33 respectively, is*

- *[A]* 240
- *[B]* 260
- *[C]* 290

- *[D]* 300
- *[E] none of the above*

Solution: Remember the relevant formula

$$S_n = \frac{n}{2}\left[2a_1 + (n-1)d\right],$$

where

$$a_7 = a_1 + 6d = 21,\ \ a_{10} = a_1 + 9d = 33$$

so $a_1 = -3$, $d = 4$ and thus

$$\text{Sum} = \frac{20}{2}\left[2 \times (-3) + (20-1) \times 4\right] = 700$$

Problem (9). *The sum of the positive integers between* 5 *and* 150*, divisible by* 9*, is*

- *[A]* 99
- *[B]* 117
- *[C]* 146
- *[D]* 155
- *[E]* 405

Solution: The numbers are

$$9, 18, 27, \ldots$$

hence, this is an arithmetic sequence.
Missing? the order of the last term:

$$a_n = 9 + (n-1)9 = 9 + 9n - 9 < 150 \Rightarrow n < \frac{150}{9} < 17 \Rightarrow n = 16$$

hence

$$S_{16} = \frac{16}{2}\left[2 \times 9 + (16-1) \times 9\right] = 405.$$

Correct answer is $[E]$.

7.1.3 Geometric sequences or Progressions

Example (1).

$$1, 2, 4, 8, 16, \ldots\ldots$$

The ratio between any two consecutive terms is always the same r*, in the example*

$$r = \frac{2}{1} = \frac{4}{2} = \frac{8}{4} = \cdots = 2$$

Generally we write:

$$a, ar, ar^2, \ldots$$

and the nth term is given by

$$a_n = a \cdot r^{n-1}$$

What is the 20th term in our case-example?

$$a_{20} = 1 \cdot 2^{19}$$

Now, a formula for the sum of the first n terms, it is:

$$S_n = \frac{a(r^n - 1)}{r - 1}, \ r \neq 1$$

In our example

$$S_n = \frac{1 \cdot (2^n - 1)}{2 - 1} = 2^n - 1$$

Be careful! *If $0 < |r| < 1$, then we derive the sum of all terms using the formula*

$$S = \frac{a}{1 - r}$$

Example (2).

$$1, \frac{1}{2}, \frac{1}{4}, \frac{1}{8}, \ldots\ldots$$

here $r = \frac{1}{2}, \quad |r| < 1$ *so*

$$S = \frac{1}{1 - \frac{1}{2}} = \frac{1}{\frac{1}{2}} = 2$$

Example. *[3] The exact value of*

$$0.222222\ldots$$

is

$$0.222\ldots = \frac{2}{10} + \frac{2}{100} + \frac{2}{1000} + \ldots$$

hence

$$0.222\ldots = \frac{\frac{2}{10}}{1 - \frac{1}{10}} = \frac{\frac{2}{10}}{\frac{9}{10}} = \frac{2}{9}$$

Example. *[4] The exact value of*

$$0.53333\ldots$$

is

$$0.533\ldots = \frac{5}{10} + \left\{ \frac{3}{100} + \frac{3}{1000} + \frac{3}{10000} + \cdots \right\}$$

so

$$0.5333\ldots = \frac{5}{10} + \frac{\frac{3}{100}}{1 - \frac{1}{10}} = \frac{5}{10} + \frac{\frac{3}{100}}{\frac{9}{100}} = \frac{5}{10} + \frac{3}{90} = \frac{8}{15}$$

7.1.4 Multiple Choice Problems

Problem (1). *The* 3rd *term of a geometric sequence with positive terms is* 4 *and the* 5th *is* 16. *The* 7th *terms is:*

- *[A]* 16
- *[B]* 32
- *[C]* 64
- *[D]* 128
- *[E]* 256

Solution - procedure

$$7^{term} \text{ is } a_7 = a \cdot r^6$$
$$\text{Missing } a \text{ and } r$$
$$\text{Use formulas}$$
$$3^{rd} \text{ term}: \quad a_3 = a \cdot r^2 \to 4 = ar^2$$
$$5^{rd} \text{ term}: \quad a_5 = a \cdot r^4 \to 16 = ar^4$$
$$\text{Solve the system: } ar^2 = 4 \text{ and } ar^4 = 16 \text{ start by dividing}$$
$$r^2 = 4 \Rightarrow r = 2 \text{ and}$$

$$a = \frac{a_3}{r^2} = \frac{4}{4} = 1$$

hence

$$a_7 = 1 \cdot 2^6 = 64$$

and the correct answer is $[C]$.

Problem (2). *If the sum of the first n terms of a geometric sequence is* $10^n - 1$, *then the eight term is*

- *[A]* 312000
- *[B]* 321500
- *[C]* 525000
- *[D]* 625000
- *[E] none of the above*

Solution: Where is the subtle point?
Let us describe

$$\frac{a_1(r^n - 1)}{r - 1} = 10^n - 1,$$

where a_1 the first term and r the ratio of the geometric sequence.

$$a_8 = S_8 - S_7 = 10^8 - 1 - (10^7 - 1) = 10^8 - 10^7 = 9 \times 10^7$$

Problem (3). *In a geometric sequence with positive terms the first term is* 1 *and the fifth is* 16. *Then, the sum of the first* 6 *terms is*

- *[A]* 32
- *[B]* 36
- *[C]* 48
- *[D]* 60
- *[E]* 63

Solution - procedure

We have to find

$$S_6 = \frac{a_1(r^6-1)}{r-1} = \frac{(r^6-1)}{r-1}$$

since $a_1 = 1$,
Missing?
r is missing
Use the data
5th term is 16 so $\to 16 = 1 \cdot r^4$
hence $r = 2$ and thus

$$S_6 = \frac{2^4-1}{2-1} = 63$$

and the correct answer is $[E]$.

Problem (4). *Three different numbers, whose product is* 125, *are consecutive terms in a geometric sequence and the same time are the first, third and the sixth terms of an arithmetic sequence. The numbers are:*

- *[A]* 1, 5, 25
- *[B]* $\frac{10}{3}$, 5, $\frac{15}{2}$
- *[C]* $\frac{1}{5}$, $\frac{24}{5}$, $\frac{30}{3}$
- *[D]* 3, 5, 7
- *[E] none of the above*

Solution-procedure:

If ω is the *ratio* of the geometric sequence
and a its first term then
an easy way is to symbolize them by

$$\{\frac{a}{\omega}, a, a \cdot \omega\}$$

and taking their product

$$a^3 = 125 \Leftrightarrow a^3 = 5^3$$

and thus the middle terms is $a = 5$. On the other hand if d is the difference of the arithmetic sequence then

$$\{5 - 2d,\ 5,\ 5 + 3d\}$$

describe them as 1st, 3rd and 5th term of this sequence. Then : since their product is 125, we find

$$25d - 30d^2 = 0 \Leftrightarrow d = 0 \text{ or } d = \frac{5}{6}$$

so

$$a_1 = \frac{10}{3},\ a_2 = 5,\ a_3 = \frac{15}{2}$$

and the correct answer is [B].

Problem (5). *If $x, 4, y$ are successive terms in an arithmetic sequence and $x, 3, y$ are successive terms in a geometric sequence then the sum $\frac{1}{x} + \frac{1}{y}$ is equal to:*

- *[A]* $\frac{7}{9}$
- *[B]* $\frac{8}{9}$
- *[C]* 1
- *[D]* $\frac{10}{9}$
- *[E]* $\frac{11}{8}$

Solution-procedure:

Observe that since $x, 4, y$ are successive terms of an arithmetic sequence then

$$x = 4 - d,\ 4,\ y = 4 + d$$

and also, using the fact that $x, 3, y$ are successive terms of a geometric sequence then

$$x = \frac{3}{\omega},\ 3,\ y = 3\omega$$

Further

$$\frac{1}{x} + \frac{1}{y} = \frac{x + y}{x \cdot y} = \frac{8}{9}$$

and the correct answer is [B].

Problem (6). *The sum of the first n terms of a sequence is*

$$S_n = 3^n - 1$$

where n is a positive integer.
If t_n represents the nth term of the sequence then

$$\frac{t_{n+1}}{t_n}$$

is equal to

- *[A]* -2
- *[B]* $\frac{2}{3}$
- *[C]* $\frac{3}{2}$
- *[D]* $+2$
- *[E]* 3

Solution-procedure:

Since

$$S_{n+1} = S_n + t_{n+1}, \; S_n = S_{n-1} + t_n$$

and

$$S_n = 3^n - 1$$

we get

$$\begin{aligned} \frac{t_{n+1}}{t_n} &= \frac{S_{n+1} - S_n}{S_n - S_{n-1}} = \frac{3^{n+1} - 1 - (3^n - 1)}{3^n - 1 - (3^{n-1} - 1)} \\ &= \frac{3^{n+1} - 3^n}{3^n - 3^{n-1}} = \frac{3^n(3-1)}{3^{n-1}(3-1)} = \frac{3^n}{3^{n-1}} \end{aligned}$$

hence

$$\frac{t_{n+1}}{t_n} = 3$$

and the correct answer is $[E]$,
i.e: (t_n) *is a geometric sequence with ratio* 3 *and* $t_1 = 3^1 - 1 = 2$

7.2 Complex Numbers

7.2.1 Basic theory

If $i \notin \mathbb{R}$ is the **imaginary unit**

$$i^2 = -1$$

then

$$\mathbb{C} = \{z = a + i \cdot b \, : \, a, b \in \mathbb{R}\}$$

is called the set of **complex numbers**.
We call $a = \Re z$ the **real part** and $b = \Im z$ the **imaginary part** of z.
On $\mathbb{C}$ we define the following operations

$$(a + ib) + (a' + ib') = (a + a') + i(b + b') \text{ addition}$$
$$((a + ib) \cdot (a' + ib')) = (aa' - bb') + i(ab' + a'b) \text{ multiplication}$$

A complex number $z = a + ib$ is represented geometrically as a point in the 2 - plane with coordinates (a, b)

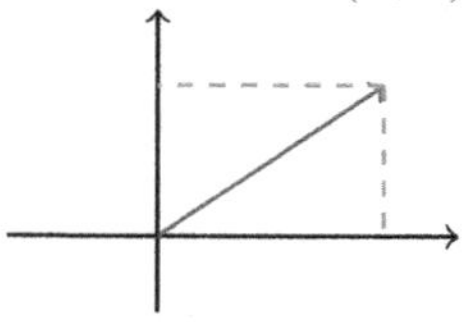

- Two complex numbers $z = x + iy$, $z' = x' + iy'$ are **equal** if and only if

$$x = x' \text{ and } y = y'$$

- The real numbers are considered as a subset of $\mathbb{C}$ by writing each $x \in \mathbb{R}$ as $x + i \cdot 0 \in \mathbb{C}$
- The operations of **addition** and **multiplication** we defined are *extensions* of the well known corresponding operations defined in $\mathbb{R}$, i.e: if we add or multiply the real numbers $x = x + i \cdot 0$, $y = y + i \cdot 0$ using the definitions given for $\mathbb{C}$ then

$$x + y = (x + i \cdot 0) + (y + i \cdot y), \;\; x \cdot y = (x + i \cdot 0) \cdot (y + i \cdot y)$$

- For each complex number $z = x + iy$ is defined its **conjugate** $\bar{z}$ by setting

$$\bar{z} = x - iy$$

and the relation

$$z\bar{z} = x^2 + y^2 = |z|$$

holds true. Then, we define

$$|z|\sqrt{x^2 + y^2} \text{ to be the modulus of } z$$

- Geometric meaning of the conjugation operation: $\bar{z}$ is the **symmetric - mirror** of z with respect to the axis $x'Ox$

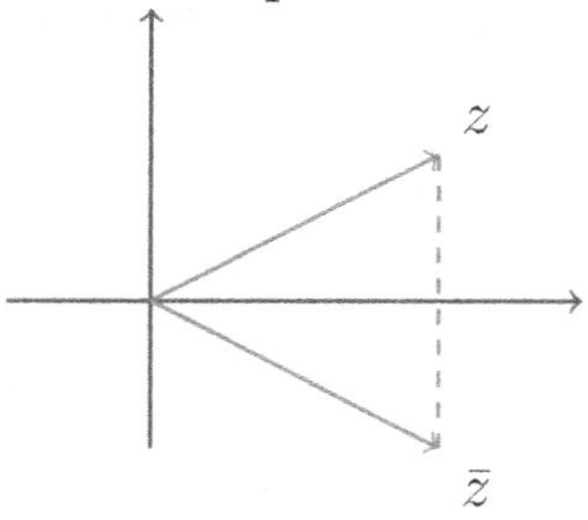

- For each complex number $z = x + i \cdot y \neq 0$ the inverse $z^{-1} \in \mathbb{C}$ is defined by the property

$$z \cdot z^{-1} = 1 \Rightarrow z^{-1} = \frac{1}{z} = \frac{\bar{z}}{|z|^2} = \frac{x}{x^2+y^2} - i\frac{y}{x^2+y^2}$$

- Geometric interpretation of the action of the **imaginary unit** i on any complex number $z = x + iy$. It is

$$iz = i(x+iy) = -y + ix$$

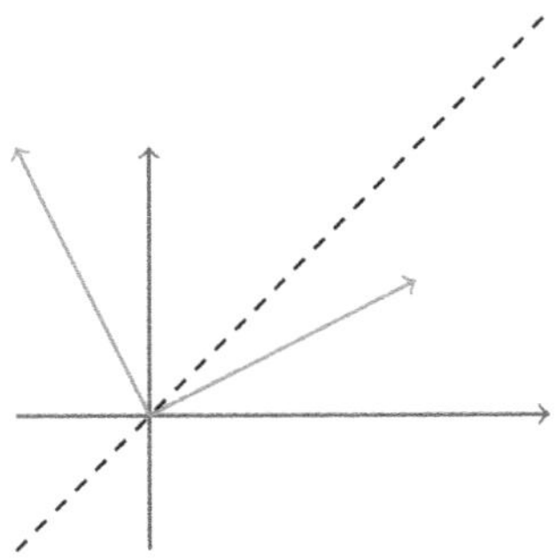

hence $i \cdot z$ represents an anticlock - wise **rotation** of z by $\frac{\pi}{2}$.

- **Powers** of i: $i^2 = -1,\ i^3 = -i,\ i^4 = 1,\ i^5 = i,\ i^6 = -1 \ldots\ldots$ hence

$$i^{2k} = (i^2)^k = (-1)^k,\ k \in \mathbb{N}$$
$$i^{2k+1} = i^{2k} \cdot i = (-1)^k \cdot i,\ k \in \mathbb{N}$$

Methodology

[1] **Transform a complex representation into the from** $a + bi$

- If there exist powers of i, we use the relations

$$i^2 = -1,\ i^3 = -i,\ i^4 = 1$$

Example (1).

$$i^{25} = i^{4\times 6+1} = (i^4)^6 \times i = i,$$
$$i^{86} = i^{4\times 21+2} = (i^4)^{21} \times i^2 = -1$$

- Powers with an exponent $n > 4$ can be reduced in the form

$$i^n = i^{4k+r} = (i^4)^k + i^i = ir,\ r \in \{0,1,2,3\}$$

- If there exist expressions of the form $(a+bi)^2$, $(c+di)^3$ and so on, we apply the standard identities.
- If there exist denominators of the form $a + bi$ we multiply both the numerator and the denominator with the conjugate $a - bi$.

Example (2). *Write into the form* $a + bi$ *the following expressions*

$$(2-i)^3 \cdot 4i = \left[2^3 - 3 \cdot 2^2 \cdot i + 3 \cdot 2 \cdot i^2 - i^3\right] \cdot 4i = [2 - 11i] \cdot 4i = 44 + 8i$$
$$\frac{2}{1+i} - \frac{3}{2+i} = \frac{2(1-i)}{(1+i)(1-i)} - \frac{3(2-i)}{(2+i)(2-i)} = \frac{2-2i}{2} - \frac{6-3i}{5} = -\frac{1}{5} - \frac{2}{5}i$$

Example (3). *Find all the $m \in \mathbb{Z}$ such that*

$$i^{5m+2} = 1.$$

Skip if it seems difficult to you.
Let $m = 4k + r$ with $k, r \in \mathbb{Z}$ and $r \in \{0, 1, 2, 3\}$. Then:

$$i^{5m+2} = i^{5\times(4k+r)+2} = i^{20k+5r+2} = (i^4)^{5k} \cdot (i^5)^r \cdot i^2 = -i^r$$

and thus

$$i^{5m+2} = 1 \Leftrightarrow -i^r = 1 \Rightarrow i^r = -1 \Rightarrow r = 2,$$

hence $m = 4k + 2$, $k \in \mathbb{Z}$.

[2] **Solving exercises including equalities of complex expressions with a, b, ... real numbers**

- We work on both members of the given equality so that it takes the form

$$A + Bi = A_1 + B_1 i$$

where A, B, A_1, B_1 are expressions non including the imaginary variable i.

- We solve the system

$$A = A_1 \text{ and } B = B_1$$

and use the deduced information to the original problem.

Example (4). *Find all $a, b \in \mathbb{R}$ such that*

$$(a - b) + (a^2 - 4b^2)i = 4.$$

Write

$$(a - b) + (a^2 - 4b^2)i = 4 + 0i \Leftrightarrow \begin{cases} a - b = 4 \\ a^2 - 4b^2 = 0 \end{cases} \Leftrightarrow \begin{cases} a = b + 4 \\ a = \pm 2b \end{cases}$$

hence $a = 8$, $b = 4$ or $a = \frac{8}{3}$, $b = \frac{-4}{3}$

Example (5). *Find all $x, y \in \mathbb{R}$ such that*

$$\frac{x}{2+i} + \frac{yi}{2-i} = \frac{1}{1-2i}$$

Multiplying numerator and denominator of each fraction with the conjugate of the corresponding denominator, we get:

$$\frac{x}{2+i} + \frac{yi}{2-i} = \frac{1}{1-2i} \Leftrightarrow \frac{x(2-i)}{(2+i)(2-i)} + \frac{yi(2+i)}{(2-i)(2+i)} = \frac{1+2i}{(1-2i)(1+2i)} \Leftrightarrow$$

$$\frac{2x - xi}{5} + \frac{-y + 2yi}{5} = \frac{1+2i}{5} \Leftrightarrow \begin{cases} 2x - y = 1 \\ -x + 2y = 2 \end{cases}$$

and thus $x = \frac{4}{3}$, $y = \frac{5}{3}$.

Example (6). *Find all x, y real numbers, such that*

$$x(1+i)^2 + y(2-i)^2 = 3 + 10i.$$

It is

$$x(1+i)^2 + y(2-i)^2 = 3 + 10i \Leftrightarrow$$
$$3y + (2x - 4y)i = 3 + 10i \Leftrightarrow \begin{cases} 3y = 3 \\ 2x - 4y = 10 \end{cases}$$

hence $x = 7$, $y = 1$.

[3] **Working with an expression of a complex variable which is equal to a real or to an imaginary number**

- Let u be the expression, then:
- If u is a real number, we consider the equality $u = \bar{u}$
- If u is an imaginary number, we consider the equality $u = -\bar{u}$.

Example (7). *If $z = x + yi$ and $u = (z-1)(\bar{z} - i)$ is a real number, prove that $x = y + 1$.*
Indeed,

$$u \in \mathbb{R} \Leftrightarrow \bar{u} = u \Leftrightarrow \overline{(z-1)(\bar{z}-i)} = (z-1)(\bar{z}-i) \Leftrightarrow$$
$$(\bar{z}-1)(z+i) = (z-1)(\bar{z}-i) \Leftrightarrow -zi - \bar{z} + i = \bar{z}i - z - i$$

and thus, for $z = x + yi$:

$$x - y - 1 = 0 \Leftrightarrow x = y + 1.$$

More Examples

Example (8). *Calculate*

$$Q(i) = i + i^2 + i^3 + i^4 + i^5$$

Solution:

$$Q(i) = i + (-1) + i^2 i + i^2 i^2 + i^4 i = i - 1 - i + 1 + i = i$$

Example (9). *Find $l \in \mathbb{R}$ such that $z = \dfrac{l+4i}{1+li}$ is a real number.*
Solution: *It is*

$$z = \frac{l+4i}{1+li} \in \mathbb{R} \Leftrightarrow \frac{(l+4i)(1-li)}{(1+li)(1-li)} \in \mathbb{R} \Leftrightarrow$$
$$\frac{5l}{l^2+1} + \frac{4-l^2}{1+l^2} \cdot i \in \mathbb{R} \Leftrightarrow \frac{4-l^2}{1+l^2} = 0 \Leftrightarrow l^2 = 2$$

hence $l = \pm 2$.

Example (10). *Let $z = x + yi$, $x, y \in \mathbb{R}$, be a complex number such that*

$$z\bar{z} + 4z + 4\bar{z} - 9 = 0$$

Prove that the points $M(x, y)$ lie on a circle and find the center and the radius of this circle.
Solution: *Using $z = x + iy$, we have $z\bar{z} = |z|^2 = x^2 + y^2$, $z + \bar{z} = 2x$, therefore:*

$$|z|^2 + 4z + 4\bar{z} - 9 = 0 \Leftrightarrow x^2 + y^2 - 8x - 9 = 0 \Leftrightarrow$$
$$(x-4)^2 + y^2 = 5^2$$

which is the equation of a circle with center $K(4, 0)$ and radius 5.

Example (11). *Let $z = x + yi$ be a complex number, if $u = \dfrac{z - 2i}{z - 1}$, $z \neq 1$ and $\mathbb{I}m(u) = 0$, prove that the image of z lies on a straight line.*
Solution: *It is*

$$u = \frac{z - 2i}{z - 1} = \frac{(z - 2i)(\bar{z} - 1)}{(z - 1)(\bar{z} - 1)} = \frac{|z|^2 - z - 2\bar{z}i + 2i}{1 + |z^2|} =$$
$$\frac{x^2 + y^2 - x - yi - 2xi - 2y + 2i}{1 + x^2 + y^2} =$$
$$\frac{x^2 + y^2 - x - 2y}{1 + x^2 + y^2} + \frac{2 - 2x - y}{1 + x^2 + y^2} \cdot i \in \mathbb{R} \Rightarrow$$
$$\frac{2 - 2x - y}{1 + x^2 + y^2} = 0 \Rightarrow y = -2x + 2$$

and the assertion has been proved.

Example (12). *In the complex plane the point M moves on the image of z such that*

$$z\bar{z} - 2(z + \bar{z}) + 3 = 0 \ (1)$$

and the point N moves on the image of z_1 such that

$$z_1\bar{z}_1 + 2(z_1 + \bar{z}_1) = 1 \ (2).$$

Find the trajectories of the points M and N and the point where these trajectories intersect, if there exist such a point.
Solution: *Let $M(z)$ with $z = x + yi$ then using $(\mathbf{1})$ we have:*

$$z\bar{z} - 2(z + \bar{z}) + 3 = 0 \Leftrightarrow x^2 + y^2 - 4x + 3 = 0$$

so

$$(x - 2)^2 + y^2 = 1 \qquad \text{(a)}$$

and thus, M moves on a circle with center $(2, 0)$ and radius 1.
Similarly working for the point $N(z_1)$, we use the same notation $z_1 = x + yi$ for the coordinates of the point:

$$z_1\bar{z}_1 + 2(z_1 + \bar{z}_1) = 1 \Leftrightarrow (x + 1)^2 + y^2 = 2 \qquad \text{(b)}$$

and thus N moves on a circle with center $(-1, 0)$ and radius $\sqrt{2}$.
It is easy to check, graphically, that the circles given by (a) and (b) do not intersect and consequently the points M and N dot not meet each other.

Problems

Problem (1). *The standard form of* $a+ib$ *of* $\dfrac{2+i}{1-i}$ *is*

- *[A]* $1+3i$
- *[B]* $\frac{1}{2}-\frac{3}{2}i$
- *[C]* $1-3i$
- *[D]* $\frac{1}{2}+\frac{3}{2}i$
- *[E]* $-1+3i$

Solution: It is

$$\frac{2+i}{1-i}=\frac{(2+i)(1+i)}{(1-i)(1+i)}=\frac{1+3i}{2}=\frac{1}{2}+\frac{3}{2}i$$

Correct answer is $[D]$.

Problem (2). *The modulus of* $3-4i$ *is*

- *[A]* 4
- *[B]* 5
- *[C]* 7
- *[D]* 8
- *[E]* 25

Solution: It is

$$|z|=\sqrt{z\cdot\bar{z}}=\sqrt{3^2+4^2}=\sqrt{25}=5$$

Correct answer is $[B]$.

Problem (3). *The roots of*

$$x^2+x+1$$

are

- *[A]* $\frac{-1}{2}\pm\frac{\sqrt{3}}{2}i$
- *[B]* $\frac{1}{2}+\frac{\sqrt{3}}{2}i$ *only*
- *[C]* $\frac{-1}{2}-\frac{\sqrt{3}}{2}i$ *only*
- *[D]* $\frac{-1}{2}+\frac{\sqrt{3}}{2}i$ *only*
- *[E] none of the above*

Solution: Use formula

$$x_{1,2} = \frac{-b \pm \sqrt{b^2 - 4ac}}{2a} = \frac{-1 \pm \sqrt{-3}}{2} \overset{\sqrt{-3}=i\sqrt{3}}{=} \frac{-1 \pm \sqrt{3}i}{2}$$

Correct answer is $[A]$.
The roots are complex conjugate numbers.

Problem (4). *If $1+i$ is a root of $x^3 + ax + b = 0$, a, b are real numbers, then $1-i$ is also a root of the same polynomial.*

Solution: Indeed, since $1+i$ is a root of $x^3 + ax + b = 0$ then:

$$(1+i)^3 + a(1+i) + b = 0 \Rightarrow \overline{(1+i)^3 + a(1+i) + b} = 0 \overset{\overline{z+w}=\bar{z}+\bar{w}}{\Rightarrow}$$
$$\overline{(1+i)^3} + a\overline{(1+i)} + b = 0 \overset{\bar{z^\nu}=(\bar{z})^\nu}{\Rightarrow} (1-i)^3 + a(1-i) + b = 0$$

hence $1 - i = \overline{1+i}$ is also a root of $x^3 + ax + b = 0$.
Remark. *The above result holds true for any polynomial of degree k, i.e: if $p(x) = a_0 + a_1 x + \cdots a_k x^k$, $a_i \in \mathbb{R}$ and $p(z_0) = 0$ for $z_0 \in \mathbb{C}$ then $p(\bar{z_0}) = 0$.*
We conclude that if a polynomial p has all its roots complex numbers in $\mathbb{C} \setminus \mathbb{R}$ then the degree of p is an even number.

Problem (5).

1. *If $z = 2ab + (a^2 - b^2)i$, $a, b \in \mathbb{R}$ then z is real if and only if*
   - *[A]* $a = 0$
   - *[B]* $b = 0$
   - *[C]* $ab = 0$
   - *[D]* $|a| = |b|$
   - *[E]* $a = b$
2. *If $z = 2ab + (a^2 - b^2)i$, $a, b \in \mathbb{R}$ then $|z|$ is*
   - *[A]* a
   - *[B]* b
   - *[C]* $a^2 + b^2$
   - *[D]* $2ab$
   - *[E]* 1

Solution: 1) It is

$$z \in \mathbb{R} \Leftrightarrow a^2 - b^2 = 0 \Leftrightarrow \sqrt{a^2} = \sqrt{b^2} \Leftrightarrow |a| = |b|$$

hence $[D]$ holds true.
2) It is

$$|z| = \sqrt{(2ab)^2 + (a^2 - b^2)^2} = \sqrt{a^4 + 2a^2b^2 + b^4}$$
$$= \sqrt{(a^2 + b^2)^2} = a^2 + b^2$$

hence, the $[C]$ holds true.

Problem (6). *If $z_1 = 6 + 8i$ and $|z_2| = 12$ then, the maximum value of $|z_1 + z_2|$ is:*

- *[A]* 18
- *[B]* 20
- *[C]* 22
- *[D]* 26
- *[E] none of the above*

Solution: **Use the triangular inequality** to get:

$$|z_1 + z_2| \leq |z_1| + |z_2|,$$

where $|z_1| = \sqrt{6^2 + 8^2} = 10$ and $|z_2| = 12$, hence

$$|z_1 + z_2| \leq |z_1| + |z_2| = 10 + 12 = 22$$

and the $[C]$ holds true.

Problem (7). *If $z = \dfrac{i}{1-i}$ then the position vector of z with the xx' axis, is*

- *[A]* 135°
- *[B]* 45°
- *[C]* 90°
- *[D]* 60°
- *[E] none of the above*

Solution: **Use inner product** to compute the angle. Let P be the image of z on the Euclidean plane. Then:

$$z = \frac{i}{1-i} = \frac{-1+i}{2}$$

therefore the coordinates of P are $(\frac{-1}{2}, \frac{1}{2})$ i.e: $\vec{OP} = (\frac{-1}{2}, \frac{1}{2})$ and since the coordinates of $\vec{i} = (1, 0)$ then the angle ϕ of $\vec{OP}$ with $\vec{i}$ is

$$\cos\phi = \vec{OP} \cdot \vec{i} = (\frac{-1}{2}, \frac{1}{2}) \cdot (1, 0) = -\frac{1}{2}$$

hence $\phi = 135^o$, i.e: $[A]$ holds true.

Problem (8). *The images of $z \in \mathbb{C}$ such that $|z - i| = 4$, lie on :*

- *[A] a circle of center $(0, 1)$ and radius 2*
- *[B] a circle of center $(1, 0)$ and radius 2*
- *[C] a circle of center $(1, 1)$ and radius 4*

- *[D] a straight line passing through the point* $0, 1$
- *[E] none of the above.*

Solution: Let $z = x + yi$, then

$$|z - i| = 4 \Leftrightarrow$$
$$|x + (y-1)i| = 4 \Leftrightarrow$$
$$(x-0)^2 + (y-1)^2 = 4^2$$

hence, z lies on a circle with center $(0, 1)$ and radius 4.

Problem (9). *If* $\omega = \dfrac{3z+2}{z-3}$ *and* $|z| = 1$ *then, the maximum value of* $|\omega|$ *is*

- *[A]* $\frac{1}{4}$
- *[B]* $\frac{1}{2}$
- *[C]* $\sqrt{\frac{13}{10}}$
- *[D]* $\frac{5}{4}$
- *[E]* $\frac{5}{2}$

Solution: **Using the triangular inequality** for both the numerator and the denominator:

$$|3z + 2| \leq 3|z| + 2 = 3 \times 1 + 2 = 5 \qquad (1)$$
$$|z - 3| \geq ||z| - |3|| = |1 - 3| = 2 \qquad (2)$$

hence, from (1) and (2), we deduce:

$$|\omega| = \left|\frac{3z+2}{z-3}\right| = \frac{|3z+2|}{|z-3|} \geq \frac{5}{2}$$

and $[E]$ holds true.

Problem (10). *If* $z = \dfrac{2}{1+i}$ *then* $\mathbb{I}m(z)$ *is*

- *[A]* 2
- *[B]* 1
- *[C]* $\sqrt{2}$
- *[D]* -1
- *[E]* -2

Solution: Use the **conjugate** of the denominator and multiply both:

$$z = \frac{2}{1+i} = \frac{2(1-i)}{(1+i)(1-i)} = \frac{2(1-i)}{2} = 1 - i$$

hence $\mathbb{I}mz = -1$ and the correct answer is $[D]$.

7.3 Vectors

7.3.1 Basic theory on vectors

- Let $\vec{i}$ be the unit vector on the Ox axis and $\vec{j}$ the unit vector on Oy axis.
- Decompose each vector $\vec{u}$ with respect to $\{\vec{i}, \vec{j}\}$ as

$$\vec{u} = u_1 \cdot \vec{i} + u_2 \cdot \vec{j}$$

in a **unique way**.

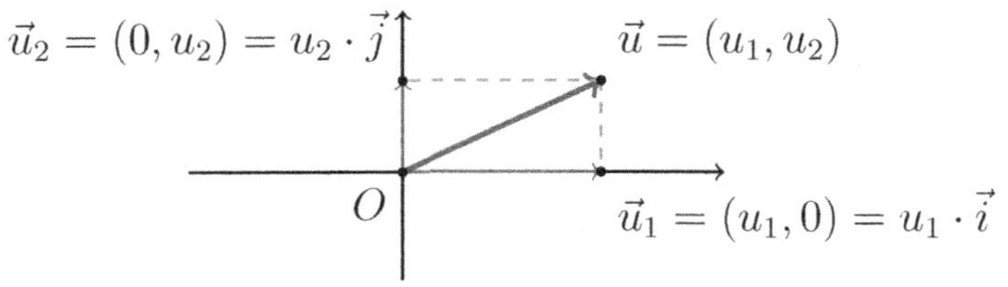

- If $\vec{u} = (u_1, u_2) = u_1 \cdot \vec{i} + u_2 \cdot \vec{j} = \vec{u}_1 + \vec{u}_2$ and $\vec{w} = (w_1, w_2) = w_1 \cdot \vec{i} + w_2 \cdot \vec{j} = \vec{w}_1 + \vec{w}_2$ and $\lambda \in \mathbb{R}$ we have the following formulas for the **sum** of two vector and the **product** of a scalar by a vector:

$$\vec{u} + \vec{w} = (u_1 + w_1, u_2 + w_2) = (u_1 + w_1)\vec{i} + (u_2 + w_2)\vec{j}$$
$$\lambda \cdot \vec{u} = \lambda \cdot (u_1, u_2) = (\lambda u_1, \lambda u_2)$$

- **Parallelism condition** (for vectors):

$$\vec{u} \parallel \vec{w} \Leftrightarrow \vec{u} = \mu \cdot \vec{w}, \ \mu \in \mathbb{R}$$

The null vector is parallel to any vector $\vec{u}$

- **Inner product:** for any $\vec{u} = (u_1, u_2)$, $\vec{w} = (w_1, w_2)$ their inner product (or scalar product) is defined by

$$\vec{u} \cdot \vec{w} = u_1 w_1 + u_2 w_2$$

- **Perpendicularity condition:**

$$\vec{u} \perp \vec{w} \Leftrightarrow \vec{u} \cdot \vec{w} = 0$$

- **Norm of a vector:** for any $\vec{u} = (u_1, u_2)$ its norm is given by

$$|\vec{u}| = \sqrt{\vec{u} \cdot \vec{u}} = \sqrt{u_1^2 + u_2^2}$$

7.3.2 Examples & problems on vectors

Example (1).

- *If $\vec{v} = (3, 4)$ then its norm is*

$$|\vec{v}| = \sqrt{3^2 + 4^2} = \sqrt{25} = 5$$

- *If* $\vec{u} = (2,3)$, $\vec{w} = (-2,4)$ *then*

$$\vec{u} + \vec{w} = (2-2, 3+4) = (0,7)$$

and

$$\vec{u} \cdot \vec{w} = 2 \times (-2) + 3 \times 4 = -4 + 12 = 10$$

- *If* $\vec{u} = (x,3)$ *and* $\vec{w} = (-1,4)$ *then,* $\vec{u} \perp \vec{w}$ *if and only if*

$$\vec{u} \cdot \vec{w} \Leftrightarrow -x + 12 = 0 \Leftrightarrow x = 12$$

- *If* $\vec{u} = (x, x-1)$ *and* $\vec{w} = (1,2)$ *then* $\vec{u} \parallel \vec{w}$ *if and only if there exists* $\mu \neq 0$ *such that*

$$\vec{u} = \mu\vec{w} \Leftrightarrow (x, x-1) = (\mu, 2\mu) \Leftrightarrow x = \mu \ \& \ x - 1 = 2\mu \Rightarrow$$
$$\mu = 1 + 2\mu \Rightarrow \mu = -1$$

Way *2:* *Use the determinant condition:*

$$\vec{u} \parallel \vec{w} \Leftrightarrow \det(\vec{u}, \vec{w}) = 0 \Leftrightarrow \begin{vmatrix} x & x-1 \\ 1 & 2 \end{vmatrix} = 0 \Leftrightarrow x = -1$$

- *If* $\vec{u} = (x,3)$, $\vec{v} = (-1,4)$ *and* $\vec{u} \perp \vec{v}$ *then*

$$\vec{u} \cdot \vec{v} = 0 \Leftrightarrow (-1)x + 4 \cdot 4 = 0 \Leftrightarrow -x + 12 = 0 \Leftrightarrow x = 12$$

Problem (1). *If* $\vec{u} = (6,8)$, $\vec{w} = (12,x)$ *and* $\vec{u} \cdot \vec{w} = 0$ *then* x *is equal to*

- *[A]* -9
- *[B]* -8
- *[C]* 0
- *[D]* 8
- *[E]* -9

Solution: Let us apply the formula $\vec{u} \cdot \vec{w} = 0$, it is

$$\vec{u} \cdot \vec{w} = 0 \Leftrightarrow 6 \cdot 12 + 8x = 0 \Leftrightarrow 72 + 8x = 0 \Leftrightarrow x = -9$$

Correct answer is $[A]$.

Problem (2). *If* $\vec{u} = (2,4)$ *and* $\vec{w} = (-1,-2)$ *then* $|3\vec{u} + 4\vec{w}|$ *is equal to*

- *[A]* $\sqrt{5}$
- *[B]* $2\sqrt{5}$
- *[C]* $3\sqrt{5}$
- *[D]* $2\sqrt{10}$
- *[E]* *none of the above*

Solution: It is

$$3\vec{u} + 4\vec{w} = (6, 12) + (-4, -8) = (2, 4) \Rightarrow$$
$$|3\vec{u} + 4\vec{w}| = \sqrt{2^2 + (4)^2} = \sqrt{20} = 2\sqrt{5}$$

Correct answer is $[B]$.

Problem (3). *A unit vector parallel to* $\vec{u} = (6, -8)$ *is*

- *[A]* $(\frac{-3}{5}, \frac{2}{5})$
- *[B]* $(\frac{-3}{5}, \frac{-4}{5})$
- *[C]* $(\frac{3}{5}, \frac{-4}{5})$
- *[D]* $(\frac{2}{5}, \frac{-4}{5})$
- *[E]* $(\frac{4}{5}, \frac{3}{5})$

Solution: It is

$$|\vec{u}| = \sqrt{6^2 + (-8)^2} = \sqrt{100} = 10$$

hence

$$\frac{\vec{u}}{|\vec{u}|} = (\frac{3}{5}, \frac{-4}{5}).$$

Correct answer is $[C]$.

Problem (4). *Which of the following vectors are parallel to the vector* $3\vec{a} - 4\vec{b}$*:*

- *[A]* $6\vec{a} - 8\vec{b}$
- *[B]* $2\vec{a} - 3\vec{b}$
- *[C]* $\frac{3}{2}\vec{a} - 2\vec{b}$
- *[D]* $4\vec{a} - 3\vec{b}$
- *[E]* $\vec{a} + \frac{4}{3}\vec{b}$

Solution: Checking each one of the given linear combinations of $\vec{a}$, $\vec{b}$, it is easy to verify that $[A]$ is the correct answer, because

$$6\vec{a} - 8\vec{b} = 2\left(3\vec{a} - 4\vec{b}\right) \Leftrightarrow 3\vec{a} - 4\vec{b} = \frac{1}{2}\left(3\vec{a} - 4\vec{b}\right)$$

Problem (5). *In the rectangle* $ABCD$ *of the figure* M *is the midpoint of* DC *and* N *is a point on* CB *such that* $\frac{NC}{NB} = \frac{1}{3}$ *and denote* $\vec{p} = \vec{AD}$, $\vec{q} = \vec{AB}$. *Then,* $\vec{MN}$ *is equal to*

- *[A]* $\frac{-1}{4}\vec{p}+\frac{1}{2}\vec{q}$
- *[B]* $\frac{5}{3}\vec{p}+\frac{1}{2}\vec{q}$
- *[C]* $4\vec{p}+2\vec{q}$
- *[D]* $\frac{7}{4}\vec{p}+\frac{3}{2}\vec{q}$
- *[E] none of the above*

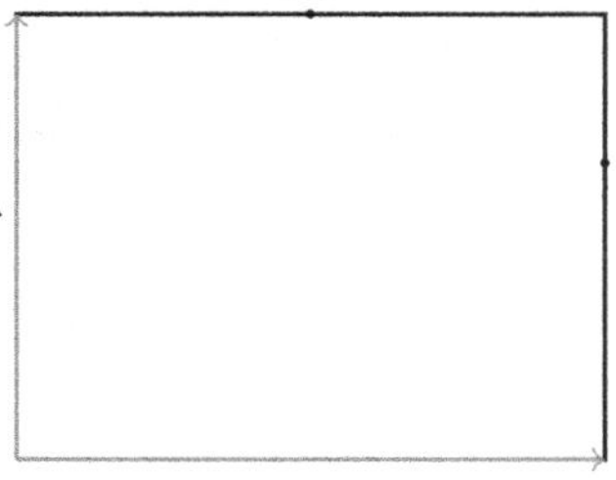

Solution: It is

$$\vec{MN} = \vec{MC} + \vec{CN} \quad (1)$$

where

$$\vec{MC} = \frac{1}{2}\vec{AB} = \frac{1}{2}\vec{q} \quad (2)$$

$$\vec{CN} = -\vec{NC} = -\frac{1}{3}\vec{BC} = -\frac{1}{3}\vec{AD} = -\frac{1}{3}\vec{p} \quad (3)$$

hence

$$\vec{MN} = \frac{1}{2}\vec{q} - \frac{1}{3}\vec{p}$$

Problem (6). *The vector* $3\vec{i} - \frac{2}{\sqrt{3}}\vec{j}$ *is parallel to*

- *[A]* $3\vec{i}-2\sqrt{3}\vec{j}$
- *[B]* $-9\vec{i}+2\sqrt{3}\vec{j}$
- *[C]* $\vec{i}+3\vec{j}$
- *[D]* $-\vec{i}+3\vec{j}$
- *[E] none of the above*

Solution: Observe that if we multiply $3\vec{i} - \frac{2}{\sqrt{3}}\vec{j}$ by -3 we get

$$3\left(3\vec{i} - \frac{2}{\sqrt{3}}\vec{j}\right) = -9\vec{i} + 2\sqrt{3}\vec{j}$$

Correct answer is $[B]$.

Problem (7). *The vectors $\vec{a}$, $\vec{b}$ are parallel if*

- *[A]* $|\vec{a}| = |\vec{b}|$
- *[B]* $|\vec{a}+\vec{b}| = |\vec{a}-\vec{b}|$
- *[C]* $|\vec{a}-\vec{b}| = |\vec{a}| + |\vec{b}|$
- *[D]* $|\vec{a}+\vec{b}| = ||\vec{a}| - |\vec{b}||$
- *[E] none of the above*

Solution: Observe that

$$\begin{aligned}
|\vec{a}-\vec{b}| &= |\vec{a}| + |\vec{b}| \Rightarrow \\
|\vec{a}|^2 + |\vec{b}|^2 - 2\vec{a}\vec{b} &= |\vec{a}|^2 + |\vec{b}|^2 + 2|\vec{a}||\vec{b}| \Rightarrow \\
-2\vec{a}\vec{b} &= 2|\vec{a}||\vec{b}| \Rightarrow \\
-|\vec{a}||\vec{b}|\cos\phi &= |\vec{a}||\vec{b}|
\end{aligned}$$

hence

$$\begin{aligned}
|\vec{a}||\vec{b}|\,(1+\cos\phi) = 0 \Rightarrow \\
\cos\phi = -1 \Rightarrow \phi = \pi
\end{aligned}$$

and thus $\vec{a} \parallel \vec{b}$. Correct answer is $[C]$.

Problem (8). *The vectors $\vec{a}$, $\vec{b}$ are equal if*

- *[A]* $\vec{a} \parallel \vec{b}$
- *[B]* $\vec{a} \uparrow\downarrow \vec{b}$ *and* $|\vec{a}| = |\vec{b}|$
- *[C]* $|\vec{a}+\vec{b}| = |\vec{a}| + \vec{b}$
- *[D]* $|\vec{a}+\vec{b}| = |\vec{a}| - |\vec{b}|$
- *[E]* $\vec{a} \parallel \vec{b}$ *and* $|\vec{a}| = |\vec{b}|$

Solution: It is obvious that $\vec{a} = \vec{b}$ if and only if $\vec{a} \parallel \vec{b}$ and $|\vec{a}| = |\vec{b}|$

Problem (9). *If $\vec{AB} = \vec{CD}$ then the following equality holds true*

- *[A]* $\vec{AC} = \vec{BD}$
- *[B]* $\vec{AD} = \vec{CB}$
- *[C]* $\vec{AC} = \vec{DB}$
- *[D]* $\vec{AD} = \vec{BD}$
- *[E] none of the above*

Solution: **Observe that**

$$\vec{AB} = \vec{CD} \Rightarrow \vec{AB} + \vec{BC} = \vec{BC} + \vec{CD} \Rightarrow \vec{AC} = \vec{BD}$$

Correct answer is $[A]$.

Problem (10). *In a regular hexagon $ABCDEZ$ with center O which of the following relations holds true?*

- *[A]* $\vec{OB} = \vec{OE}$
- *[B]* $\vec{OB} = \vec{ZA}$
- *[C]* $\vec{OC} = \vec{OB}$
- *[D]* $\vec{OD} = \vec{AZ}$
- *[E]* $\vec{AC} = 2\vec{AB}$

Solution: It is easy to check that

$$\vec{OB} = \vec{ZA}$$

Correct answer is $[B]$.

Problem (11). *In the parallelogram $ABCD$ let M be the midpoint of the side BC. Then $\vec{DM}$ is equal to*

- *[A]* $-\vec{b} + \vec{a}$
- *[B]* $\dfrac{\vec{a} + \vec{b}}{2}$
- *[C]* $\dfrac{\vec{a} - \vec{b}}{2}$
- *[D]* $\dfrac{2\vec{a} - \vec{b}}{2}$
- *[E]* *none of the above*

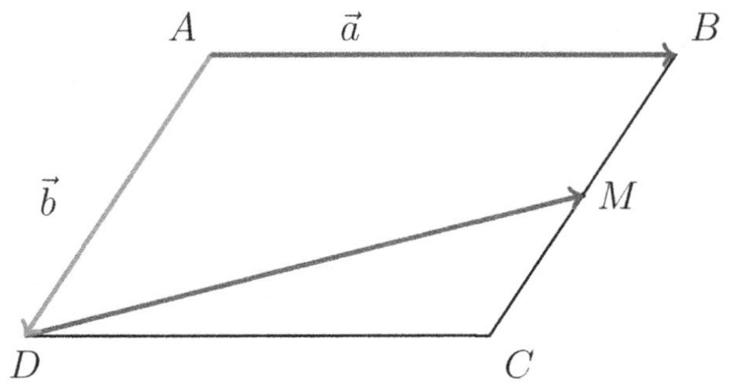

Solution: It is

$$\begin{aligned}\vec{DM} &= \vec{DC} + \vec{CM} = \vec{DC} + \frac{\vec{CB}}{2} \\ &= \vec{AB} + \frac{\vec{DA}}{2} = \vec{AB} - \frac{\vec{AD}}{2} \\ &= \vec{a} - \frac{\vec{b}}{2} = \frac{2\vec{a} - \vec{b}}{2}\end{aligned}$$

Correct answer is $[D]$.

Problem (12). *If $\vec{a} \uparrow\uparrow \vec{b}$ non null vectors and $k, l \in \mathbb{R}^{\star}$ such that $k\vec{a} + l\vec{b} = \vec{0}$. Then:*

- *[A] k, l are negative*
- *[B] k, l are positive*
- *[C] k, l are inverse one each other*
- *[D] k, l have different signature*
- *[E] none of the above*

Solution: **Where is the subtle point?**

$$k\vec{a} + l\vec{b} = \vec{0} \Rightarrow \vec{a} = -\frac{l}{k}\vec{b} \Rightarrow k \cdot l < 0$$

hence $[D]$ holds true.

Problem (13). *The vector $\vec{a} = (2 - 3l + l^2, 0)$ is the null vector if and only if*

- *[A] $l = 0$*
- *[B] $l = 1$ or $l = 2$*
- *[C] $l = -1$*
- *[D] $l = -2$*
- *[E] none of the above*

Solution: Note that

$$\begin{gathered}\vec{a} = \vec{0} \Leftrightarrow (2 - 3l + l^2, 0) = (0, 0) \Leftrightarrow \\ 2 - 3l + l^2 = 0 \Leftrightarrow (l - 1)(l - 2) = 0\end{gathered}$$

hence $[B]$ holds true.

Problem (14). *If the vectors $\vec{a} = (1 + l, n - 2)$ and $\vec{b} = (2, 2)$ are equal then*

- *[A] $l = -1$ and $n = 3$*

- *[B]* $l = 1$ *and* $n = 4$
- *[C]* $l = 1$ *and* $n = -4$
- *[D]* $l = -2$ *and* $n = 0$
- *[E]* $l = 0$ *and* $n = 4$

Solution: Where is the subtle point?
Recall that
Equality of vectors is equivalent to
equality of corresponding coordinates, so

$$\vec{a} = \vec{b} \Leftrightarrow (1+l, n-2) = (2,2) \Leftrightarrow$$
$$1 + l = 2 \text{ and } n - 2 = 2 \Leftrightarrow l = 1 \text{ and } n = 4$$

and $[B]$ holds true.

Problem (15). *If* $\vec{u} = \left(\frac{1}{2}, \frac{3}{4}\right)$, $\vec{v} = \left(x, \frac{\sqrt{5}}{4}\right)$ *and* $|\vec{u}| = |\vec{v}|$ *then* x *is equal to*

- *[A]* $\frac{1}{4}$ *or* $\frac{-1}{4}$
- *[B]* $\frac{1}{2}$
- *[C]* 0
- *[D]* $\frac{5}{16}$
- *[E]* $\frac{\sqrt{2}}{2}$ *or* $\frac{-\sqrt{2}}{2}$

Solution: Use the formula for the norm of the vector with coordinates $\vec{a} = (x, y)$: $|\vec{a}| = \sqrt{x^2 + y^2}$:

$$|\vec{u}| = \sqrt{\frac{1}{4} + \frac{9}{16}}, \ |\vec{v}| = \sqrt{x^2 + \frac{5}{16}} \overset{|\vec{u}|=|\vec{v}|}{\Longrightarrow}$$
$$\frac{1}{4} + \frac{9}{16} = x^2 + \frac{5}{16} \Rightarrow x^2 = \frac{1}{2} \Rightarrow x \pm \frac{\sqrt{2}}{2}$$

and $[E]$ holds true.

Problem (16). *If* $\vec{a} = 5\vec{\imath} + 3\vec{\jmath}$, $\vec{b} = 5\vec{\imath} - 3\vec{\jmath}$, $\vec{c} = 2\vec{\imath} + 4\vec{\jmath}$ *and* $\vec{d} = 4\vec{\imath} - 2\vec{\jmath}$ *which of the following relations holds true?*

- *[A]* $\vec{a} \perp \vec{b}$
- *[B]* $\vec{b} \perp \vec{c}$

- *[C]* $\vec{a} \perp \vec{d}$
- *[D]* $\vec{c} \perp \vec{d}$
- *[E]* $\vec{b} \perp \vec{d}$

Solution: Where is the subtle point?
Use inner product to characterize **perpendicular** vectors, i.e:
if $\vec{u} = (x_1, y_2)$ and $\vec{v} = (x_2, y_2)$, then:
$\vec{u} \perp \vec{b} \Leftrightarrow \vec{u} \cdot \vec{v} = x_1 x_2 + y_2 y_2 = 0$.
It is easy to check that

$$\vec{c} \cdot \vec{d} = 2 \times 4 + 4 \times (-2) = 0$$

hence $\vec{c} \perp \vec{d}$ and $[D]$ is true.

Problem (17). *Given the vector $\vec{a} = (5, 3)$ the x - coordinate of a vector $\vec{b}$ normal to $\vec{a}$, is equal to 8. Then, the y - coordinate of $\vec{b}$ is:*

- *[A]* $\dfrac{-8}{5}$
- *[B]* $\dfrac{15}{8}$
- *[C]* $\dfrac{-40}{3}$
- *[D]* $\dfrac{40}{3}$
- *[E]* $\dfrac{24}{5}$

Solution: Use the characterization of perpendicularity in terms of the **inner product**.
Let $\vec{b} = (8, y)$ and since $\vec{b} \perp \vec{a} = (5, 3)$ we deduce:

$$\vec{a} \cdot \vec{b} = 0 \Leftrightarrow (5, 3) \cdot (8, y) = 0 \Leftrightarrow 40 + 3y = 0$$

hence $[C]$ holds true.

Problem (18). *If $\vec{a} \cdot \vec{b} = 0$ find which of the following relations does not hold:*

- *[A]* $\vec{a} \perp \vec{b}$
- *[B] the angle* $\widehat{(\vec{a}, \vec{b})}$ *is* $\dfrac{\pi}{2}$
- *[C]* $|\vec{a}| = |\vec{b}|$ *and* $\widehat{(\vec{a}, \vec{b})} = \dfrac{\pi}{2}$
- *[D]* $\vec{b} = \vec{0}$

- *[E]* $\widehat{(\vec{a},\vec{b})} = \frac{\pi}{3}$ *and* $|\vec{a}| = |\vec{b}|$

Solution: Where is the subtle point?
If the **inner product** vanishes then $\vec{a} \perp \vec{b}$, so
$\widehat{(\vec{a},\vec{b})}$ is $\frac{\pi}{2}$ or one of the vectors is $\vec{0}$.
Hence, $[E]$ cannot hold.

Problem (19). *According to the figure below the inner product* $\vec{a} \cdot \vec{b}$ *is equal to*

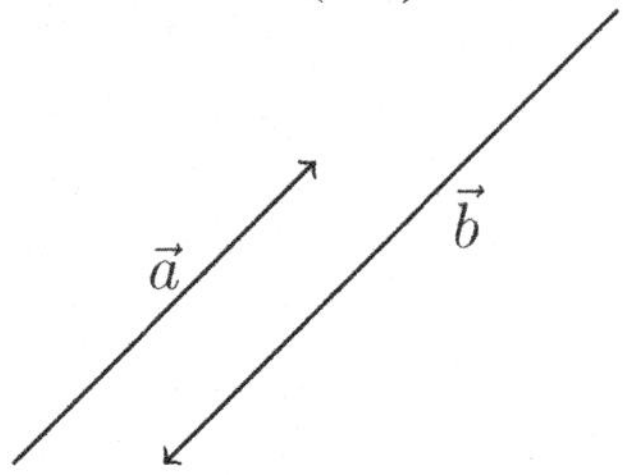

- *[A]* $-|\vec{a}|\vec{b}$
- *[B]* $\frac{1}{2}|\vec{a}|\vec{b}$
- *[C]* 0
- *[D]* $-\frac{1}{2}|\vec{a}|\vec{b}$
- *[E]* $|\vec{a}|\vec{b}$

Solution: Where is the subtle point?
Angle of $\vec{a}$**,** $\vec{b}$ is π, hence

$$\vec{a} \cdot \vec{b} = |\vec{a}||\vec{b}| \cos \pi = -|\vec{a}||\vec{b}|$$

and $[A]$ is the correct answer.

Problem (20). *The non -null vectors* $\vec{a} = (m^2, 2m)$ *and* $\vec{b} = (1, 2)$ *are parallel. Then* m *is equal to*

- *[A]* 1
- *[B]* -1
- *[C]* 2
- *[D]* -2
- *[E]* $\frac{1}{2}$

Solution: Where is the subtle point?
Use the determinant to characterize parallelism

$$\vec{a} \parallel \vec{b} \Leftrightarrow \begin{vmatrix} m^2 & 1 \\ 2m & 2 \end{vmatrix} = 0 \Leftrightarrow 2m^2 - 2m = 0 \Leftrightarrow$$

$$2m(m-1) = 0 \Rightarrow m = 0 \text{ or } m = 1$$

and since $\vec{a} \neq \vec{0}$ we get $m = 1$. So, $[A]$ is the correct answer.

Problem (21). *If M is the midpoint of the side AB of the quadrilateral $ABCD$ then $\vec{MD} + \vec{MC}$ is equal to*

- *[A]* $\vec{DA} + \vec{BC}$
- *[B]* $\vec{AB} + \vec{DC}$
- *[C]* $\vec{AD} + \vec{BC}$
- *[D]* $\vec{DA} + \vec{CB}$
- *[E] none of the above*

Solution: Where is the subtle point?
Use the fact that $\vec{MA} + \vec{MB} = \vec{0}$, then:

$$\vec{MD} + \vec{MC} = \vec{MA} + \vec{AD} + \vec{MB} + \vec{BC} = \vec{AD} + \vec{BC}$$

Correct answer is $[C]$.

Problem (22). *The point P with position vector $\vec{r} = t\vec{i} + (2t-1)\vec{j}$, $t \in \mathbb{R}$, belongs to the straight line with equation*

- *[A]* $x + y = -1$
- *[B]* $2x - y = 1$
- *[C]* $2x + y = 1$
- *[D]* $x - y = 1$
- *[E]* $y = 2$

Solution: Let $x = t$ and $y = 2t - 1$ for the coordinates of the point P, then:

$$y = 2x - 1$$

and thus $[B]$ holds true.

Problem (23). *The vector $\vec{u} = (3, -2)$ is parallel to the straight line with analytical equation*

- *[A]* $y = 3x - 2$
- *[B]* $2x + 3y + 1 = 0$
- *[C]* $x + y = 1$
- *[D]* $3x - 2y = 2$
- *[E]* $x - y - 2 = 0$

Solution: The direction of the vector $\vec{u}$ is $\lambda_{\vec{u}} = \dfrac{-2}{3}$ and is equal to the direction of the straight line with equation $2x + 3y + 1 = 0$. Correct answer is $[B]$.

Problem (24). *We are given the vectors* $\vec{a} = (5k, 3 - l)$, $\vec{b} = (4 - l, 4k)$ *with* $k, l \in \mathbb{R}$. *The vectors* $\vec{a}$ *and* $\vec{b}$ *are* ***equal*** *if*

- *[A]* $k = 1,\ l = 2$
- *[B]* $k = 2,\ l = 2$
- *[C]* $k = 2,\ l = 1$
- *[D]* $k = 2,\ l = -2$
- *[E] none of the above*

Solution: **Recall that equality of vectors** is equivalent to the **equality of coordinates**, hence:

$$\vec{a} = \vec{b} \Leftrightarrow (5k, 3 - l) = (4 - l, 4k) \Leftrightarrow$$
$$5k = 4 - l \text{ and } 3 - l = 4k \Rightarrow k = 1,\ l = -1$$

Problem (25). *Given the vectors* $\vec{a} = (5k, 3 - l)$, $\vec{b} = (4 - l, 4k)$ *with* $k, l \in \mathbb{R}$, *if* $l = 8$, $k > 0$ *and* $\vec{a} \parallel \vec{b}$, *then* k *is equal to*

- *[A]* 4
- *[B]* 1
- *[C]* 2
- *[D]* 3
- *[E]* 5

Solution: Where is the subtle point?
Parallelism of $\vec{a}$ and $\vec{b}$ is equivalent to the **vanishing of the determinant** of $\vec{a}$, $\vec{b}$. So

$$\begin{vmatrix} 5k & -4 \\ -5 & 4k \end{vmatrix} = 0 \Leftrightarrow k^2 - 1 = 0 \Rightarrow$$
$$k = \pm 1 \text{ and since } k > 0 \Rightarrow k = 1$$

and thus, $[B]$ holds true.

Problem (26). *Let $\vec{a}$, $\vec{b}$ be given with $|\vec{a}| = 2$, $|\vec{b}| = 4$. If the vectors $3\vec{a} + l\vec{b}$ and $3\vec{a} - l\vec{b}$ are perpendicular, then l is equal to*

- *[A]* $\frac{3}{2}$
- *[B]* $-\frac{3}{2}$
- *[C]* $\frac{3}{2}$ *or* $-\frac{3}{2}$
- *[D]* $-2, +2$
- *[E] none of the above*

Solution: Where is the subtle point?
Use inner product

$$3\vec{a} + l\vec{b} \perp 3\vec{a} - l\vec{b} \Leftrightarrow \left(3\vec{a} + l\vec{b}\right) \cdot \left(3\vec{a} - l\vec{b}\right) = 0 \Leftrightarrow$$
$$9|\vec{a}|^2 - l^2|\vec{b}|^2 = 0 \Leftrightarrow 36 - 16l^2 = 0$$

hence

$$l = \pm\frac{3}{2}$$

and $[C]$ is the correct answer.

Problem (27). *The vector $\vec{a} = (5, -3)$ has as origin the point $(-2, 4)$, then its end point is*

- *[A]* $(-3, 1)$
- *[B]* $(3, 7)$
- *[C]* $(-3, -1)$
- *[D]* $(3, -1)$
- *[E]* $(7, 1)$

Solution: Where is the subtle point?
If $A(x_1, y_1)$ and $B(x_2, y_2)$ are given points then

$$\vec{AB} = (x_2 - x_1, y_2 - y_1)$$

hence, if (x, y) is the end point of $\vec{a}$:

$$(x+2, y-4) = (5, -3) \Leftrightarrow x+2 = 5 \text{ and } y-4 = 3$$

and thus $x = 3$, $y = 7$. The correct answer is [B].

7.4 Matrices

Examples

a. A **matrix** of type 2×3 with 2 rows and 3 columns is an object of the form

$$P = \begin{bmatrix} a & b & c \\ x & y & z \end{bmatrix}$$

Similarly we define a matrix of type $m \times n$ with m - rows and n columns.

b. A **matrix** of type 2×2 with 2 rows and 2 columns is an object of the form

$$Q = \begin{bmatrix} a & b \\ x & y \end{bmatrix}$$

Any matrix of type $m \times m$, with m - rows and m - columns, is called a square matrix.

c. **Addition - sum** of matrices:

$$\begin{bmatrix} a & b & c \\ x & y & z \end{bmatrix} + \begin{bmatrix} a' & b' & c' \\ x' & y' & z' \end{bmatrix} = \begin{bmatrix} a+a' & b+b' & c+c' \\ x+x' & y+y' & z+z' \end{bmatrix} \tag{1}$$

Remark. *The operation of the addition of matrices is defined only between of matrices of the same type.*
Carefull!!! *We cannot add squares matrices of different type, that is: there is no meaning in adding a 2×2 matrix with a matrix of type 3×3 ....*

d. **Multiplication of a scalar with a matrix:** Let $\lambda \in \mathbb{R}$: then for P a type 2×3 matrix

$$\lambda \cdot P = \lambda \cdot \begin{bmatrix} a' & b' & c' \\ x' & y' & z' \end{bmatrix} = \begin{bmatrix} \lambda \cdot a' & \lambda \cdot b' & \lambda \cdot c' \\ \lambda \cdot x' & \lambda \cdot y' & \lambda \cdot z' \end{bmatrix} \tag{2}$$

e. **Mutliplication of two matrices:**

- **Square matrices** 2×2: Let $A = \begin{bmatrix} a & b \\ c & d \end{bmatrix}$ and $B = \begin{bmatrix} x & y \\ z & w \end{bmatrix}$, then

$$A \cdot B = \begin{bmatrix} ax+bz & ay+bw \\ cx+dz & cy+dw \end{bmatrix} \tag{3}$$

Mnemonic Rule:
first row of A × first column of B → first row-first column element of $A \cdot B$
first row of A × second column of B → first row-second column element of $A \cdot B$
second row of A × first column of B → second row-first column element of $A \cdot B$
second row of A × second column of B → second row-second column element of $A \cdot B$

- **Product of a 2×3 with a 3×2 matrix:** Let $P = \begin{bmatrix} a & b & c \\ d & e & f \end{bmatrix}$ and $Q = \begin{bmatrix} x & y \\ z & w \\ l & m \end{bmatrix}$, then

$$P \cdot Q = \begin{bmatrix} a & b & c \\ d & e & f \end{bmatrix} \cdot \begin{bmatrix} x & y \\ z & w \\ l & m \end{bmatrix} = \begin{bmatrix} ax + bz + cl & ay + bw + cm \\ dx + cz + fl & dy + cw + fm \end{bmatrix} \qquad (4)$$

Mnemonic Rule:
first row of P × first column of Q → first row-first column element of $P \cdot Q$
first row of P × second column of Q → first row-second column element of $P \cdot Q$
second row of P × first column of Q → second row-first column element of $P \cdot Q$
second row of P × second column of Q → second row-second column element of $P \cdot Q$

Remark. ***Attention!!!!*** :
THE PRODUCT $A \cdot B$ OF TWO MATRICES A, B IS DEFINED IF AND ONLY IF THE NUMBER OF COLUMNS OF THE FIRST FACTOR-MATRIX A OF THE PRODUCT IS EQUAL TO THE NUMBER OF THE ROWS OF THE SECOND FACTOR-MATRIX B OF THE PRODUCT.
If the product $A \cdot B$ is well defined the, in general, $B \cdot A$ has no meaning-it is not defined.
Only in the case of square matrices A, B of the same type, both $A \cdot B$ and $B \cdot A$ are well defined.
Remark. ***In general the product of square matrices is non-commutative, i.e:***

$$A \cdot B \neq B \cdot A \qquad (5)$$

Example (1). *Let* $P = \begin{bmatrix} 1 & 2 & -3 \\ 4 & 0 & 6 \end{bmatrix}$, $Q = \begin{bmatrix} -1 & 0 & 2 \\ -3 & 2 & -1 \end{bmatrix}$ *and* $\lambda \in \mathbb{R}$, *then:*

$$\begin{aligned} P + Q &= \begin{bmatrix} 1 & 2 & -3 \\ 4 & 0 & 6 \end{bmatrix} + \begin{bmatrix} -1 & 0 & 2 \\ -3 & 2 & -1 \end{bmatrix} \\ &= \begin{bmatrix} 1+(-1) & 2+0 & -3+2 \\ 4+(-3) & 0+2 & 6+(-1) \end{bmatrix} = \begin{bmatrix} 0 & 2 & -1 \\ 1 & 2 & 5 \end{bmatrix} && \text{(i)} \\ \lambda \cdot P = \lambda \cdot \begin{bmatrix} 1 & 2 & -3 \\ 4 & 0 & 6 \end{bmatrix} &= \begin{bmatrix} \lambda & 2\lambda & -3\lambda \\ 4\lambda & 0 & 6\lambda \end{bmatrix} && \text{(ii)} \end{aligned}$$

Example (2).

- ***Product of square*** 2×2 ***matrices:*** *Let* $A = \begin{bmatrix} 1 & 2 \\ -1 & 3 \end{bmatrix}$, $B = \begin{bmatrix} 1 & 0 \\ 2 & -1 \end{bmatrix}$, *then:*

$$A \cdot B = \begin{bmatrix} 1 & 2 \\ -1 & 3 \end{bmatrix} \cdot \begin{bmatrix} 1 & 0 \\ 2 & -1 \end{bmatrix}$$

$$= \begin{bmatrix} 1 \times 1 + 2 \times 2 & 1 \times 0 + 2 \times (-1) \\ -1 \times 1 + 3 \times 2 & -1 \times 0 + 3 \times (-1) \end{bmatrix} = \begin{bmatrix} 5 & -2 \\ 5 & -3 \end{bmatrix} \quad \text{(i)}$$

$$B \cdot A = \begin{bmatrix} 1 & 0 \\ 2 & -1 \end{bmatrix} \cdot \begin{bmatrix} 1 & 2 \\ -1 & 3 \end{bmatrix}$$

$$= \begin{bmatrix} 1 \times 1 + 0 \times (-1) & 1 \times 2 + 0 \times 3 \\ 2 \times 1 + (-1) \times (-1) & 2 \times 2 + (-1) \times 3 \end{bmatrix} = \begin{bmatrix} 1 & 2 \\ 3 & 1 \end{bmatrix} \quad \text{(ii)}$$

It is clear that: $A \cdot B \neq B \cdot A$ (iii)

- ***Product of a*** 2×3 ***with a*** 3×2 ***matrix:*** *Let* $P = \begin{bmatrix} 1 & 0 & 2 \\ 0 & 1 & 3 \end{bmatrix}$, $Q = \begin{bmatrix} 2 & 0 \\ 0 & 1 \\ 3 & 1 \end{bmatrix}$,

then, $P \cdot Q$ *is well defined and*

$$P \cdot Q = \begin{bmatrix} 1 & 0 & 2 \\ 0 & 1 & 3 \end{bmatrix} \cdot \begin{bmatrix} 2 & 0 \\ 0 & 1 \\ 3 & 1 \end{bmatrix}$$

$$= \begin{bmatrix} 1 \times 2 + 0 \times 0 + 2 \times 3 & 1 \times 0 + 0 \times 1 + 2 \times 1 \\ 0 \times 2 + 1 \times 0 + 3 \times 3 & 0 \times 0 + 1 \times 1 + 3 \times 1 \end{bmatrix} = \begin{bmatrix} 8 & 2 \\ 9 & 4 \end{bmatrix} \quad \text{(iv)}$$

The product $Q \cdot P$ *is also defined since the number* 3 *of columns of* Q *is equal to the number of rows* 3 *of* P*, but this is the exeption.*
For example, if P *is a* 2×3 *matrix and* Q *is a* 3×4 *matrix, then* $P \cdot Q$ *is well defined as* 2×4 *matrix but the product* $Q \cdot P$ *has no meaning.*

Let $A = \begin{bmatrix} a_{11} & a_{12} \\ a_{21} & a_{22} \end{bmatrix}$ be a square matrix. Its **determinant** is defined as the number

$$\det A = a_{11}a_{22} - a_{12}a_{21}$$

Mnemonic rule for the calculation of 2×2 determinants: Take the product of the first diagonal and subtract the product of the second diagonal.

The matrix

$$I_{2\times 2} = \begin{bmatrix} 1 & 0 \\ 0 & 1 \end{bmatrix}$$

is called the **identity** matrix.

The **inverse** of the square matrix $A = \begin{bmatrix} a & b \\ c & d \end{bmatrix}$ is a matrix B of the same type that has the following property

$$A \cdot B = B \cdot A = I.$$

Calculation of the inverse matrix - Mnemonic Rule: Let $A = \begin{bmatrix} a & b \\ c & d \end{bmatrix}$ be square 2×2 matrix with $\det A = ad - bc \neq 0$, then its inverse A^{-1} is given by:

$$A^{-1} = \begin{bmatrix} a & b \\ c & d \end{bmatrix}^{-1} = \frac{1}{ad - bc} \begin{bmatrix} d & -b \\ -c & a \end{bmatrix} \tag{6}$$

where $\det A = ad - bc \neq 0$.

7.4.1 Problems

Problem (1). *If* $A = (1\ 2)$, $\begin{pmatrix} -2 \\ 3 \end{pmatrix}$ *then the determinant of* BA *is*

- *[A]* -6
- *[B]* -4
- *[C]* -2
- *[D]* 0
- *[E]* 1

Solution: It is

$$BA = \begin{pmatrix} -2 \\ 3 \end{pmatrix} \cdot \begin{pmatrix} 1 & 2 \end{pmatrix} = \begin{pmatrix} -2 & -4 \\ 3 & 6 \end{pmatrix} = -12 - (-12) = 0.$$

Correct answer is $[D]$.

Problem (2). *Let*

$$A = \begin{bmatrix} 2 & 3 \\ 4 & x \end{bmatrix}, \quad x \in \mathbb{R}.$$

If A *is not invertible then* x *is equal to*

- *[A]* 1
- *[B]* 2
- *[C]* 5
- *[D]* 6
- *[E]* 8

Solution: Where is the subtle point?
Invertible means that $\det A \neq 0$ so **non invertible means**

$$\det A = 0 \Leftrightarrow 2x - 12 = 0 \Leftrightarrow x = 6,$$

hence $[D]$ holds true.

Problem (3). *Let*

$$A = \begin{bmatrix} 4 & a \\ b & 5 \end{bmatrix}, \quad a, b \in \mathbb{N}^\star.$$

If A is not invertible then (a, b) is equal to

- *[A]* $(2, 5)$
- *[B]* $(2, 10)$
- *[C]* $(3, 10)$
- *[D]* $(4, 10)$
- *[E]* *none of the above*

Solution: Non invertible is equivalent to

$$\det A = 0 \Leftrightarrow 20 - ab = 0 \Leftrightarrow ab = 20,$$

and since $a, b \in \mathbb{N}^\star$ we get

$$(a, b) \in \{(1, 20), (20, 1), (2, 10), (10, 2), (4, 5), (5, 4)\},$$

hence $[B]$ holds true.

Problem (4). *If* $W = \begin{bmatrix} -2 & 4 \\ 3 & 5 \end{bmatrix}$, $S = \begin{bmatrix} -2 & 0 \\ -2 & 2 \end{bmatrix}$ *then* $W + \frac{1}{2}S = \ldots$

- *[A]* $\begin{bmatrix} -2\frac{1}{2} & 2 \\ 2 & 6 \end{bmatrix}$
- *[B]* $\begin{bmatrix} -\frac{1}{2} & 2 \\ \frac{1}{2} & 6 \end{bmatrix}$
- *[C]* $\begin{bmatrix} -3 & 4 \\ 2 & 6 \end{bmatrix}$
- *[D]* $\begin{bmatrix} 2\frac{1}{2} & -4 \\ 2 & -1 \end{bmatrix}$
- *[E]* $\begin{bmatrix} 3\frac{1}{2} & -4 \\ 1 & 1 \end{bmatrix}$

Solution: Use the definition of the **addition** of matrices and of **multiplication** of a scalar times a matrix:

$$W + \frac{1}{2}S = \begin{bmatrix} -2 & 4 \\ 3 & 5 \end{bmatrix} + \frac{1}{2} \cdot \begin{bmatrix} -2 & 0 \\ -2 & 2 \end{bmatrix} = \begin{bmatrix} -2 & 4 \\ 3 & 5 \end{bmatrix} + \begin{bmatrix} -1 & 0 \\ -1 & 1 \end{bmatrix}$$
$$= \begin{bmatrix} -2-1 & 4+0 \\ 3-1 & 5+1 \end{bmatrix} = \begin{bmatrix} -3 & 4 \\ 2 & 6 \end{bmatrix}$$

Correct answer is $[C]$.

Problem (5). *The solution of the equation*

$$2\begin{bmatrix} \cos^2 x & \sin x \cos x \\ 1 & 0 \end{bmatrix} - \begin{bmatrix} 1 & 0 \\ 1 & 0 \end{bmatrix} = \begin{bmatrix} 0 & 2\sin^2 x \\ 1 & 0 \end{bmatrix}$$

is

- *[A]* $\frac{\pi}{4}$
- *[B]* $\frac{\pi}{2}$
- *[C]* $\frac{3\pi}{2}$
- *[D]* 2π
- *[E] none of the above*

Solution: **Use trigonometry formulas** : $\cos 2x = 2\cos^2 x - 1$, $2\sin x \cos x = \sin 2x$, then:

$$2\begin{bmatrix} \cos^2 x & \sin x \cos x \\ 1 & 0 \end{bmatrix} - \begin{bmatrix} 1 & 0 \\ 1 & 0 \end{bmatrix} = \begin{bmatrix} 0 & 2\sin^2 x \\ 1 & 0 \end{bmatrix} \Leftrightarrow$$
$$\begin{bmatrix} 2\cos^2 x - 1 & 2\sin x \cos x \\ 1 & 0 \end{bmatrix} = \begin{bmatrix} 0 & 2\sin^2 x \\ 1 & 0 \end{bmatrix} \Leftrightarrow$$
$$\begin{bmatrix} \cos 2x & \sin 2x \\ 1 & 0 \end{bmatrix} = \begin{bmatrix} 0 & 2\sin^2 x \\ 1 & 0 \end{bmatrix}$$

and thus we are led to the system

$$\left.\begin{aligned} \cos 2x &= 0 \\ \sin 2x &= 2\sin^2 x \end{aligned}\right\} \Rightarrow x = \frac{\pi}{4}$$

hence $[A]$ holds true.

Problem (6). *If* $A = \begin{bmatrix} 6 & 10 \\ 2 & 4 \end{bmatrix}$ *then* A^{-1} *is*

- *[A]* $\begin{bmatrix} 1 & -\frac{1}{2} \\ -\frac{1}{2} & \frac{3}{2} \end{bmatrix}$
- *[B]* $\begin{bmatrix} -4 & 10 \\ -2 & 6 \end{bmatrix}$
- *[C]* $\begin{bmatrix} -2 & 6 \\ -10 & 4 \end{bmatrix}$
- *[D]* $\begin{bmatrix} 1 & 2\frac{1}{2} \\ \frac{1}{2} & \frac{1}{2} \end{bmatrix}$

- *[E]* $\begin{bmatrix} \frac{1}{6} & \frac{1}{10} \\ \frac{1}{2} & \frac{1}{4} \end{bmatrix}$

Solution: First: check that A is invertible: $\det A = 4 \neq 0$ and **use the technique described in the theory section:**

$$A^{-1} = \frac{1}{\det A}\begin{bmatrix} 4 & -10 \\ -2 & 6 \end{bmatrix}^t, \text{ where } \det A = 4 \Rightarrow$$

$$A^{-1} = \begin{bmatrix} 1 & -\frac{5}{2} \\ -\frac{1}{2} & \frac{3}{2} \end{bmatrix}$$

and [A] holds true.

Problem (7). *If*

$$\begin{bmatrix} 3 & -2 \\ 4 & 1 \end{bmatrix} \cdot \begin{bmatrix} 6 & x \\ 2 & 2 \end{bmatrix} = \begin{bmatrix} 14 & -13 \\ 26 & -10 \end{bmatrix}$$

then x is equal to

- *[A]*-3
- *[B]*-1
- *[C]*0
- *[D]*2
- *[E]*3

Solution: It is

$$\begin{bmatrix} 3 & -2 \\ 4 & 1 \end{bmatrix} \cdot \begin{bmatrix} 6 & x \\ 2 & 2 \end{bmatrix} = \begin{bmatrix} 14 & -13 \\ 26 & -10 \end{bmatrix} \Leftrightarrow \begin{bmatrix} 14 & 3x-4 \\ 26 & 4x+2 \end{bmatrix} = \begin{bmatrix} 14 & -13 \\ 26 & -10 \end{bmatrix}$$

and thus we get the system

$$\left.\begin{aligned} 3x - 4 &= -13 \\ 4x + 2 &= -10 \end{aligned}\right\} \Rightarrow x = -3$$

hence [A] holds true.

Problem (8). *If* $A = \begin{bmatrix} 2 & 3 \\ 4 & 5 \end{bmatrix}$, $B = \begin{bmatrix} 3 & 1 \\ 2 & 2 \end{bmatrix}$ *then $AB = \ldots$ equal to*

- *[A]* $\begin{bmatrix} 5 & -4 \\ 6 & 7 \end{bmatrix}$
- *[B]* $\begin{bmatrix} 10 & 14 \\ 12 & 16 \end{bmatrix}$

- *[C]* $\begin{bmatrix} 6 & 3 \\ 8 & 10 \end{bmatrix}$
- *[D]* $\begin{bmatrix} 12 & 8 \\ 22 & 14 \end{bmatrix}$
- *[E]* $\begin{bmatrix} 1 & 3 \\ 3 & 10 \end{bmatrix}$

Solution: **Use the definition of the multiplication of matrices:**

$$\begin{aligned} A \cdot B &= \begin{bmatrix} 2 & 3 \\ 4 & 5 \end{bmatrix} \cdot \begin{bmatrix} 3 & 1 \\ 2 & 2 \end{bmatrix} \\ &= \begin{bmatrix} 2 \times 3 + 3 \times 2 & 2 \times 1 + 3 \times 2 \\ 4 \times 3 + 5 \times 2 & 4 \times 1 + 5 \times 2 \end{bmatrix} \\ &= \begin{bmatrix} 12 & 8 \\ 22 & 14 \end{bmatrix} \end{aligned}$$

and $[D]$ holds true.

Problem (9). *If* $K = \begin{bmatrix} 2 & 4 \\ -1 & -2 \end{bmatrix}$ *and* $L = \begin{bmatrix} -2 & 2 & 4 \\ 1 & -1 & -2 \end{bmatrix}$ *then* $KL = \ldots$

- *[A]* $\begin{bmatrix} -2 & 2 & 0 \\ 0 & 0 & 0 \end{bmatrix}$
- *[B]* $\begin{bmatrix} -4 & 4 & 16 \\ -1 & -1 & 4 \end{bmatrix}$
- *[C]* $\begin{bmatrix} 0 & 0 & 0 \\ 4 & -4 & 1 \end{bmatrix}$
- *[D]* $\begin{bmatrix} 0 & 0 & 0 \\ 0 & 0 & 0 \end{bmatrix}$
- *[E]* $\begin{bmatrix} 0 & -2 & 1 \\ 0 & 0 & 0 \end{bmatrix}$

Solution: **Check that the product has meaning:**
Indeed, K is of type 2×2 and L is 2×3, so KL has meaning and is of type $2 \times 2 \times 2 \times 3 = 2 \times 3$
Use definition of multiplication

$$\begin{aligned} KL &= \begin{bmatrix} 2 & 4 \\ -1 & -2 \end{bmatrix} \cdot \begin{bmatrix} -2 & 2 & 4 \\ 1 & -1 & -2 \end{bmatrix} \\ &= \begin{bmatrix} 2 \times (-2) + 4 \times 1 & 2 \times 2 + 4 \times (-1) & 2 \times 4 + 4 \times (-2) \\ (-1) \times (-2) + (-2) \times 1 & (-1) \times 2 + (-2) \times (-1) & (-1) \times 4 + (-2) \times (-2) \end{bmatrix} \\ &= \begin{bmatrix} 0 & 0 & 0 \\ 0 & 0 & 0 \end{bmatrix} \end{aligned}$$

and $[D]$ holds true.

Problem (10). *If* $K = \begin{bmatrix} 1 & 0 & -1 \\ 0 & 5 & 3 \end{bmatrix}$ *and* $L = \begin{bmatrix} 1 & 3 \\ -2 & 0 \\ 4 & 6 \end{bmatrix}$*, then* $K \cdot L$ *is equal to*

- *[A]* $\begin{bmatrix} -3 & -3 \\ 2 & 18 \end{bmatrix}$
- *[B]* $\begin{bmatrix} 3 & 3 \\ -2 & 18 \end{bmatrix}$
- *[C]* $\begin{bmatrix} -3 & -3 \\ -2 & 18 \end{bmatrix}$
- *[D]* $\begin{bmatrix} -3 & -3 \\ 2 & -18 \end{bmatrix}$
- *[E]none of the above*

Solution: Check that KL **has meaning**
Indeed, K is of type 2×3 and L is of type 3×2, so
KL is of type $2 \times 3 \times 3 \times 2 = 2 \times 2$, it is well defined.

$$\begin{aligned} KL &= \begin{bmatrix} 1 & 0 & -1 \\ 0 & 5 & 3 \end{bmatrix} \cdot \begin{bmatrix} 1 & 3 \\ -2 & 0 \\ 4 & 6 \end{bmatrix} \\ &= \begin{bmatrix} 1 \times 1 + 0 \times (-2) + (-1) \times 4 & 1 \times 3 + 0 \times 0 + (-1) \times 6 \\ 0 \times 1 + 5 \times (-3) + 3 \times 4 & 0 \times 3 + 5 \times 0 + 3 \times 6 \end{bmatrix} \\ &= \begin{bmatrix} -3 & -3 \\ 2 & 18 \end{bmatrix} \end{aligned}$$

so $[A]$ holds true.

Problem (11). *If* $X = \begin{bmatrix} 1 & -2 & 0 \end{bmatrix}$ *and* $\Psi = \begin{bmatrix} 2 \\ -1 \\ 0 \end{bmatrix}$ *then* $X \cdot \Psi$ *is equal to*

- *[A]*$[-4]$
- *[B]*$[4]$
- *[C]*$[0]$
- *[D]* $\begin{bmatrix} 3 \\ -2 \\ 0 \end{bmatrix}$
- *[E] none of the above*

Solution: Check the meaning of the product:
X is of type 1×3 and Ψ is of type 3×1, so
$X\Psi$ is of type $1 \times 3 \times 3 \times 1 = 1 \times 1$, i.e: is a scalar.

$$X \cdot \Psi = \begin{bmatrix} 1 & -2 & 0 \end{bmatrix} \cdot \begin{bmatrix} 2 \\ -1 \\ 0 \end{bmatrix} = [1 \times 2 + (-2) \times (-1) + 0 \times 0] = [4]$$

and [B] holds true.

Problem (12). *The inverse of the matrix* $A = \begin{bmatrix} 2 & -1 \\ 3 & 4 \end{bmatrix}$

- *[A]* $\begin{bmatrix} \frac{-4}{11} & \frac{1}{11} \\ \frac{3}{11} & \frac{2}{11} \end{bmatrix}$
- *[B]* $\begin{bmatrix} \frac{4}{11} & \frac{1}{11} \\ \frac{3}{11} & \frac{2}{11} \end{bmatrix}$
- *[C]* $\begin{bmatrix} \frac{4}{11} & \frac{1}{11} \\ -\frac{3}{11} & \frac{2}{11} \end{bmatrix}$
- *[D]* $\begin{bmatrix} -\frac{4}{11} & -\frac{1}{11} \\ -\frac{3}{11} & \frac{2}{11} \end{bmatrix}$
- *[E] none of the above*

Solution: Check that is invertible:

$$\det A = 2 \times 4 - (-1) \times 3 = 11 \neq 0$$

so A is indeed invertible.

Recall formula for the inverse of a 2×2 **matrix** M**:** If $M = \begin{bmatrix} a & b \\ c & d \end{bmatrix}$ with $\det M \neq 0$ then

$$M^{-1} = \frac{1}{\det M} \begin{bmatrix} d & -b \\ -c & a \end{bmatrix}$$

and thus, in our case

$$A^{-1} = \frac{1}{11} \begin{bmatrix} 4 & 1 \\ -3 & 2 \end{bmatrix}$$

so [C] holds true.

Problem (13). *The inverse of the matrix* $A = \begin{bmatrix} -1 & 3 \\ 2 & -5 \end{bmatrix}$ *is*

- *[A]* $\begin{bmatrix} 3 & 6 \\ -2 & -4 \end{bmatrix}$
- *[B]* $\begin{bmatrix} 5 & 3 \\ 2 & 1 \end{bmatrix}$
- *[C]* $\begin{bmatrix} -3 & 6 \\ 2 & 4 \end{bmatrix}$

- [D] $\begin{bmatrix} 3 & 6 \\ -2 & 4 \end{bmatrix}$

- [E] $\begin{bmatrix} -3 & -6 \\ -2 & 4 \end{bmatrix}$

Solution: **Check if A is invertible:**

$$\det A = (-1) \times (-5) - 2 \times 3 = 5 - 6 = -1 \neq 0$$

so, A is invertible.

Recall the formula for the inverse of a non - singular 2×2 matrix

$$A^{-1} = \frac{1}{\det A} \begin{bmatrix} -5 & -3 \\ -2 & -1 \end{bmatrix} = \begin{bmatrix} 5 & 3 \\ 2 & 1 \end{bmatrix}$$

Chapter 8

Topic 5 - Data Analysis - Combining & Probabilities

8.1 Statistics

Mean, Median

[a] If $\{x_1, x_2, \ldots\ldots, x_m\}$ are m observations, their **mean** $\bar{x}$ defined by

$$\overline{x} = \frac{x_1 + x_2 + \cdots + x_m}{m} = \frac{\sum_{i=1}^{k} x_i}{m} \tag{1}$$

[b] The **median** is defined, in the case of non-grouped values, as follows:

- For an odd number $(= 2m + 1)$ of observations the median[1] is

$$\delta = \frac{x_{m+1}}{2} \tag{2}$$

- If we have an even number of observations the median is

$$\delta = \frac{x_{\frac{m}{2}} + x_{\frac{m}{2}+1}}{2} \tag{3}$$

Dispersion measures

[c] **Range** R:

$$R = \text{maximum value} - \text{minimum value} \tag{4}$$

[d] **Interquartile Range** Q:

$$Q = Q_3 - Q_1 \tag{5}$$

where Q_1, Q_3 the first and the third quartile, which they are calculated in the following way:

- In non-grouped samples, the median divides the sample into two equal parts. Then, Q_1 is the median of the left part an Q_3 is the median of the right part of the sample.

[1] See Example 1

Examples in Statistics

Example (1). *Consider the following set of data*

7	9	10	13	20	11	6	7	12

Find
a) The mean b) The mode c) The median d) The range
Solution:

- *The mean is*

$$Mean = \frac{7+9+10+13+20+11+6+7+12}{9} = \frac{95}{9} = 10.6$$

- *The mode is the most common value, i.e: 7*
- *To find the median, the first step consists into ordering the data in ascending order:*

6	7	7	9	10	11	12	13	20

and here 10 *is the middle value, hence median* = 10.

- *For the range*

$$Range = max - min = 20 - 6 = 14$$

- *Also,* Q_1 *is the median of* $6, 7, 7, 9$ *and* Q_3 *is the median of* $11, 12, 13, 20$*, that is:* $Q_1 = 7$ *and* $Q_3 = 12.5$

Remark. *If there are two middle values, as in the set*

$$8, 9, 10, 13, 14, 15, 16, 17$$

then, the median is their average, here

$$mean = \frac{13+14}{2} = 13.5$$

Also, Q_1 *is the median of the set* $8, 9, 10, 13$ *and* Q_3 *is the median of* $14, 15, 16, 17$*, so:*

$$Q_1 = 9.5,\ Q_3 = 15.5$$

Example (2). *Find the mean value of the variable* x *when their values are presented in the following table:*

x_i	ν_i
100	7
150	10
200	15
250	6
300	2

Solution: *It is*

$$\bar{x} = \frac{100 \times 7 + 150 \times 10 + 200 \times 15 + 250 \times 6 + 300 \times 2}{7 + 10 + 15 + 6 + 2} = \frac{7.300}{40} = 182,5$$

Example (3). *For the entrance examination in a university school the minimum grade is* 16. *The subjects that are examined have the following weight coefficients:*

Mathematics: 3, *Chemistry:* 2, *Physics:* 2, *Essay:* 1
The grades that 2 *students - candidates have obtained are:*

	Mathematics	Chemistry	Physics	Essay
A	14	16	16	19
B	18	16	15	13

Solution: ***We calculate the mean for each student:***

$$\bar{x}_A = \frac{14\times 3+16\times 2+16\times 2+19\times 1}{3+2+2+1} = \frac{125}{8} = 15,625 \text{ rejected}$$
$$\bar{x}_B = \frac{18\times 3+16\times 2+15\times 2+13\times 1}{3+2+2+1} = \frac{129}{8} = 16,125 \text{ accepted}$$

hence, the first candidate failed and the second one has been accepted.

Problems in statistics

Problem (1). *The mean of the first* 100 *terms of the arithmetic progress* $-10, -7, -4, \ldots\ldots$ *is*

- *[A]* 138
- *[B]* 138.5
- *[C]* 139
- *[D]* 140
- *[E]* 277

Solution: Let us find the sum of the first 100 terms:

$$S_{100} = \frac{100}{2}[2\times(-10)+99\times 3] = 50\times 277 = 13850$$

hence

$$\text{Mean} = \frac{13850}{100} = 138.5$$

and the correct answer is [B].

Problem (2). *If the mean of the numbers* $12, 13, 15, 20, 30, x$ *is* 17 *then the median is*

- *[A]* 12
- *[B]* 14
- *[C]* 14.5
- *[D]* 15
- *[E]* 16

Solution: Let us apply the formula for the mean:

$$\bar{x} = \frac{12 + 13 + 15 + 20 + 30 + x}{6} = 17 \Rightarrow x = 12,$$

hence, in ascending order we have:

$$12, 12, 13, 15, 20, 30$$

so the median is

$$\text{Median} = \frac{13 + 15}{2} = 14.$$

Correct answer is $[B]$.

Problem (3). *In the set of data* $1, 5, 2, 3, 3, 4, 6, 2, 5, 2$ *the interquartile range is*

- *[A]* 1
- *[B]* 2
- *[C]* 3
- *[D]* 4
- *[E] none of the above*

Solution: Arrange the data in ascending order:

$$1, 2, 2, 2, 3, 3, 4, 5, 5, 6$$

hence

$$\text{Median} = \frac{3 + 3}{2} = 3 \tag{a}$$
$$Q_1 = 2,\ Q_3 = 5$$

and thus

$$Q_3 - Q_1 = 5 - 2 = 3.$$

Correct answer is $[C]$.

Problem (4). *In a sample of* 30 *students of a school, the* 16 *boys have a mean weight of* 72*kgr. Find the mean weight of the girls if the mean weight of all the students is* 65*kgr.*

- *[A]* 55
- *[B]* 56
- *[C]* 57
- *[D]* 59
- *[E] none of the above.*

Solution: If the mean weight of the girls is $\bar{x}$, then

$$65 = \frac{16 \times 72 + 14 \times \bar{x}}{30},$$

solving in $\bar{x}$ we get:

$$\bar{x} = \frac{30 \times 65 - 16 \times 72}{14} = 57,$$

so the correct answer is $[C]$.

8.2 Probability

Summary

1. **Probability**
$$p = \frac{\text{Number of successful events}}{\text{Number of all possible events}}$$

2. If two events A and B have no effect on one another, then
$$p(A \cap B) = p(A) \cdot p(B)$$
and they are called **independent events**.

3. If
$$p(A \cap B) = 0$$
then
$$p(A \cup B) = p(A) + p(B)$$
and A, B are called **mutually exclusive events**.

4. For any two events A and B we have:
$$p(A \cup B) = p(A) + p(B) - p(A \cap B)$$

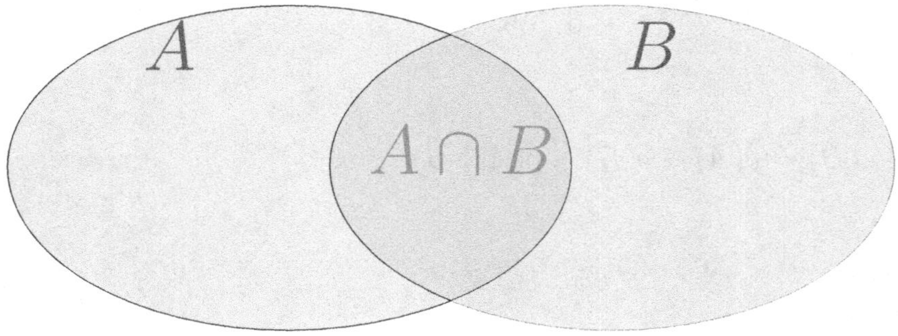

Probability

The probability of getting a 4 when a die is thrown, is
$$p = \frac{1}{6}$$
The probability of getting a 4 or 5 when a die is thrown ifs
$$p = \frac{2}{6}$$
and generally, probability is given by
$$p = \frac{\text{Number of successful events}}{\text{Number of all possible events}}$$

Independent Events

Two independent effects do not have effect one on each other

$$p(A \cap B) = \text{ probability that both events A and B happen}$$

If A and B are **independent** then

$$p(A \cap B) = p(A) \cdot p(B)$$

Example (1). *I throw two dice, the probability of getting two asses is*

$$p(\textit{two asses}) = \frac{1}{6} \cdot \frac{1}{6} = \frac{1}{36}$$

8.2.1 Mutually exclusive events

If $p(A \cap B) = 0$ then the events A, B are said to be **mutually exclusive**.

Example (2). *I throw a die. The event A of getting 1 and the event B of getting 3 cannot happen both in one throw. We shall have either 1 or 3 or one of the numbers 2, 4, 5, 6. We thus infer that*

$$p(A \cap B) = 0$$

*The probability to get either 1 or 3 is (**Careful now!**)*

$$p(A \cup B) = p(A) + p(B) = \frac{1}{6} + \frac{1}{6} = \frac{2}{6} = \frac{1}{3}$$

Generally speaking

$$p(A \cup B) = p(A) + p(B) - p(A \cap B)$$

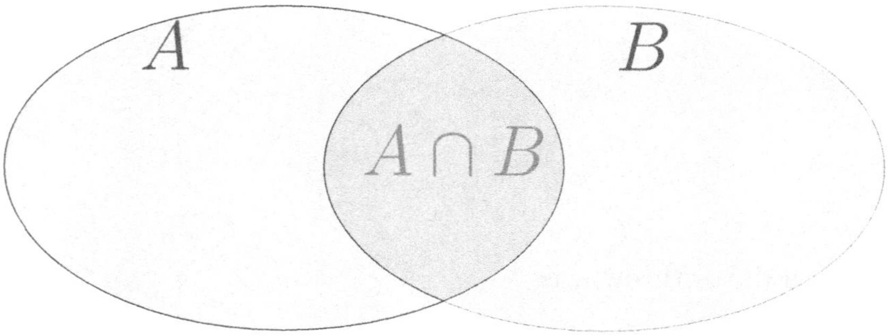

Problem Solving Now

Be Careful here!

In probability problems, the first step should be to make clear and explicit all the given information. Sometimes we see subtle points, implicit information

Example (3). *I throw a die, find the probability of getting a number greater than 4 or an even number.*

Try it for a while before you go on reading

Solution: Let A be the event *greater than* 4, then

$$p(A) = \frac{2}{6} = \frac{1}{3}$$

(two numbers 5 or 6).
Let B be the event *even number*, then

$$p(B) = \frac{3}{6} = \frac{1}{2}$$

(three even numbers $\{2, 4, 6\}$).
Let C be the event *greater than* 4 *and be an even number* then

$$p(C) = \frac{1}{6}$$

and also

$$C = A \cap B$$

therefore if we define the event $A \cup B$ *greater than* 4 *or an even number* we get:

$$\begin{aligned} p(A \cup B) &= p(A) + p(B) - p(A \cap B) \\ &= \frac{1}{3} + \frac{1}{2} - \frac{1}{6} \\ &= \frac{4}{6} = \frac{2}{3}. \end{aligned}$$

8.3 Some practice now

Problem (1). *Throw two dice, then the probability of getting a product of* 20 *or* 36 *is*

- *[A]* $\frac{1}{12}$
- *[B]* $\frac{3}{12}$
- *[C]* $\frac{5}{12}$
- *[D]* $\frac{16}{36}$
- *[E]* $\frac{2}{3}$

Solution: Denote by (x, y) what we get by throwing the two dies.
Let A be the event (x, y) *have a product of* 20
Let B be the event (x, y) *have a product of* 36
Then

$$A = \{(4,5),(5,4)\}$$

so

$$p(A) = \frac{2}{36}$$

and

$$B = \{(6,6)\}$$

so

$$p(B) = \frac{1}{36}$$

Further,

$$A \cap B = \emptyset \Rightarrow p(A \cap B) = 0$$

hence

$$p(A \cup B) = p(A) + p(B) - p(A \cap B) = \frac{2}{36} + \frac{1}{36} = \frac{3}{36} = \frac{1}{12}$$

and the correct answer is [A].

Problem (2). *A box, let us called it A, contains* 5 *red balls and* 3 *white balls. Another box, say it B, contains* 6 *red balls and* 4 *white balls. Draw an item at random from each box. Then, the probability that at least one ball is white is*

- *[A]* $\frac{1}{8}$
- *[B]* $\frac{3}{8}$
- *[C]* $\frac{3}{10}$
- *[D]* $\frac{5}{8}$
- *[E]* $\frac{7}{8}$

Solution: **Try to solve for a while before reading on**
First way:

- Since the result of the draw from box A is independent of the corresponding result from the box B we deduce that the probability of the event $C =$ *both the balls are red* is given by

$$p(C) = p(\{\text{red ball from} A\}) \cdot (\{\text{red ball from} B\}) = \frac{5}{8}\frac{6}{10} = \frac{3}{8}$$

- Clearly, the event *At least one of the balls is white* has as its *complementary* event C

- Since for complementary events X, Y the

$$p(X) + p(Y) = 1$$

holds true, then:

$$p(\text{At least one of the balls is white}) = 1 - p(C) = 1 - \frac{3}{8} = \frac{5}{8}.$$

The correct answer is [D]

Second way:

- Let us use the rule for *mutually exclusive events*
- Define the events $X_1 = \{\text{red}_\text{A} - \text{white}_\text{B}\}$, $X_2 = \{\text{white}_\text{A} - \text{red}_\text{B}\}$, $X_3 = \{\text{white}_\text{A} - \text{white}_\text{B}\}$ with

$$p(X_1) = \frac{5}{8} \cdot \frac{4}{10} = \frac{1}{4},\ p(X_2) = \frac{3}{8} \cdot \frac{6}{10} = \frac{9}{40},\ p(X_3) = \frac{3}{8} \cdot \frac{4}{10} = \frac{3}{20},$$

- hence

$$\begin{aligned} p(\text{At least one of the balls is white}) &= p(X_1) + p(X_2) + p(X_3) \\ &= \frac{1}{4} + \frac{9}{40} + \frac{3}{20} = \frac{5}{8}. \end{aligned}$$

Problem (3). *An integer is chosen at random from the first* 1000 *positive integers. The probability that the number is a multiple of* 3 *is*

- *[A]*0.25
- *[B]*0.3
- *[C]*0.31
- *[D]*0.32
- *[E]*0.333

Try for a while to solve it before reading on
Solution:

- It is enough to count the multiples of 3 between 1 and 1000.
- **Way** 1: The multiples of 3 are $3, 6, 9, \ldots, \ldots 999$. This is an arithmetic sequence with $a_1 = 3$, $a_n = 999$ and $d = 3$. Let us use the formula

$$a_n = a_1 + (n-1)d$$

hence

$$999 = 3 + 3(n-1) \Rightarrow n = 333$$

and thus

$$p = \frac{333}{1000}$$

- **Way 2:** The positive multiples of 3 are

$$3 \cdot \mathbb{N} = \{3 \cdot k \, : \, k \in \mathbb{N}\}$$

- So it is enough to determine bounds on $k \in \mathbb{N}$ such that

$$1 \leq 3 \cdot k \leq 1000$$

hence

$$\frac{1}{3} \leq k \leq \frac{1000}{3}, \quad k \in \mathbb{N}$$

hence

$$p(3 \cdot \mathbb{N}) = \frac{333}{1000} = 0.333$$

and the correct answer is $[E]$.

Sample Space & Probability

1 **Sample Space:**

Example (1). *If we toss a die, the sample space consists of the six positive numbers*

$$\{1, 2, 3, 4, 5, 6\}$$

Example (2). *If we toss two coins then the sequence of heads (H) and tails (T) is*

$$\{HH, HT, TH, TT\}$$

This is the sample space.

2 **Probability:** In Example 1, what is the probability of getting a number greater that 4?

Solution: If $X > 4$ then $X = 5$ or $X = 6$, so
Probability:

$$p(X > 4) = \frac{\text{number of } X > 4}{\text{number of } X} = \frac{2}{6} = \frac{1}{3}.$$

Also

$$p(X < 3) = p(X = 1 \,\text{or}\, X = 2) = \frac{1}{6} + \frac{1}{6} = \frac{2}{6} = \frac{1}{3}$$

3 **Generally** For the values X, $p(X)$ we use a table (**distribution table**) as

X:	X_1	X_2	X_3	X_4	X_5	X_6
$p(X)$:	$p(X_1)$	$p(X_2)$	$p(X_3)$	$p(X_4)$	$p(X_5)$	$p(X_6)$

It is always

$$p(X_1) + \cdots + p(X_6) = 1,$$

then, the **mean** or **expected value** of X is

$$E(X) = X_1 p(X_1) + \cdots X_6 p(X_6) = \sum_i X_i p(X_i) = \mu \tag{1}$$

4 **Variance** of X:

$$\mathrm{Var}(X) = E(X-\mu)^2 = E(X^2) - \mu^2 \tag{2}$$

5 **Standard Deviation**

$$\sigma_X = \sqrt{\mathrm{Var}(X)} \tag{3}$$

8.4 Counting rule

If an event A can happen in m ways and thereafter an event B can happen in n ways - and so on - all together can happen, in this order, in

$$m \times n \times \cdots\cdots$$

ways

Examples (1). *Basics on the counting rule to get familiar with the methods and techniques used:*

[a]: ***How many*** 5 ***- figure numbers can be formed with the digits***

$$1, 2, 3, 4, 5, 6, 7, 8, 9$$

all different?
Answer: *The first figure can be chosen in* 9 *ways,*
The second figure can be chosen in 8 *ways*
and so on up to the fifth figure which can be chosen in 5 *ways, hence, using the counting rule, we have*

$$9 \times 8 \times 7 \times 6 \times 5 = 15.120$$

such numbers.

[b]: ***In the previous example, if the figures can be repeated, find the number of such numbers.***
Answer: *Each digit can be chosen in* 9 *ways since they can be repeated, hence*

$$9 \times 9 \times 9 \times 9 \times 9 = 9^5 = 59.049$$

such numbers

[c]: ***In how many ways can the letters of the word*** *pencil* ***be arranged on a line?***
Answer: *The first letter* p *can be placed in* 5 *different ways*
the second letter can be placed in the remaining 4 *places and so on up to the sixth letter, hence*

$$6 \times 5 \times 4 \times 3 \times 2 \times 1 = 720$$

different words

[d]: ***In how many different ways can the numbers***

$$1, 2, 3, 4, 5, 6$$

be arranged around the circumference of a circle?
Answer: *Fixing the first digit on the circumference the remaining* 5 *five digits can be arranged in*

$$5 \times 4 \times 3 \times 2 \times 1 = 5! = 120$$

different ways.

Careful!!!! In this topic we shall present everything through examples & problems - This is the easiest way to learn about it.

Example (2).

[i] ***In how many ways can we select*** 2 ***students out of a group of*** 12***?***
Solution: *We use the formula*

$$x = C_2^{12} = \frac{12!}{2! \times (12-2)!} = \frac{12!}{2! \times 10!} = \frac{11 \times 12}{2} = 66$$

[ii] ***In a class there are*** 14 ***girls, in how many ways can we select a group of*** 8 ***girls?***
Solution: *It is*

$$y = C_8^{14} = \frac{14!}{8! \times (14-8)!} = \frac{14!}{8! \times 6!} = 3 \times 7 \times 11 \times 13 = 3003$$

Remark. *Generally, the number of combinations of n objects, k at at a time, $n \geq k$, is*

$$C_k^n = \frac{n!}{k! \times (n-k)!}$$

Example (3). ***In how many ways can we select*** 3 ***boys from a group of*** 9 ***boys?***
Solution: We shall use the formula for the combination of m objects in subsets of n:

$$C_n^m = \binom{m}{n} = \frac{m!}{n! \cdot (m-n)!},$$

where

$$m! = 1 \times 2 \times \cdots \times (m-1) \times m$$

hence

$$C_3^9 = \binom{9}{3} = \frac{9!}{3!(9-3)!} = \frac{9!}{3! \cdot 6!} = \frac{7 \times 8 \times 9}{1 \times 2 \times 3} = 84$$

Example (4). ***How many*** 3 ***- digit numbers can be formed, using once each digit*** 1, 2, 3, 4, 5, 6, 7, 8, 9***?***

Solution: We shall use the formula for the number $\mathcal{P}_k^m$ of **permutations** of m objects, by k each time: it is

$$\mathcal{P}_k^m = m \times (m-1) \times (m-2) \times \cdots\cdots \times (m-k+1),$$

hence, given that in our case $m = 9$ and $k = 3$, we get

$$\mathcal{P}_3^9 = 9 \times 8 \times 7 = 504$$

Remark. *The counting rule is equivalent to this formula.*

Equivalently we have that the required number is

$$\binom{9}{3} \times 3! = 504$$

Example (5). *In how many ways can we select 3 digits from the set* $\{1, 2, 3, 4, 5, 6, 7, 8, 9\}$

Solution: The required number, since we do not care about the order of the digits, is

$$C_3^9 = \binom{9}{3} = \frac{9!}{3! \times (9-3)!} = \left(\frac{9!}{3! \times 6!}\right) = 84$$

Counting Rule & Combinations - Problems

Problem (1). *We have 10 different books, if we can select 3 books in x ways, then x is equal to*

- *[A]* 30
- *[B]* 60
- *[C]* 90
- *[D]* 120
- *[E]* 240

Solution: It is

$$x = \binom{10}{3} = \frac{10!}{3! \times (10-3)!} = \frac{10!}{3! \times 7!} = \frac{10 \times 9 \times 8}{1 \times 2 \times 3} = 120.$$

Correct answer is $[D]$.

Problem (2). *We have 11 different books, in how many ways can we select 5 books, when a certain book is already included?*

- *[A]* 70
- *[B]* 90
- *[C]* 120
- *[D]* 150

- *[E]* 210

Solution: It is

$$x = \binom{10}{4} = \frac{10!}{4! \times (10-4)!} = \frac{10!}{4! \times 6!} = 210$$

Correct answer is $[E]$.

Problem (3). *In how many ways can we select 5 books from 14 if a certain book is already excluded?*

- *[A]* 70
- *[B]* 90
- *[C]* 120
- *[D]* 150
- *[E]* 1287

Solution: It like selecting 5 books from a collection of 13, hence the required number is

$$C_5^{13} = \binom{13}{5} = \frac{13!}{5! \times (13-5)!} = \frac{13!}{5! \times 8!} = 1287.$$

Correct answer is $[E]$.

Problem (4). *A boy has 5 friends, in how many ways can he invite one or more of his friends?*

- *[A]* 15
- *[B]* 16
- *[C]* 21
- *[D]* 30
- *[E]* 31

Solution: He can invite them one by one, or in pairs, or in triads, or in groups of four or, finally, all the five toghether. Hence the required number is

$$\binom{5}{1} + \binom{5}{2} + \binom{5}{3} + \binom{5}{4} + \binom{5}{5} = 31.$$

Correct answer is $[E]$.

Problem (5). *In how many ways can 10 books be divided into two subsets each one containing 5 books:*

- *[A]* 72

- *[B]* 126
- *[C]* 144
- *[D]* 252
- *[E]* *none of the above*

Solution: First we observe that there are C_5^{10} subsets each one containg 5 books but we also have to take care about the $2! = 1 \times 2 = 2$ different ways the subsets do appear and to disgard this ordering. So the number we are asking for is

$$\frac{1}{2}\binom{10}{5} = \frac{1}{2}\frac{10!}{5! \times 5!} = 126.$$

Correct answer is $[B]$.

Problem (6). *We have* 10 *different books and* 2 *boxes* A *and* B*. In how many ways can we divide the books in two subsets and put them into the boxes* A *and* B

- *[A]* 72
- *[B]* 126
- *[C]* 252
- *[D]* 256
- *[E]* *none of the above*

Solution: It is natural to accept that we put 5 books in each of the boxes, so the question is equivalent to the following :
find the number of subsets of 5 books we can get from the given set of 10 different books hence the required number is

$$x = C_5^{10} = 252$$

Correct answer is $[C]$.

Problem (7). *In how many ways six passengers can seat themselves in a railway compartement with eight seats (four a side), if two of them insist on having back to the engine.*

- *[A]* 1440
- *[B]* 2160
- *[C]* 4300
- *[D]* 4320
- *[E]* 8640

Solution: Think for a while before reading on
Let us use the counting rule:
There are

$$4 \times 3$$

ways to sit 2 passengers with their back to the engine,
Hence for the other 4 passengers there are

$$6 \times 5 \times 4 \times 3$$

ways to sit and thus, in total:

$$(4 \times 3) \times (6 \times 5 \times 4 \times 3) = 4320$$

different ways, hence the correct answer is $[D]$.

Problem (8). *How many 3 - digit numbers, divisible by 5, can be formed using the digits* $0, 1, 2, 3, 4, 5$*?*

- *[A]* 33
- *[B]* 34
- *[C]* 35
- *[D]* 36
- *[E]* 40

Solution: Think for a while before reading on
Such a number has the form abc where :
c is equal either to 0 or 5
if $c = 0$ then a is one of $1, 2, 3, 4, 5$ and for b is the remaing 4 choices hence
for $c = 0$ we have $5 \times 4 = 20$ different numbers
if $c = 5$ then
a is one of the remaing $1, 2, 3, 4$ digits and b is one the other 4 possible digits, thus
for $c = 5$ we have $4 \times 4 = 16$ possible numbers
In total we have

$$5 \times 4 + 4 \times 4 = 36$$

different 3-digits divisible by 5.
Correct answer is D. .

Problem (9). *In how many ways can* 5 *men and* 5 *women be placed around a table so that each woman is sitting between* 2 *men?*

- *[A]* 256
- *[B]* 512
- *[C]* 2880
- *[D]* 14400

- *[E]* 28800

Solution: Think for a while before reading on
Let the men sit first then there are 4! ways (circular permutations)
Then, there are 5 places for the women so 5! ways,
hence, we get

$$4! \times 5! = 2880$$

ways, hence the correct answer is $[C]$.

Problem (10). *In how many ways can sit on a row* 5 *men and* 5 *women not two men or two women sit together?*

- *[A]* 512
- *[B]* 2880
- *[C]* 14400
- *[D]* 28800
- *[E]* *none of the above*

Solution: Think for a while before reading on
Let the men take their seat first. Then the women can seat either this way

$$\text{mwmwmwmwmw}$$

or that way

$$\text{wmwmwmwmwm}.$$

So, there are

$$2 \times 5! \times 5! = 28.800$$

different ways.
Correct answer is $[D]$.

Problem (11). *In how many ways can seven boys be selected from twelve so that a certain boy is always one of the seven selected?*

- *[A]* 462
- *[B]* 924
- *[C]* 1848
- *[D]* 5544
- *[E]* 33264

Solution: Think for a while before reading on
Since the one boy has to be always selected then the required number is

$$x = \binom{11}{6} = \frac{11!}{6! \times (11-6)!} = \frac{11!}{6! \times 5!} = 462.$$

Correct answer is $[A]$.

Problem (12). *In how many ways can we give six different history books and six different science books to 2 students, each student to take 3 books of each kind?*

- *[A]* 120
- *[B]* 400
- *[C]* 800
- *[D]* 2160
- *[E] none of the above*

Solution: Think for a while before reading on
The number of different triads of history books is $\binom{6}{3}$
The number of different triads of science books is $\binom{6}{3}$
The number of students is $2 = \binom{2}{1}$, hence the required number is

$$\binom{6}{3} \times \binom{6}{3} \times \binom{2}{1} = 800.$$

Correct answer is $[C]$.

Problem (13). *In how many ways can 8 different objects be divided in 4 pairs?*

- *[A]* 105
- *[B]* 210
- *[C]* 256
- *[D]* 420
- *[E]* 2520

Solution: Think for a while before reading on
The first pair can be chosen in $\binom{8}{2}$,
The second pair can be chosen in $\binom{6}{2}$,
The third pair can be chosen in $\binom{4}{2}$,
The fourth pair can be chosen in $\binom{2}{2}$,
The fourth pairs can be arranged in 4! different ways
Hence, the required number is

$$x = \frac{\binom{8}{2} \times \binom{6}{2} \times \binom{4}{2} \times \binom{2}{2}}{4!} = 105.$$

Correct answer is $[A]$.

Problem (14). *In how many ways can we divide a group of 10 students into two equal groups?*

- *[A]* 32

- *[B]* 64
- *[C]* 126
- *[D]* 128
- *[E]* 252

Solution: Think for a while before reading on
Since the two partitions are counted twice we have

$$x = \frac{\binom{10}{5}}{2} = \frac{1}{2} \times \frac{10!}{5! \times 5!} = 126.$$

Correct answer is $[C]$.

Problem (15). *Eight (8) boys went on an excursion and at the evening they came to a hotel. In the hotel there were only 4 rooms with 2 beds each. In how many ways can the students stay in the hotel?*

- *[A]* 2510
- *[B]* 2515
- *[C]* 2520
- *[D]* 2525
- *[E]* 2530

Solution: The number of different ways the students can stay in the hotel can be calculated as follows:

- In the first room we have $\binom{8}{2}$ different choices
- In the second room we shall have $\binom{6}{2}$ choices
- In the third room $\binom{4}{2}$ choices and finally
- In the fourth room $\binom{2}{2} = 1$ choice

hence, by the multiplicative principle:

$$\binom{8}{2} \times \binom{6}{2} \times \binom{4}{2} \times \binom{2}{2} = \frac{8!}{6! \times 2!} \times \frac{6!}{4! \times 2!} \times \frac{4!}{2! \times 2!} \times \frac{2!}{2!} = 28 \times 15 \times 6 = 2520.$$

Correct answer is $[C]$.

Problem (16). *How many 3 - digit numbers can be formed using the numbers* $0, 1, 2, 3, 4, 5$ *without repetitions?*

- *[A]* 100
- *[B]* 150

- *[C]* 155
- *[D]* 200
- *[E]* 300

Solution: For the first digit there are 5 different ways-choices (since 0 is not allowable to be the first digit), for the second we have 5 possible choices and for the third digit there are 4 different ways to choose it. In total there are

$$5 \times 5 \times 4 = 100$$

possible 3 - digit numbers. Correct answer is $[A]$.

Problem (17). *How many* 3 *- digit numbers,* divisible *by* 10*, can be formed using the numbers* $0, 1, 2, 3, 4, 5, 6, 7$*, without repetitions?*

- *[A]* 42
- *[B]* 44
- *[C]* 46
- *[D]* 48
- *[E]* 49

Solution: Where is the subtle point?
The number is divisible by 10 if and only if the last digit is 0, hence:
The first digit can be chosen in 7 ways,
The second digit can be chosen in 6 ways since repetitions are not allowed and the third digit must be zero.
It follows that the required number of possible 3 - digit divisible with 10 is

$$7 \times 6 = 42.$$

Correct answer is $[A]$.

Problem (18). *In how many ways can a group of* 5 *persons arrange themselves in a row of* 7 *chairs?*

- *[A]* 2500
- *[B]* 2510
- *[C]* 2515
- *[D]* 2520
- *[E]* 2525

Solution: The first person has 7 different choices to occupy one of the chairs, the second has 6 different chairs to choose among them and so on

$$\text{Ways} = 7 \times 6 \times 5 \times 4 \times 3 = 2520.$$

Correct answer is $[D]$.

Problem (19). *In how many ways can 7 persons sit around a circular table?*

- *[A]* 710
- *[B]* 720
- *[C]* 730
- *[D]* 740
- *[E]* 750

Solution: The first person has no limitation because they do not give us any information on the number of possible choices of seats around the table ... **For the remaining 6 persons there are 6! different permutations** and thus

$$\text{Ways} \ = 6! = 1 \times 2 \times 3 \times 4 \times 5 \times 6 = 720$$

different ways, hence the correct answer is $[B]$.

Problem (20). *In how many ways can 5 boys and 2 girls sit in a row with 7 seats if the girls are going to sit together?*

- *[A]* 120
- *[B]* 720
- *[C]* 840
- *[D]* 1440
- *[E]* *none of the above*

Solution: **Look at the row**

1	2	3	4	5	6	7

The girls can sit either at the places $1, 2$, or at $2, 3$, or at $3, 4$, or at $4, 5$, or at $5, 6$ or $6, 7$, so they do have 6 different choices. But, since the girls can change their *internal* order we have it it total $6 \times 2 = 12$ different choices for the girls.
For the 5 boys and the remaining 5 seats they are allowed $5! = 1 \times 2 \times 3 \times 4 \times 5 = 120$ different ways and thus

$$\text{Total number} \ = 12 \times 5! = 12 \times 120 = 1440$$

ways, hence the correct answer is $[D]$.

8.5 Sample Space & Probabilities

Problems on probabilities

Skip any problem that seems difficult for you.

Problem (1). *In a class there are 12 boys and 10 girls, 6 boys and 4 girls are taller than 1.70m. If we select one student at random the probability that the student is a girl or taller that 1.70m is equal to*

- *[A]* $\frac{10}{22}$
- *[B]* $\frac{12}{22}$
- *[C]* $\frac{14}{22}$
- *[D]* $\frac{16}{22}$
- *[E]* *none of the above*

Solution: Define the events

$$T = \{\text{taller than } 1.70m\}$$

and

$$G = \{\text{the student is a girl }\}$$

By the hypothesis

$$p(T) = \frac{10}{22} \qquad \text{(i)}$$

$$p(G) = \frac{10}{22} \qquad \text{(ii)}$$

$$p(G \cap T) = \frac{4}{22} \qquad \text{(iii)}$$

hence, since we are asked about the probability of the event

$$G \cup T = \{ \text{ is a girl or is taller than } 1.70m\}$$

we have:

$$p(G \cup T) = p(G) + p(T) - p(G \cap T) = \frac{10}{22} + \frac{10}{22} - \frac{4}{22} = \frac{16}{22} = \frac{8}{11},$$

the correct answer is $[D]$.

Problem (2). *In a class there are 28 students, 18 of them are boys. The height of 10 boys is less than 1.70m and the height of 16 students is less than 1.70m. We choose a student at random, what is the probability to be a girl taller than 1.70m ?*

- *[A]* $\frac{1}{28}$
- *[B]* $\frac{2}{28}$
- *[C]* $\frac{3}{28}$
- *[D]* $\frac{4}{28}$
- *[E]* $\frac{16}{28}$

Solution: Where is the subtle point?
Based on the data define the event:

$$T = \{\text{taller than } 1.70 \text{ m}\},\ p(T) = 1 - \frac{16}{28} = \frac{12}{28} = \frac{3}{7}$$

There are $28 - 18 = 10$ girls among the students and there are 4 girls taller than 1.70 m, hence

$$p(\{\text{the student is a girl taller than} 1.70\,\text{m}\}) = \frac{4}{28} = \frac{1}{7}$$

and the correct answer is $[D]$.

Problem (3). *We toss a coin three times, find the probability of the event*

$$A = \{\ \textit{get a head at least twice}\}$$

,

- *[A]* $\frac{3}{8}$
- *[B]* $\frac{4}{8}$
- *[C]* $\frac{6}{8}$
- *[D]* $\frac{7}{8}$
- *[E]* *none of the above*

Solution: Define the event

$$A = \{\text{HHT}, \text{HTH}, \text{THH}, \text{HHH}\}$$

Usin the multiplication principle we easily get that the sample space Ω contains

$$2^3 = 2 \times 2 \times 2 = 8$$

elements, therfore the probability of the event A is given by

$$p(A) = \frac{N(A)}{N(\Omega)} = \frac{4}{8} = \frac{1}{2}$$

and thus $[B]$ is the correct answer.

Problem (4). *We toss a coin three times, find the probability of the event*

$$A = \{ \text{ get both head and tail } \}$$

,

- *[A]* $\frac{1}{8}$
- *[B]* $\frac{2}{8}$
- *[C]* $\frac{3}{8}$
- *[D]* $\frac{4}{8}$
- *[E]* $\frac{3}{4}$

Solution: **Define the event**

$$A_{H\&K} = \{HHK, HKH, HK, KHH, KHK, KKH\}$$

As we have already seen, in problem 3, the sample space Ω of the experiment, has

$$N(\Omega) = 2^3 = 8$$

elements and thus the probability of the event $A_{H\&K}$ is given by

$$p(A_{H\&K}) = \frac{N(A_{H\&K})}{N(\Omega)} = \frac{6}{8} = \frac{3}{4}$$

and thus $[E]$ is the correct answer.
Way 2: Observe that

$$p\left(A_{H\&K}\right) = 1 - p\left(\{HHH, KKK\}\right) = 1 - \frac{N\left(\{HHH, KKK\}\right)}{N(\Omega)} = 1 - \frac{2}{8} = \frac{3}{4}.$$

Problem (5). *Forty students have been asked if they play soccer or basketball.* 25 *of them have answered positively for soccer,* 18 *yes for basketball and* 5 *answered negatively to both sports. If we denote by $p(S)$ the probability that a student, chosen at random, plays soccer and with $p(B)$ the probability that practises basketball, calculate the probability of the event: a student chosen at random, is* ***enrolled in both****, soccer and basketball.*

- *[A]* $\frac{2}{40}$
- *[B]* $\frac{8}{40}$
- *[C]* $\frac{16}{40}$

- *[D]* $\frac{20}{40}$
- *[E]* $\frac{32}{40}$

Solution: **Define the proper events:**

$$S = \{\text{the student is a soccer player}\}, \; p(S) = \frac{25}{40} \quad \text{(i)}$$
$$B = \{\text{the student is a basketball player}\}, \; p(B) = \frac{18}{40}$$

and since 5 students do not play either soccer nor basketball

$$p\left((S \cup B)'\right) = \frac{5}{40} \Rightarrow p(S \cup B) = 1 - \frac{5}{40} = \frac{35}{40} \; (iii)$$

Now, we are asked to compute the probability of $S \cap B$, where, using (i)-(iii):

$$\begin{aligned} p(S \cup B) &= p(S) + p(B) - p(S \cap B) \Rightarrow \\ p(S \cap B) &= p(S) + p(B) - p(S \cup B) \\ &= \frac{25}{40} + \frac{18}{40} - \frac{35}{40} = \frac{8}{40}, \end{aligned}$$

hence $[B]$ holds true.

Problem (6). *Under the assumptions of problem* (5), *calculate the probability of the event: a student chosen at random, is enrolled* **at least in one of the two**, *soccer or basketball.*

- *[A]* $\frac{14}{40}$
- *[B]* $\frac{17}{40}$
- *[C]* $\frac{20}{40}$
- *[D]* $\frac{23}{40}$
- *[E]* $\frac{35}{40}$

Solution: **Where is the subtle point?**
Express the event

$$E_{s_b} = \{\text{a randomnly chosed student practises at least soccer or basketball}\}$$

in terms of the events S and T:

$$\begin{aligned} E_{s_b} &= S \cup B \Rightarrow \\ p(E_{s_b}) &= p(S) + p(B) - p(S \cap B) \Rightarrow \\ p(E_{s_b}) &= \frac{25}{40} + \frac{18}{40} - \frac{8}{40} = \frac{35}{40}. \end{aligned}$$

Problem (7). *Under the assumptions of problem (5), calculate the probability of the event:* *a student chosen at random, is enrolled* **only** *in soccer.*

- *[A]* $\frac{14}{40}$
- *[B]* $\frac{15}{40}$
- *[C]* $\frac{16}{40}$
- *[D]* $\frac{17}{40}$
- *[E]* $\frac{18}{40}$

Solution: Where is the subtle point?
Enrolling only in soccer is equivalent to: $E_{S\cap B'}$, hence we have to compute

$$p(S \cap B') \tag{i}$$

where

$$S = (S \cap B') \cup (S \cap B) \tag{ii}$$
$$(S \cap B') \cap (S \cap B) = \emptyset \tag{iii}$$

Using (i), (ii) and taking into account (iii) we can write:

$$\begin{aligned} p(S \cap B') &= p(S) - p(S \cap B) \\ &= \frac{25}{40} - \frac{8}{40} = \frac{17}{40} \end{aligned}$$

Problem (8). *In a game - competition they participate four players, A, B, C and D. Suppose that the probability that A wins is the double of the probability that B wins, the probability that B wins is the double of the the probability C wins and finally the probability C wins is equal to the probability of winning for the player D, calculate the probability of the event*
wins either A or B.

- *[A]* $\frac{1}{8}$
- *[B]* $\frac{3}{8}$
- *[C]* $\frac{1}{4}$
- *[D]* $\frac{2}{4}$
- *[E]* $\frac{3}{4}$

Solution: Where is the subtle point?
Define the events properly and relate them: let

$$A = \{\text{player A wins}\}\,,\ p(A) = x \qquad \text{(i)}$$

$$B = \{\text{player B wins}\}\,,\ p(B) = \frac{p(A)}{2} = \frac{x}{2} \qquad \text{(ii)}$$

$$C = \{\text{player C wins}\}\,,\ p(C) = \frac{p(B)}{2} = \frac{x}{4} \qquad \text{(ii)}$$

$$D = \{\text{player D wins}\}\,,\ p(D) = p(C) = \frac{x}{4} \qquad \text{(iv)}$$

Now, use the fact that:

$$p(A) + p(B) + p(C) + p(D) = 1 \Leftrightarrow x + \frac{x}{2} + \frac{x}{4} + \frac{x}{4} = 1 \Leftrightarrow x = \frac{1}{2} = p(A) \qquad \text{(v)}$$

Now, by the definition of the game

$$p(A \cap B) = 0$$

hence

$$p(A \cup B) = p(A) + p(B) - p(A \cap B) = p(A) + p(B) = \frac{1}{2} + \frac{1}{2} = \frac{3}{4}$$

and thus, $[E]$ is the correct answer.

Problem (9). *In a box there are* 20 *balls,* 10 *are white, numbered from* 1 *to* 10 *and the remaining* 10 *are in color blue and numbered from* 1 *to* 10 *as well. If we choose at random one ball from the box, calculate the probability of the event*
the ball is white or the number appearing is greater that 6.

- *[A]* $\frac{6}{20}$
- *[B]* $\frac{7}{20}$
- *[C]* $\frac{9}{20}$
- *[D]* $\frac{13}{20}$
- *[E]* $\frac{14}{20}$

Solution: Where is the subtle point?
Define the events properly:

$$W = \{\text{the ball is colored white}\} \qquad \text{(i)}$$

$$N_{>6} = \{\text{number of the ball is greater than } 6\} \qquad \text{(ii)}$$

with

$$p(W) = \frac{10}{20} \tag{iii}$$
$$p(N_{>6}) = \frac{4+4}{20} = \frac{8}{20} \tag{iv}$$
$$p(W \cap N_{>6}) = \frac{4}{20} \tag{v}$$

hence, the probability of the event the ball is colored white or its number is greater than 6 is:

$$p(W \cup N_{>6}) = p(W) + p(N_{>6}) - p(W \cap N_{>6}) = \frac{10}{20} + \frac{8}{20} - \frac{4}{20} = \frac{14}{20},$$

so $[E]$ is the correct answer.

Problem (10). *In Ω they are given two events A and B such that*

$$p(A \cup B) = \frac{3}{4},\ p(A \cap B) = \frac{1}{4},\ p(A') = \frac{1}{2}.$$

Then, the probabilities $p(A)$ and $p(B)$ are:

- *[A]* $\frac{1}{8}$
- *[B]* $\frac{1}{4}$
- *[C]* $\frac{2}{4}$
- *[D]* $\frac{3}{4}$
- *[E]* *none of the above*

Solution: **Where is the subtle point?**
Use $p(A \cup B)$ formula and the fact that $p(A') = 1 - p(A)$:

$$\left.\begin{array}{l} p(A \cup B) = p(A) + p(B) - p(A \cap B) \\ p(A) = 1 - p(A) = 1 - \frac{1}{2} = \frac{1}{2} \end{array}\right\} \Rightarrow \frac{3}{4} = \frac{1}{2} + p(B) - \frac{1}{4} \Rightarrow p(B) = \frac{2}{4}$$

hence $[C]$ holds true.

Problem (11). *Let A, B be events in Ω such that*

$$p(A) = \frac{1}{2},\ p(B) = x \text{ and } p(A \cup B) = x^2 + \frac{3}{4},$$

then x is equal to:

- *[A]* $\frac{1}{8}$

- *[B]* $\frac{2}{8}$
- *[C]* $\frac{3}{8}$
- *[D]* $\frac{4}{8}$
- *[E]* *none of the above*

Solution: Where is the subtle point?
Use the formula for $p(A \cup B)$ and that $p(A) \geq 0$ for any event A:

$$p(A \cup B) = p(A) + p(B) - p(A \cap B) \Rightarrow$$
$$p(A \cup B) \leq p(A) + p(B) \Rightarrow$$
$$x^2 + \frac{3}{4} \leq \frac{1}{2} + x \Rightarrow x^2 - x + \frac{1}{4} \leq 0$$

and thus we conclude that

$$\left(x - \frac{1}{2}\right)^2 \leq 0 \Rightarrow x = \frac{1}{2}$$

and the correct answer is $[D]$.

Problem (12). *Let $\Omega = \{1, 2, 3, \ldots, 2000\}$ and the probability function*

$$p(\omega) = \begin{cases} a^2, & \text{if } 1 \leq \omega \leq 3 \\ 2a, & \text{if } \omega = 4, \ a \in \mathbb{R} \\ 0, & \text{if } \omega > 4 \end{cases}$$

Then, a is equal to

- *[A]* $\frac{1}{9}$
- *[B]* $\frac{2}{9}$
- *[C]* $\frac{2}{8}$
- *[D]* $\frac{1}{3}$
- *[E]* *none of the above*

The probability of the event

$$A = \{\omega \in \Omega \,|\, 2 \leq \omega \leq 5\}$$

is

- *[A]* $\frac{1}{9}$

- *[B]* $\frac{2}{9}$
- *[C]* $\frac{4}{9}$
- *[D]* $\frac{7}{9}$
- *[E]* $\frac{8}{9}$

Solution:

- **Where is the subtle point?**
  Use the defining property of the probability function:

$$\sum_{i=1}^{2} p(\omega_i) = 1 \Leftrightarrow p(1) + p(2) + p(3) + p(4) + p(5) + \cdots + p(2000) = 1 \Leftrightarrow$$

$$3a^2 + 2a = 1 \Leftrightarrow 3a^2 + 2a - 1 = 0 \Rightarrow a = -1 \text{ or } a = \frac{1}{3}$$

  and $a = -1$ is rejected (is negative). In conclusion $a = \frac{1}{3}$.

- The probability of $A = \{\omega \in \Omega \,|\, 2 \leq \omega \leq 5\}$ is

$$p(A) = p(2) + p(3) + p(4) + p(5) = 2a^2 + 2a \text{ and since } a = \frac{1}{3} \Rightarrow$$

$$p(A) = 2\left(\frac{1}{3}\right)^2 + 2\frac{1}{3} = \frac{8}{9}.$$

  Hence, the correct answer is $[E]$.

Summary of elementary combinatorics		
Description	Additional Details	Formula
Combinations of n objects by k or k	$\|A\| = n$ and $x_i \neq x_j$ for any $i \neq j$	$\binom{n}{k} = \frac{n!}{(n-k)! \times k!} = \frac{1 \times 2 \times \cdots \times (n-k+1)}{k!}$
Ordering n objects by k, k are similar	$n > k$	$\frac{n!}{k!} = n \times (n-1) \times \cdots \times (n-k+1)$
Permutations of n objects	Repetitions **not allowed**	$n! = 1 \times 2 \times \cdots \times n$
Permutations of k out of n objects	Repetitions **allowed**	n^k
Cyclic permutations of n objects	Fix an object & apply permutations of $(n-1)$ objects	$(n-1)! = 1 \times 2 \times \cdots \times (n-1)$
Combinations of n objects by k when k	Repetitions **allowed**	$\binom{n+k-1}{k}$

Chapter 9

Practice Tests

9.1 Practice Test A

Problem (1). If

$$x^2 - 3x + 1 = 0$$

and r_1, r_2 are the roots then, the value of $r_1^2 + r_2^2$ is
A. 4, **B.** 5, **C.** 6, **D.** 7, **E.** 8,

Problem (2). The remainder when $p(x) = x^3 + x^2 + 2x - 1$ is divided by $x + 1$ is
A. -4, **B.** -3, **C.** -2 , **D.** 1, **E.** 2,

Problem (3). If $x + 1$, $x - 2$ are factors of

$$q(x) = x^3 + 3x^2 + ax + b$$

then, the values of a, b are:
A. -6 **and** -1, **B.** -6 **and** -8, **C.** -4 **and** -2, **D.** 4 **and** -2, **E.** 4 **and** 2,

Problem (4). If the equation

$$3x^2 - 4x + a = 0$$

has two equal real roots, then the value of a is
A. $-\frac{4}{3}$, **B.** $-\frac{2}{3}$, **C.** 1, **D.** $\frac{4}{3}$, **E. none of the above,**

Problem (5). The solutions of the inequality

$$\frac{1}{x-1} < \frac{2}{x+3}$$

are
A. $x < -3$, **B.** $x < -1$, **C.** $x > -3$, **D.** $-3 < x < 1$ **or** $x > 5$, **E.** $x > 1$,

Problem (6). If

$$f(x) = \begin{cases} \dfrac{x^2 - 4}{x + 2}, & \text{if } x \neq -2 \\ a & \text{if } x = -2 \end{cases}$$

is continuous, then a is equal to
A. -4, **B.** -2, **C.** 0, **D.** 1, **E.** 2,

Problem (7). If a triangle $\triangle AOB$, formed by the two coordinate axis and the straight line with equation $y = x + 3$, is rotated around the x- axis, then the volume of the figure formed is equal to
A. 3π, **B.** 4π, **C.** 6π, **D.** 9π, **E.** 10π,

Problem (8). The domain of the function

$$f(x) = \sqrt{x - x^3}$$

is
A. $x < -1$, **B.** $x > -1$, **C.** $x < -1$ **or** $x > 1$, **D.** $x \leq -1$ **or** $0 \leq x \leq 1$, **E.** $x > 1$,

Problem (9). If

$$f(x) = \frac{5x^3 - 7x^2 + 6x + 1}{x^3 - 1}$$

the asymptotes of the graph $\mathcal{G}_f$ of f, are
A. $y = 5,\ x = 1$, **B.** $y = -5,\ x = 1$, **C.** $y = 5$, **D.** $x = -1$, **E.** $y = 1,\ x = -1$,

Problem (10). The sum of all the terms of the geometric sequence

$$10, \frac{10}{2}, \frac{10}{4}, \frac{10}{8}, \ldots, \frac{10}{2^n}, \ldots$$

is
A. 10, **B.** 12, **C.** 15, **D.** 20, **E.** 24,

Problem (11). The reciprocal of $5 - 3i$ is
A. $-\frac{5}{34} + \frac{3}{34}i$, **B.** $\frac{5}{34} - \frac{3}{34}i$, **C.** $\frac{5}{34} + \frac{3}{34}i$, **D.** $5 + 3i$, **E.** $\frac{1}{5} + \frac{1}{3}i$,

Problem (12). In the figure there are two concentric circles with arc lengths $s_1 = 5$ and $s = 1$. Then, the ratio of the areas of the larger over the smaller sector is
A. 5, **B.** 10, **C.** 15, **D.** 20, **E.** 25,

Problem (13). If the graph of the function $f(x) = -x^2 + 4$ is translated 4 units lefts and 3 units down, then the coordinates of the point $(1, f(1)) = (1, 3)$ will be
A. $(1, 3)$, **B.** $(1, 15)$, **C.** $(4, 3)$, **D.** $(4, 15)$, **E.** $(4, -15)$,

Problem (14). The solution of

$$|x + 1| < x$$

is
A. $x < \frac{-1}{2}$, **B.** $0 < x < \frac{1}{2}$, **C.** $\frac{-1}{2} < x < 1$, **D.** $\frac{-1}{2} < x < 0$,
E. no values of x satisfy the given inequality,

Problem (15). In a triangle $\triangle ABC$ it is $\sin A = \frac{1}{4}$, $\cos B = \frac{-1}{3}$, then C is equal to

A. 35°, B. 40°, C. 45°, D. 56°, E. 60°,

Problem (16). **The graph of the function represented parametrically by**

$$x = t^2,\ y = \frac{1}{t^2}$$

is a (an)
A. line segment, B. hyperbola, C. ellipse, D. line, E. none of the above,

Problem (17). **If the vectors** $\vec{v} = 2\vec{i} - 5\vec{j}$ **and** $\vec{u} = m\vec{i} + \vec{j}$ **are perpendicular, then** m **is equal to**
A. 2, B. 3, C. $\frac{5}{2}$, D. $\frac{1}{5}$, E. 10,

Problem (18). **Six coins are flipped, the probability that** 5 **tails appears is**
A. $\frac{1}{32}$, B. $\frac{3}{32}$, C. $\frac{3}{16}$, D. $\frac{11}{16}$, E. $\frac{15}{16}$,

Problem (19). **What is the range of the function**

$$f(x) = 7 - |5\cos 5x|$$

A. $[-5, 7]$, B. $[0, 7]$, C. $(2, 5)$, D. $[2, 7]$, E. $[5, 7]$,

Problem (20). **The range of the function**

$$f(x) = 2x^2 - 4x + 15$$

is
A. $y \geq 2$, B. $y \geq 4$, C. $y \geq 6$, D. $y \geq 13$, E. $y \geq 15$,

Problem (21). **If the graph of the function** $y^2 = x$ **is rotated by** 90° **around the origin, clockwise, the equation of the resulting curve is**
A. $y^2 = -x$, B. $x = y^2$, C. $y = -x^2$, D. $y = x^2$, E. **none of the above,**

Problem (22). **The limit**

$$\lim_{x \to -\infty} \frac{2x^6 - 3x^4 + 4x + 1}{-3x^5 + 7x^2 - 8x - 2}$$

is
A. $-\infty$, B. $-\frac{2}{3}$, C. $\frac{2}{3}$, D. 2, E. $+\infty$,

Problem (23). **The set of points** $(x, 2x, 3)$ **is**
A. line, B. plane, C. circle, D. point, E. cylinder,

Problem (24). **Let**

$$f(x) = \ln \tan 2x,$$

which of the following intervals belongs to the domain of f

A. $0 < x < 2\pi$, **B.** $0 < x < \pi$, **C.** $0 < x \leq \frac{\pi}{2}$, **D.** $0 < x < \frac{\pi}{4}$, **E.** $-\frac{\pi}{2} < x\frac{-\pi}{2}$,

Problem (25)**. What is the volume of a sphere is its surface is** 16π**?**
A. 32π, **B.** 16π, **C.** $\frac{32\pi}{3}$, **D.** 32π, **E.** 10π,

Problem (26)**. If**

$$f(x) = 3e^{2x} - 4,\ g(x) = 2\ln x + 1$$

then

$$(f \circ g)(1)$$

is
A. $3e^2 + 4$, **B.** $3e^2 - 2$, **C.** $3e^2 - 3$, **D.** $3e^2 - 4$, **E.** $e^2 + 7$,

Problem (27)**. If**

$$a^2x - a = b^2x + b$$

then x **is equal to**
A. $a + b$, **B.** $a - b$, **C.** $\frac{1}{a-b}$, **D.** $\frac{1}{a+b}$, **E.** $\frac{1}{b-a}$,

Problem (28)**. Given that** $f(x) = \frac{x-1}{x+3}$ **and** $g(x) = f\left(\frac{x+1}{2}\right)$ **the domain of** g **is**
A. $\mathbb{R}$, **B.** $x > 0$, **C.** $x \neq -7$, **D.** $x \neq -3$, **E.** $x \neq 7$,

Problem (29)**. If** a **is the inverse of** b **then**
A. $a = b$, **B.** $a + b = 0$, **C.** $ab = 1$, **D.** $ab = 0$, **E.** $a - b = 1$,

Problem (30)**. If** $a, b \in \mathbb{Z}$ **and** $b \neq 0$**, then** $\frac{a}{b}$ **is**
A. an integer, B. positive, C. negative, D. a rational number,
E. none of the above,

Problem (31)**.**

$$\frac{x^4 - y^4}{x^3 - y^3} : \frac{x^2 + y^2}{x^2 + xy + y^2}$$

is equal to
A. $x^2 - y^2$, **B.** $x - y$, **C.** $x + y$, **D.** $x^2 + y^2$, **E. none of the above,**

Problem (32)**. The fraction**

$$\frac{a^4 - a}{a^2 + a + 1}$$

is equal to
A. $a^2 + a + 1$, **B.** $a^2 - a + 1$, **C.** $a^2 + a$, **D.** $a^2 - a$, **E. none of the above,**

Problem (33)**. If**

$$f(x) = \sqrt{x - |x|}$$

then $f(x)$ is equal to
A. 0, **B.** $\sqrt{x}$, **C.** $2\sqrt{x}$, **D.** $\sqrt{2x}$, **E. none of the above,**

Problem (34). **If the lines**

$$\epsilon : y = -mx + 1,\ \zeta : y = (m^2 - 3m + 3)x + 4$$

are perpendicular then the value of m is
A. -1, **B.** 0, **C.** 1, **D.** 2, **E. none of the above,**
Problem (35). **The value of**

$$9^{\frac{1}{2}\log_3 18}$$

is
A. 1, B. 3, C. 9, D. 18, E. 27,

Problem (36). **If**

$$9^x - 8 \cdot 3^x - 9 = 0,$$

then x is
A. -2, B. -1, C. 1, D. 2, E. 3,

Problem (37). **In the figure below we are given**

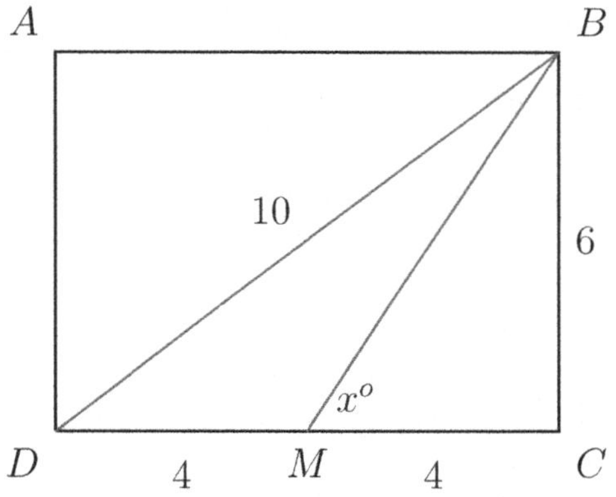

then x^o is
A. 50°, **B.** 51°, **C.** 56°, **D.** 56.3^o, **E.** 57°,

Problem (38). **The sum of four consecutive odd numbers is** 56, **then the greatest number is**
A. 12, B. 13, C. 14, D. 16, E. 17,

Problem (39). **If**

$$2^{2x} - 2^{x+1} - 8 = 0,$$

then x is equal to
A. 2, B. 4, C. 6, D. 8, E. none of the above,

Problem (40). **Let the orthogonal in the figure below be given, then x^o is equal to**

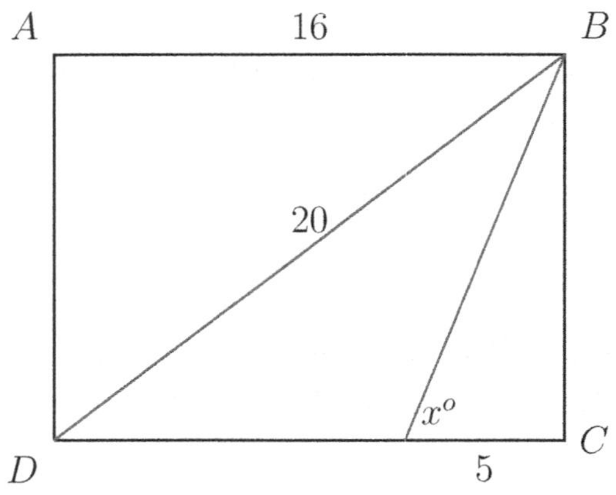

A. 60°, **B**. 62.5^o, **C**. 65.5^o, **D**. 67.2^o, **E**. 67.4^o,

Problem (41). **The smallest possible value of the angle** $\hat{B}$ **is**

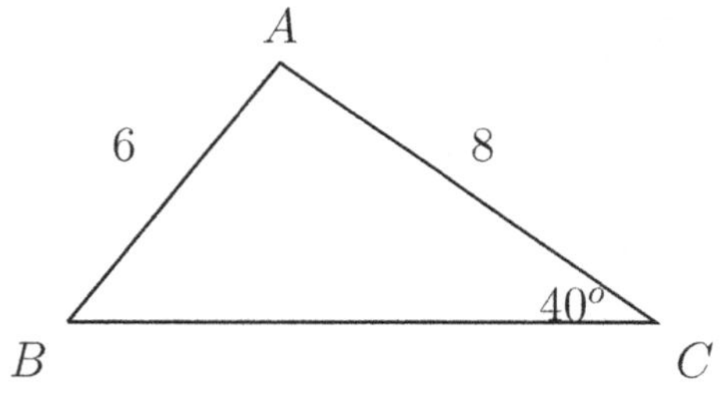

A. 59° , **B**. 62°, **C**. 80°, **D**. 88°, **E**. 89°,

Problem (42). **The smallest possible value of the area of triangle is**

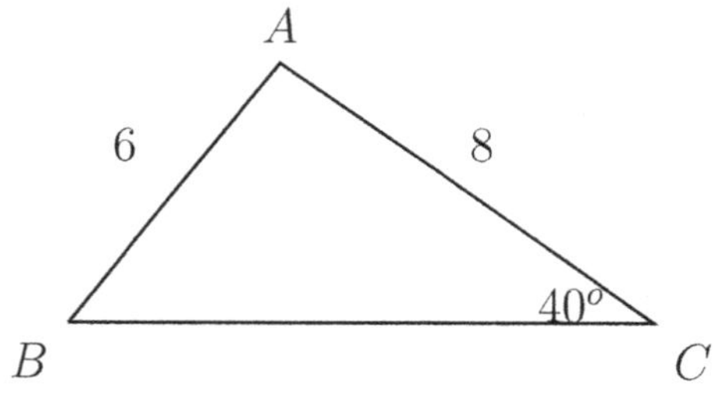

A. 20, **B**. 22, **C**. 23, **D**. 23.7, **E**. 24,

Problem (43). **If** a, b **are real numbers and one root of** $x^2 + ax + b = 0$ **is** $1 - i$ **then the values of** a **and** b **are**

A. $a = -2,\ b = -2$, **B**. $a = -2,\ b = 2$, **C**. $a = 2,\ b = -2$, **D**. $a = 2,\ b = 2$, **E**. $a = 2,\ b = 1$,

Problem (44). **If** $A = [1\ \ 2]$ **and** $B = \begin{bmatrix} -2 \\ 3 \end{bmatrix}$, **then the determinant of** BA **is**

A. -6 , **B**. -4, **C**. -2 , **D**. 0, **E**. 1,
Problem (45). **The area of the shaded figure below is**

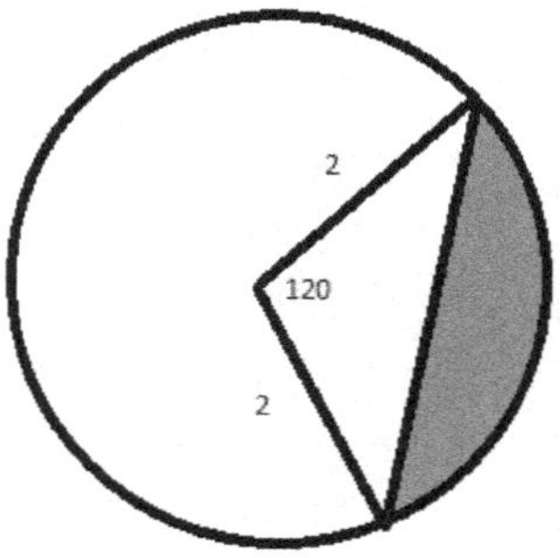

A. $\frac{4\pi}{3}-\sqrt{2}$, B. $\frac{4\pi}{3}-\sqrt{3}$, C. $\frac{4\pi}{3}+\sqrt{2}$, D. $\frac{4\pi}{3}+3$, E. $\frac{4\pi}{3}+1$,

Problem (46). If

$$\sin x + cosx = m,$$

then $\sin x \cos x$ **is**
A. $1-m^2$, **B.** $\frac{1-m^2}{2}$, **C.** m^2-1, **D.** $\frac{m^2-1}{2}$, **E. none of the above,**

Problem (47). If

$$f(x) = 2a \sin \frac{ax}{2}, \ a > 0$$

and the amplitude is 1 then, the period of this function is
A. 2π, **B.** 4π, **C.** 6π, **D.** 8π, **E. none of the above,**

Problem (48).

$$\cos \frac{\pi}{12} \cos \frac{\pi}{4} + \sin \frac{\pi}{12} \sin \frac{\pi}{4}$$

is
A. $\frac{-\sqrt{3}}{2}$, **B.** $\frac{-\sqrt{2}}{2}$, **C.** $\frac{1}{2}$, **D.** $\sqrt{2}$, **E.** $\frac{\sqrt{3}}{2}$,

Problem (49).

$$\sin 170^o \cos 10^o + \cos 170^o \sin 10^o$$

is
A.-1 , **B.** $\frac{-\sqrt{3}}{2}$, **C.**$\frac{-1}{2}$, **D.** 0, **E.** $\frac{\sqrt{3}}{2}$,

Problem (50). The side a **of the triangle** $\triangle ABC$ **is the figure below is**

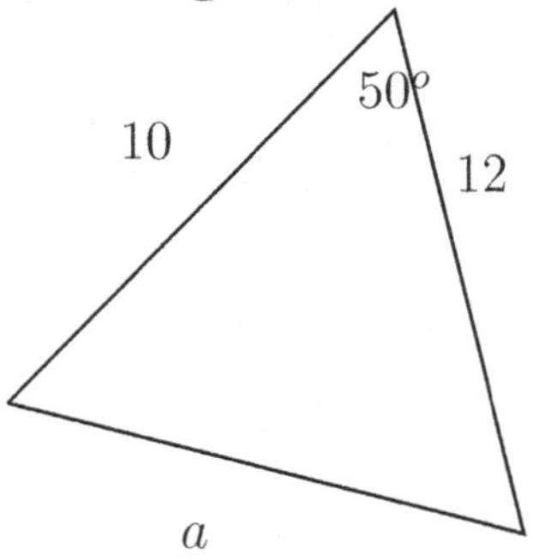

A. 20, B. 23, C. 24, D. 25, E. 26,

9.1.1 Solutions of Practice Test A

Problem (1). If $x^2-3x+1=0$ and r_1, r_2 are the roots then, the value of $r_1^2+r_2^2$ is
A. 4, **B.** 5, **C.** 6, **D.** 7, **E.** 8,
Since $r_1+r_2=3,\ r_1r_2=1$ *it follows*

$$r_1^2+r_2^2=(r_1+r_2)^2-2r_1r_2=3^2-2\cdot 1=7$$

and the correct answer is $[D]$.

Problem (2). *The remainder when* $p(x)=x^3+x^2+2x-1$ *is divided by* $x+1$ *is*
A. -4, *B.* -3, *C.* -2 , *D.* 1, *E.* 2,
Given that $x+1=0$ *for* $x=-1$ *it follows that the remainder of the division* $\dfrac{p(x)}{x+1}$, *we are asking for, is*

$$p(-1)=-3$$

and the correct answer is $[B]$.

Problem (3). *If* $x+1$, $x-2$ *are factors of*

$$q(x)=x^3+3x^2+ax+b$$

then, the values of a, b *are:*
A. -6 *and* -1, *B.* -6 *and* -8, *C.* -4 *and* -2, *D.* 4 *and* -2, *E.* 4 *and* 2,
Being $x+1$, $x-2$ *factors of* $q(x)$, *we get:*

$$\left.\begin{array}{l} q(-1)=0 \\ q(2)=0 \end{array}\right\} \Rightarrow \left.\begin{array}{c} -1+3-a+b=0 \\ +8+12+2a+b=0 \end{array}\right\} \Rightarrow a=-6,\ b=-8$$

and the correct answer is $[B]$.

Problem (4). *If the equation*

$$3x^2-4x+a=0$$

has two equal real roots, then the value of a *is*
A. $-\frac{4}{3}$, *B.* $-\frac{2}{3}$, *C.* 1, *D.* $\frac{4}{3}$, *E. none of the above,*
The given equation has two equal real roots, therefore

$$\Delta=0 \Rightarrow 16-12a=0 \Rightarrow a=\frac{4}{3}$$

the correct answer is $[D]$.

Problem (5). *The solutions of the inequality*

$$\frac{1}{x-1}<\frac{2}{x+3}$$

are [1]
A. $x<-3$, *B.* $x<-1$, *C.* $x>-3$, *D.* $-3<x<1$ *or* $x>5$, *E.* $x>1$,

[1] In inequalities we do not eliminate the denominators when are expressed in terms of the variable x

Remark. *In inequalities we* **do not eliminate denominators** *if they are expressions with* x *...*

Follow this very easy technique

$$\frac{1}{x-1} < \frac{2}{x+3} \Leftrightarrow \frac{5-x}{(x-1)(x+3)} < 0$$

Now, write on a line the roots of the numerator and of the denominator, take any value x_0 *in one of the intervals between these roots and check the sign of the quantity*

$$\frac{5-x}{(x-1)(x+3)}$$

for this particular value of the variable x.
Let $x_0 = 0 \in (-3, 1)$, *then*

$$\frac{(5-0)(0-1)}{0+3} = \frac{-5}{3} < 0$$

and we should have that the inequality holds true for $x \in (-3, 1)$, *or for* $x > 5$, *hence the correct answer is* $[D]$.
Note 9.1.1. *We repeat : if an interval, here* $-3 < x < 1$, *is accepted, then the next one is rejected, either on the left or on the right and so on.*

Problem (6). *If*

$$f(x) = \begin{cases} \dfrac{x^2-4}{x+2}, & \text{if } x \neq -2 \\ a & \text{if } x = -2 \end{cases}$$

is continuous, then a *is equal to*
A. -4, *B.* -2, *C.* 0, *D.* 1, *E.* 2,

First, for $x_0 \neq -2$ *the function* f *is continuous. At* $x_0 = -2$ *we shall have*

$$\lim_{x\to-2} \frac{x^2-4}{x+2} = f(-2) \Leftrightarrow$$
$$\lim_{x\to-2} \frac{(x-2)(x+2)}{x+2} = a \Leftrightarrow$$
$$\lim_{x\to-2} (x-2) = a \Leftrightarrow a = -4$$

and the correct answer is $[A]$.

Problem (7). *If a triangle* $\triangle AOB$, *formed by the two coordinate axis and the straight line with equation* $y = x + 3$, *is rotated around the* x*- axis, then the volume of the figure formed is equal to*
A. 3π, *B.* 4π, *C.* 6π, *D.* 9π, *E.* 10π,

$$V = \frac{1}{3}\pi r^2 h = \frac{1}{3}\pi 3^2 \cdot 3 = 9\pi$$

and the correct answer is $[D]$.

Problem (8). *The domain of the function*

$$f(x) = \sqrt{x - x^3}$$

is

A. $x < -1$, *B.* $x > -1$, *C.* $x < -1$ *or* $x > 1$, *D.* $x \leq -1$ *or* $0 \leq x \leq 1$, *E.* $x > 1$,

$$A_f = \{x \in \mathbb{R} \,|\, x - x^3 \geq 0\}$$

where

$$x - x^3 \geq 0 \Leftrightarrow x(1 - x^2) \geq 0 \Leftrightarrow x(1 - x)(1 + x) \geq 0$$

the roots are -1, 0, 1 *and for* $x = 2 \in (1, +\infty)$ *we have* $2(2 - 1)(2 + 1) = -6 < 0$:

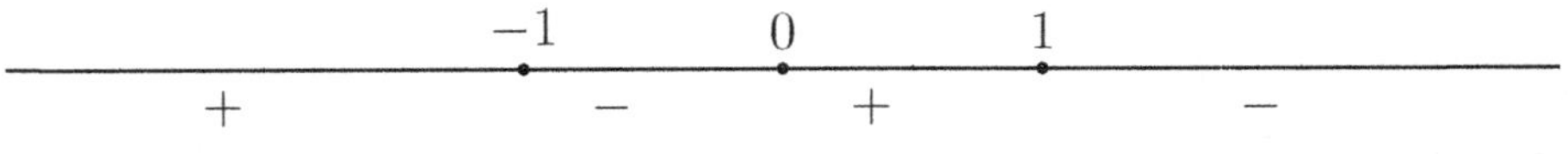

hence

$$x \in (-\infty, -1] \cup [0, 1]$$

and the correct answer is $[D]$.

Problem (9). *If*

$$f(x) = \frac{5x^3 - 7x^2 + 6x + 1}{x^3 - 1}$$

the asymptotes of the graph $\mathcal{G}_f$ *of* f, *are*

A. $y = 5$, $x = 1$, *B.* $y = -5$, $x = 1$, *C.* $y = 5$, *D.* $x = -1$, *E.* $y = 1$, $x = -1$,

Remark. *In order to determine the asymptotes of the graph of a function we need to study its limit at the points that belong to the boundary*[2] *of its domain.*

Start with the domain of f, *it is:*

$$x^3 - 1 \neq 0 \Leftrightarrow (x - 1)(x^2 + x + 1) \neq 0 \Leftrightarrow x \neq 1$$

because $x^2 + x + 1 \geq 1$ *for any* $x \in \mathbb{R}$. *We compute the limits* $\lim_{x \to 1} f(x)$, $\lim_{x \to \pm\infty} f(x)$:

$$\lim_{x \to 1} \frac{5x^3 - 7x^2 + 6x + 1}{x^3 - 1} = \frac{5}{0^{\pm}} = \pm\infty$$

and

$$\lim_{x \to \pm\infty} \frac{5x^3 - 7x^2 + 6x + 1}{x^3 - 1} = \lim_{x \to \pm\infty} \frac{5x^3}{x^3} = 5$$

hence $x = 1$ *is the vertical asymptote and* $y = 5$ *is the horizontal asymptote of* f, *i.e: the correct answer is* $[A]$.

[2] As a **boundary** point of a subset of the real line we can characterize each point that has the following property: any open interval, as small as possible, with center at this point has common points with the subset and the complementary of the subset, without taking under consideration this certain point

Problem (10). *The sum of all the terms of the geometric sequence*

$$10, \frac{10}{2}, \frac{10}{4}, \frac{10}{8}, \ldots, \frac{10}{2^n}, \ldots$$

is

A. 10, *B.* 12, *C.* 15, *D.* 20, *E.* 24,

The ratio of the geometric sequence is $\frac{1}{2}$ *and thus*

$$S = \frac{10}{1 - \frac{1}{2}} = 20,$$

hence the correct answer is [*D*].

Problem (11). *The reciprocal of* $5 - 3i$ *is*

A. $-\frac{5}{34} + \frac{3}{34}i$, *B.* $\frac{5}{34} - \frac{3}{34}i$, *C.* $\frac{5}{34} + \frac{3}{34}i$, *D.* $5 + 3i$, *E.* $\frac{1}{5} + \frac{1}{3}i$,

$$\frac{1}{5 - 3i} = \frac{5 + 3i}{25 + 9} = \frac{5 + 3i}{34}$$

and the correct answer is [*C*].

Problem (12). *In the figure there are two concentric circles with arc lengths* $s_1 = 5$ *and* $s = 1$. *Then, the ratio of the areas of the larger over the smaller sector is*

A. 5, *B.* 10, *C.* 15, *D.* 20, *E.* 25,

It is

$$\left.\begin{aligned} s &= r\theta \\ s_1 &= r_1\theta \\ s_1 &= 5s \end{aligned}\right\} \Rightarrow \frac{r}{r_1} = \frac{1}{5}$$

hence

$$\frac{A_1}{A} = \frac{\pi r_1^2}{\pi r^2} = \frac{r_1^2}{r^2} = 25$$

and the correct answer is [*E*].

Problem (13). *If the graph of the function* $f(x) = -x^2 + 4$ *is translated 4 units left and 3 units down, then the coordinates of the point* $(1, f(1)) = (1, 3)$ *will be*

A. $(1, 3)$, *B.* $(1, 15)$, *C.* $(4, 3)$, *D.* $(4, 15)$, *E.* $(4, -15)$,

We have

$$\begin{aligned} x \to x + 3 \Rightarrow 1 \to 4 \\ y \to y - 3 \end{aligned}$$

and thus

$$3 = f(1) \to f(1 + 3) - 3 = f(4) - 3 = -15$$

so, the correct answer is [*E*].

Problem (14). *The solution of*

$$|x+1| < x$$

is

A. $x < \frac{-1}{2}$, *B.* $0 < x < \frac{1}{2}$, *C.* $\frac{-1}{2} < x < 1$, *D.* $\frac{-1}{2} < x < 0$,
E. no values of x *satisfy the given inequality,*

Note that for $x \geq 0$ *the inequality becomes*

$$x + 1 < x \Leftrightarrow 1 < 0,$$

i.e: it is impossible. In the rest of the cases the assumption $x \leq 0$ *contradicts the inequality, hence the correct answer is* $[E]$.

Problem (15). *In a triangle* ΔABC *it is* $\sin A = \frac{1}{4}$, $\cos B = \frac{-1}{3}$, *then* C *is equal to*
A. 35°, *B.* 40°, *C.* 45°, *D.* 56°, *E.* 60°,

$$\sin A = \frac{1}{4} \Rightarrow A = 14.5^o$$
$$\cos B = \frac{-1}{3} \Rightarrow B = 109.5^o$$

so

$$C = 180^o - (14.5^o + 109.5^o) = 56^o$$

and the correct answer is $[D]$.

Problem (16). *The graph of the function represented parametrically by*

$$x = t^2, \ y = \frac{1}{t^2}$$

is a (an)

A. line segment, B. hyperbola, C. ellipse, D. line, E. none of the above,

Observe that

$$xy = t^2\frac{1}{t^2} \Rightarrow xy = 1$$

and thus we get a portion of a hyperbola. The correct answer is $[B]$.

Problem (17). *If the vectors* $\vec{v} = 2\vec{i} - 5\vec{j}$ *and* $\vec{u} = m\vec{i} + \vec{j}$ *are perpendicular, then* m *is equal to*
A. 2, *B.* 3, *C.* $\frac{5}{2}$, *D.* $\frac{1}{5}$, *E.* 10,

Using the characterization of perpendicularity by means of the inner product:

$$\vec{v} \perp \vec{u} \Leftrightarrow \vec{v} \cdot \vec{u} = 0 \Leftrightarrow 2m - 5 = 0 \Leftrightarrow m = \frac{5}{2}$$

and the correct answer is $[C]$.

Problem (18). *Six coins are flipped, the probability that 5 tails appear is*
A. $\frac{1}{32}$, *B.* $\frac{3}{32}$, *C.* $\frac{3}{16}$, *D.* $\frac{11}{16}$, *E.* $\frac{15}{16}$,

It is

$$probability = \frac{\binom{6}{5}}{2^6} = \frac{6}{64} = \frac{3}{32}$$

hence, the correct answer is $[B]$.

Problem (19). *What is the range of the function*

$$f(x) = 7 - |5\cos 5x|$$

A. $[-5, 7]$, *B.* $[0, 7]$, *C.* $(2, 5)$, *D.* $[2, 7]$, *E.* $[5, 7]$,
It is

$$\begin{gathered} 0 \le |5\cos 5x| \le 5 \Rightarrow \\ -5 \le -|5\cos 5x| \le 0 \Rightarrow \\ 2 \le 7 - |5\cos 5x| \le 7 \end{gathered}$$

and thus, the correct answer is $[D]$.

Problem (20). *The range of the function*

$$f(x) = 2x^2 - 4x + 15$$

is

A. $y \ge 2$, *B.* $y \ge 4$, *C.* $y \ge 6$, *D.* $y \ge 13$, *E.* $y \ge 15$,
The coefficient of x^2 *is* $2 > 0$ *therefore* $f(x)$ *attains its minimum at*

$$x = \frac{-b}{2a} = 1 \textit{ with } f(1) = 13$$

so the correct answer is $[D]$.

Problem (21). *If the graph of the function* $y^2 = x$ *is rotated by* 90° *around the origin, clockwise, the equation of the resulting curve is*
A. $y^2 = -x$, *B.* $x = y^2$, *C.* $y = -x^2$, *D.* $y = x^2$, *E. none of the above,*
The correct answer is $[C]$ *since by the reflections about the* x *-axis and* y *-axis, we should have*

$$\begin{gathered} (x, y) \to (x, -y), \textit{ if } y \ge 0 \\ (x, y) \to (-x, y) \textit{ if } y < 0 \end{gathered}$$

hence the equation of the rotated curve is $y = -x^2$.

Problem (22). *The limit*

$$\lim_{x \to -\infty} \frac{2x^6 - 3x^4 + 4x + 1}{-3x^5 + 7x^2 - 8x - 2}$$

is

A. $-\infty$, *B.* $-\frac{2}{3}$, *C.* $\frac{2}{3}$, *D.* 2, *E.* $+\infty$,

It is

$$lim_{x\to-\infty}\frac{2x^6-3x^4+4x+1}{-3x^5+7x^2-8x-2}=\lim_{x\to-\infty}\frac{2x^6}{-3x^5}=\lim_{x\to-\infty}\frac{-2}{3}x=+\infty$$

and the correct answer is [E].

Problem (23). *The set of points $(x, 2x, 3)$ is*
A. line, B. plane, C. circle, D. point, E. cylinder,
It is the intersection of the plane

$$y = 2x$$

with the plane

$$z = 3$$

and thus a straight line. Hence, the correct answer is [A].

Problem (24). *Let*

$$f(x) = \ln\tan 2x,$$

which of the following intervals belongs to the domain of f
A. $0 < x < 2\pi$, *B.* $0 < x < \pi$, *C.* $0 < x \le \frac{\pi}{2}$, *D.* $0 < x < \frac{\pi}{4}$, *E.* $-\frac{\pi}{2} < x\frac{-\pi}{2}$,
The function f is defined for

$$\tan 2x > 0 \Leftrightarrow 0 < 2x < \frac{\pi}{2} \Leftrightarrow 0 < x < \frac{\pi}{4}$$

and the correct answer is [D].

Problem (25). *What is the volume of a sphere if its surface is 16π?*
A. 32π, *B.* 16π, *C.* 12π, *D.* $\frac{32}{3}\pi$, *E.* 10π,
It is

$$4\pi r^2 = 16\pi \Rightarrow r = 2 \Rightarrow V = \frac{4}{3}\pi r^3 = \frac{32}{3}\pi$$

and the correct answer is [C].

Problem (26). *If*

$$f(x) = 3e^{2x} - 4,\ g(x) = 2\ln x + 1$$

then

$$(f \circ g)(1)$$

is

A. $3e^2 + 4$, *B.* $3e^2 - 2$, *C.* $3e^2 - 3$, *D.* $3e^2 - 4$, *E.* $e^2 + 7$,

$$\begin{gathered} g(x) = \ln(ex^2) \Rightarrow \\ (f \circ g)(x) = f\left(\ln(ex^2)\right) = 3(x^2 e)^2 - 4 \Rightarrow \\ (f \circ g)(1) = 3e^2 - 4 \end{gathered}$$

i.e: the correct answer is D.

Problem (27). *If $a \neq b$ and*

$$a^2x - a = b^2x + b$$

then x is equal to

A. $a+b$, *B.* $a-b$, *C.* $\frac{1}{a-b}$, *D.* $\frac{1}{a+b}$, *E.* $\frac{1}{b-a}$,

$$\begin{aligned} a^2x - a &= b^2x + b \Leftrightarrow \\ x(a^2 - b^2) &= a + b \Rightarrow \\ x &= \frac{1}{a-b} \end{aligned}$$

and the correct answer is [*C*].

Problem (28). *Given that* $f(x) = \frac{x-1}{x+3}$ *and* $g(x) = f\left(\frac{x+1}{2}\right)$ *the domain of* g *is*

A. $\mathbb{R}$, *B.* $x > 0$, *C.* $x \neq -7$, *D.* $x \neq -3$, *E.* $x \neq 7$,

$$g(x) = f\left(\frac{x+1}{2}\right) = \frac{x-1}{x+7}$$

hence g is defined for $x \neq -7$ and the correct answer is [*C*].

Problem (29). *If a is the inverse of b then*

A. $a = b$, *B.* $a + b = 0$, *C.* $ab = 1$, *D.* $ab = 0$, *E.* $a - b = 1$,

The correct answer is [*C*].

Problem (30). *If $a, b \in \mathbb{Z}$ and $b \neq 0$, then $\frac{a}{b}$ is*

A. an integer, B. positive, C. negative, D. a rational number, E. none of the above,

The correct answer is D.

Problem (31).

$$\frac{x^4 - y^4}{x^3 - y^3} : \frac{x^2 + y^2}{x^2 + xy + y^2}$$

is equal to

A. $x^2 - y^2$, *B.* $x - y$, *C.* $x + y$, *D.* $x^2 + y^2$, *E. none of the above,*

The correct answer is C.

Problem (32). *The fraction*

$$\frac{a^4 - a}{a^2 + a + 1}$$

is equal to

A. $a^2 + a + 1$, *B.* $a^2 - a + 1$, *C.* $a^2 + a$, *D.* $a^2 - a$, *E. none of the above,*

$$\frac{a^4 - a}{a^2 + a + 1} = \frac{a(a^3 - 1)}{a^2 + a + 1} = \frac{a(a-1)(a^2 + a + 1)}{a - 1} = a^2 - a$$

and the correct answer is [*D*].

Problem (33). *If*

$$f(x) = \sqrt{x - |x|}$$

then $f(x)$ *is equal to*

A. 0, *B.* $\sqrt{x}$, *C.* $2\sqrt{x}$, *D.* $\sqrt{2x}$, *E. none of the above,*

$$x - |x| \geq 0 \Rightarrow x = 0$$

hence, the correct answer is [*A*].

Problem (34). *If the lines*

$$\epsilon : y = -mx + 1, \ \zeta : y = (m^2 - 3m + 3)x + 4$$

are perpendicular then the value of m is

A. −1, *B.* 0, *C.* 1, *D.* 2, *E. none of the above,*

$$\epsilon \perp \zeta \Leftrightarrow -m(m^2 - 3m + 3) = -1 \Rightarrow -(m-1)^3 = 0 \Rightarrow m = 1$$

and the correct answer is [*C*].

Problem (35). *The value of*

$$9^{\frac{1}{2}\log_3 18}$$

is

A. 1, B. 3, C. 9, D. 18, E. 27,

It is

$$9^{\frac{1}{2}\log_3 18} = \left(\sqrt{9}\right)^{\log_3 18} = 3^{\log_3 18} = 18$$

hence the correct answer is [*D*].

Problem (36). *If*

$$9^x - 8 \cdot 3^x - 9 = 0,$$

then x is

A. -2, B. -1, C. 1, D. 2, E. 3,

It is

$$9^x - 8 \cdot 3^x - 9 = 0 \Rightarrow 3^{2x} - 8 \cdot 3^x - 9 = 0 \Rightarrow 3^x = 9$$

hence $x = 2$ *and the correct answer is* [*D*].

Problem (37). *In the figure below we are given*

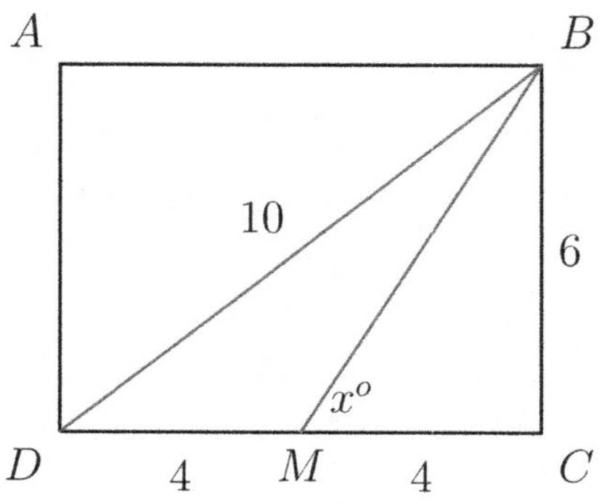

then x^o is

A. 50°, *B.* 51°, *C.* 56°, *D.* 56.3^o, *E.* 57°,
Using the Pythagorean theorem

$$DC^2 = 10^2 - 6^2 = 8^2 \Rightarrow MC = 4$$

hence

$$\tan x^o = \frac{6}{4} = 1.5$$

so

$$x^o = 56.3^o$$

and the correct answer is [*D*].

Problem (38). *The sum of four consecutive odd numbers is 56, then the greatest number is*
A. 12, B. 13, C. 14, D. 16, E. 17,
It is

$$x + (x+2) + (x+4) + (x+6) = 56 \Rightarrow 4x = 44$$

and the correct answer is [*E*].

Problem (39). *If*

$$2^{2x} - 2^{x+1} - 8 = 0,$$

then x is equal to
A. 2, B. 4, C. 6, D. 8, E. none of the above,
Let $2^x = w$ *then*

$$w^2 - 2w - 8 = 0 \Rightarrow w = \frac{2 \pm \sqrt{36}}{2} = \frac{2 \pm 6}{2} \in \{4, -2\}$$

so $2^x = -2$ *(which is* ***rejected****) or* $2^x = 4$ *i.e:* $x = 2$ *and the correct answer is* [*A*].

Problem (40). *Let the orthogonal in the figure below be given, then x^o is equal to*

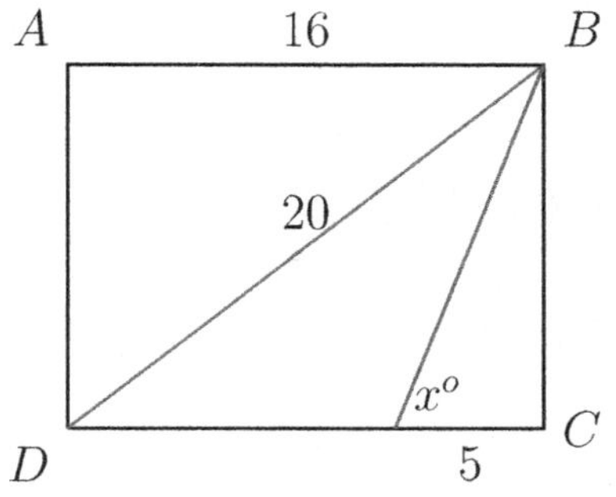

A. 60°, *B.* 62.5^o, *C.* 65.5^o, *D.* 67.2^o, *E.* 67.4^o,
It is

$$AD^2 = BC^2 = 20^2 - 16^2 = 144 \Rightarrow BC = 12$$

hence

$$\tan x = \frac{12}{5} \Rightarrow x^o = 67.4^o$$

and the correct answer is [*E*].

Problem (41). *The smallest possible value of the angle B is*

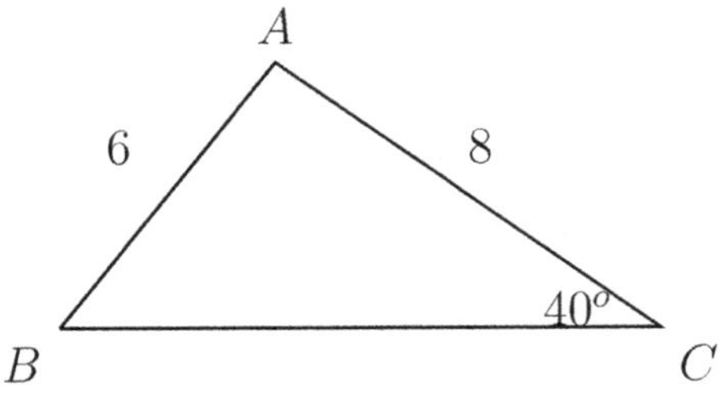

A. 59° , *B.* 62°, *C.* 80°, *D.* 88°, *E.* 89°,

It is

$$\frac{\sin 40^o}{6} = \frac{\sin B}{8} \Rightarrow \sin B = 0.64 \times \frac{8}{6} = 0.86$$

so $B = 59^o$ *or* $B = 121^o$ *and the correct answer is* [*A*].

Problem (42). *The smallest possible value of the area of triangle is*

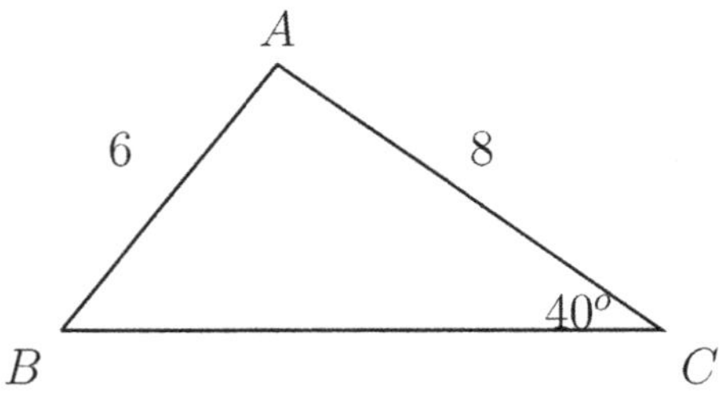

A. 20, *B.* 22, *C.* 23, *D.* 23.7, *E.* 24,
It is

$$\frac{\sin 40^o}{6} = \frac{\sin B}{8} \Rightarrow \sin B = 0.64 \times \frac{8}{6} = 0.86 \Rightarrow B \in \{59^o, 121^o\} \Rightarrow$$
$$B = 59^o \text{ and } A = 81^o$$

hence, the area is given by

$$A = \frac{1}{2} 6 \times 8 \times \sin A = \frac{1}{2} 6 \times 8 \times \sin 81^o = 23.7$$

and the correct answer is [*D*].

Problem (43). *If a, b are real numbers and one root of $x^2 + ax + b = 0$ is $1 - i$ then the values of a and b are*
A. $a = -2,\ b = -2$, *B.* $a = -2,\ b = 2$, *C.* $a = 2,\ b = -2$, *D.* $a = 2,\ b = 2$,
E. $a = 2,\ b = 1$,
The other root is $1 + i$, *i.e: the* ***conjugate*** *of* $1 - i$, *and thus*

$$\left.\begin{array}{l} r_1 + r_2 = 2 = -a \\ r_1 r_2 = 2 = b \end{array}\right\}$$

and the correct answer is [*B*].

Problem (44). *If* $A = [1\ 2]$ *and* $B = \begin{bmatrix} -2 \\ 3 \end{bmatrix}$, *then the determinant of* BA *is*

A. -6 , *B.* -4, *C.* -2 , *D.* 0, *E.* 1,
It is

$$BA = \begin{bmatrix} -2 \\ 3 \end{bmatrix} \cdot [1\ 2] = \begin{bmatrix} -2 & -4 \\ 3 & 6 \end{bmatrix}$$

and thus $\det BA = -12 + 12 = 0$, *the correct answer is* [D].

Problem (45). *The area of the shaded figure below is*

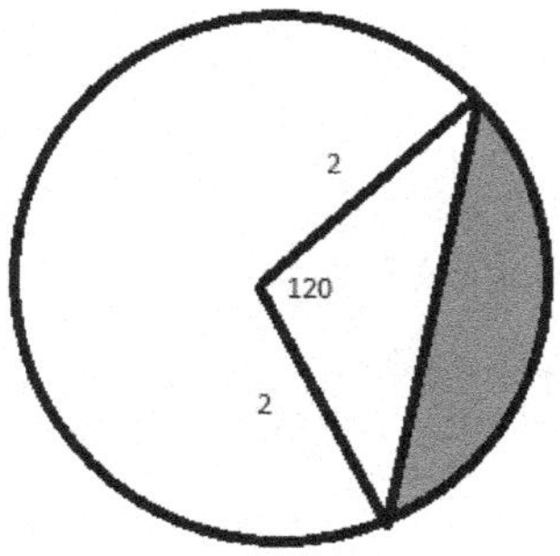

A. $\frac{4\pi}{3} - \sqrt{2}$, *B.* $\frac{4\pi}{3} - \sqrt{3}$, *C.* $\frac{4\pi}{3} + \sqrt{2}$, *D.* $\frac{4\pi}{3} + 3$, *E.* $\frac{4\pi}{3} + 1$,
We have

$$\begin{aligned} A_{shaded} &= \frac{1}{2}\theta r^2 - \frac{1}{2} r \times r \times \sin 120^o \\ &= \frac{1}{2}\frac{2\pi}{3}2^2 - \frac{1}{2}2 \times 2 \times \frac{\sqrt{3}}{2} = \frac{4\pi}{3} - \sqrt{3} \end{aligned}$$

so, the correct answer is [B].
Remark. *Remember*

$$s = \theta R, \ Area = \frac{1}{2}\theta R^2$$

Problem (46). *If*

$$\sin x + cosx = m.$$

then $\sin x \cos x$ *is*

A. $1 - m^2$, *B.* $\frac{1-m^2}{2}$, *C.* $m^2 - 1$, *D.* $\frac{m^2-1}{2}$, *E. none of the above,*

$$\sin x + cosx = m \Rightarrow (\sin x + \cos x)^2 = m^2 \Rightarrow \sin x \cos x = \frac{m^2-1}{2}$$

and the correct answer is [D].

Problem (47). *If*

$$f(x) = 2a \sin \frac{ax}{2}, \ a > 0$$

and the amplitude is 1 *then, the period of this function is*

A. 2π, *B.* 4π, *C.* 6π, *D.* 8π, *E. none of the above,*
Since amplitude is 1, *then*

$$2a = 1 \Rightarrow a = \frac{1}{2} \Rightarrow \frac{ax}{2} = \frac{x}{4}$$

and for the period T_f *of the function* f *we shall have*

$$T_f = \frac{2\pi}{\frac{1}{4}} = 8\pi$$

so, the correct answer is [*E*].

Problem (48).

$$\cos\frac{\pi}{12}\cos\frac{\pi}{4} + \sin\frac{\pi}{12}\sin\frac{\pi}{4}$$

is

A. $\frac{-\sqrt{3}}{2}$, *B.* $\frac{-\sqrt{2}}{2}$, *C.* $\frac{1}{2}$, *D.* $\sqrt{2}$, *E.* $\frac{\sqrt{3}}{2}$,

$$\begin{aligned}\cos\frac{\pi}{12}\cos\frac{\pi}{4} + \sin\frac{\pi}{12}\sin\frac{\pi}{4} &= \cos\left(\frac{\pi}{12} - \frac{\pi}{4}\right)\\ &= \cos\left(\frac{\pi}{4} - \frac{\pi}{12}\right) = \cos 30^o = \frac{\sqrt{3}}{2}\end{aligned}$$

so, the correct answer is [*D*].

Problem (49).

$$\sin 170^o \cos 10^o + \cos 170^o \sin 10^o$$

is

A.-1 , *B.* $\frac{-\sqrt{3}}{2}$, *C.*$\frac{-1}{2}$, *D.* 0, *E.* $\frac{\sqrt{3}}{2}$,

$$\sin 170^o \cos 10^o + \cos 170^o \sin 10^o = \sin(170^o + 10^o) = \sin 180^o = 0$$

so, the correct answer is D.

Problem (50). *The side* a *of the triangle* $\triangle ABC$ *in the figure below is*

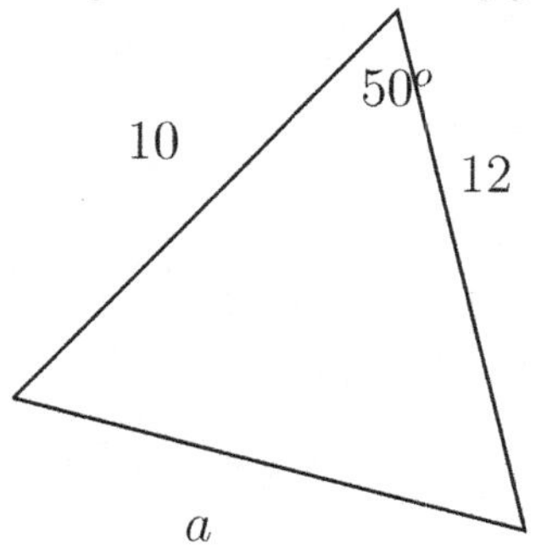

A. 20, B. 23, C. 24, D. 25, E. 26,
It is

$$a^2 = 10^2 + 12^2 - 210 \times 12 \times \cos 50^o \Rightarrow a = 19.46 \sim 20$$

and thus the correct answer is [*A*].

9.2 Practice Test B

Problem (1). **If**

$$x^2 + y^2 - 2x + 4y + 1 = 0$$

is the equation of a circle, then the coordinates of the center and the length of the radius are
A. $(1, 2)$ $r = 2$, **B.** $(-1, 2)$ $r = 1$, **C.** $(1, 2)$ $r = 4$, **D.** $(1, -2)$ $r = 2$,
E. none of these,

Problem (2). **If**

$$x^2 + y^2 + 2ax + 2by + 2c = 0$$

is the equation of a circle, then
A. $a^2 + b^2 = 4c^2$, **B.** $a^2 + b^2 = 2c$, **C.** $a^2 + b^2 + 2c > 0$, **D.** $a^2 + b^2 > 2c$,
E. none of these,

Problem (3). **If**

$$y^2 - x + 2y = 4$$

is the equation of a parabola, then

A. vertex is $(-5, -1)$ and focus $(\frac{1}{4}, 1)$,

B. vertex is $(-5, -1)$ and focus $(\frac{-19}{4}, 1)$,

C. vertex is $(5, -1)$ and focus $(\frac{-19}{4}, 1)$,

D. vertex is $(-5, -1)$ and focus $(\frac{-19}{4}, -1)$, E. none of these,

Problem (4). **One focus of the ellipse**

$$\frac{x^2}{2} + \frac{y^2}{3} = 1,$$

is
A. $(2, 3)$, **B.** $(0, \sqrt{2})$, **C.** $(\sqrt{3}, 0)$, **D.** $(-1, 0)$, **E.** $(0, -1)$.

Problem (5). **Three dice are tossed and it is given that at least one** 4 **appears. The probability that** 1 **will appear twice is**
A. $\frac{1}{72}$, **B.** $\frac{1}{36}$, **C.** $\frac{1}{24}$, **D.** $\frac{1}{12}$, **E.** $\frac{1}{6}$,

Problem (6).

- **Six coins are flipped. The probability that** 5 **or** 6 **tails appear is**

  A. $\frac{7}{64}$, B. $\frac{1}{8}$, C. $\frac{5}{32}$, D. $\frac{7}{32}$, E. none of these,

- **A pair of dice is tossed. The probability that the sum of the sides facing up is an odd number is**

  A. 0.15, **B.** 0.25, **C.** 0.30, **D.** 0.40, **E.** 0.50,

Problem (7).

- **A coin is tossed three times, the probability that exactly two heads will appear is**

  A. 0.300, **B.** 0.325, **C.** 0.345, **D.** 0.365, **E.** 0.375,

- **How many** 5 **- digit numbers can be formed from the numbers** $1, 2, 3, 5, 0$ **if no digit is repeated?**

  A. 72, **B.** 76, **C.** 80, **D.** 90, **E.** 96,

Problem (8).

- **We flip** 8 **coins, the first three all come up tails. The probability that at least** 1 **more tail will be flipped is**

  A. $\frac{25}{32}$, **B.** $\frac{27}{32}$, **C.** $\frac{29}{32}$, **D.** $\frac{30}{32}$, **E.** $\frac{31}{32}$,

- **Given that a multiple of** 4 **is less that** 100**, what is the probability that it is between** 30 **and** 60**?**

  A. $\frac{4}{25}$, **B.** $\frac{5}{25}$, **C.** $\frac{6}{25}$, **D.** $\frac{7}{25}$, **E.** $\frac{8}{25}$,

Problem (9). If

$$f(x) = \frac{2}{x-2}, \quad g(x) = \sqrt{4-x^2}$$

the domain of the function $f \circ g$ **is**
A. $x > 2$, **B.** $x < 2$, **C.** $-2 \leq x < 0$, **D.** $0 < x \leq 2$, **E.** $-2 \leq x < 0$ **or** $0 < x \leq 2$,

Problem (10). If the remainder of the division of

$$p(x) = x^4 + 3x^2 + x + a^4 - 1$$

by $x + 1$**, is** 83**, the value of** a **is**
A. 3, **B.** 5, **C.** 6, **D.** 8, **E.** 30,

Problem (11). If the function

$$f(x) = \begin{cases} \frac{x^3-8}{x^2-4}, & \text{if } x \neq 2 \\ m, & \text{if } x = 2 \end{cases}$$

is continuous, the value of m **is**
A. 0, **B.** 1, **C.** 2, **D.** 3, **E.** 4,

Problem (12). **The value of**

$$\lim_{x\to 3}\frac{x^2-9}{(x-3)^3}$$

is

A. $-\infty$, **B.** -3, **C.** 3, **D.** $+\infty$, **E.** none of these,

Problem (13). **If** M **is a point on the axis** xx', $\widehat{AMB}=90^o$, $A(-1,7)$, $B(7,1)$ **the coordinates of** M **are**

A. $(-6,0)$, $(0,0)$, **B.** $(-2,0)$, $(0,0)$, **C.** $(0,0)$, $(2,0)$, **D.** $(0,0)$, $(6,0)$,
E. none of the above,

Problem (14). **The points** $A(3,4)$ **and** $B(4,3)$ **are symmetric with respect to**

A. the point $(0,0)$, **B.** straight line $x'x$, **C.** straight line $y'y$,
D. bisector of the angle $x\hat{O}y$, **E.** none of the above,

Problem (15). **If the straight lines**

$$y=(m^3-8)x+3,\ y=-2(m^2-2m)x+5$$

are parallel then m **is equal to**

A. -2 or -1, **B.** -2 or 2, **C.** 0, **D.** 2 or 0, **E.** 2 or 1,

Problem (16). **If the distance** OM, $O(0,0)$ **and** $M(x+1,x-1)$ **is minimum then the value of** x **is**

A. -2, **B.** -1, **C.** 0, **D.** 1, **E.** 2,

Problem (17). **If** $A(1,9)$, $B(2,14)$ **are two points of**

$$f(x)=x^2+ax+b$$

then $b-a$ **is**

A. -4, **B.** -2, **C.** 0, **D.** 2, **E.** 4,

Problem (18). **If** r_1, r_2 **are roots of** $x^2-x-1=0$, **then the value**

$$\frac{r_1}{r_2}+\frac{r_2}{r_1}$$

is

A. -3, **B.** -2, **C.** -1, **D.** 0, **E.** 3,

Problem (19). **The third term of a geometric sequence is** 12 **and he sixth is** 96. **Then, the sum of the first** 10 **terms is**

A. $2^{10}-1$, **B.** $2^{11}-2$, **C.** $2^{11}+1$, **D.** $3(2^{10}-1)$, **E.** $3(2^{10}+1)$,

Problem (20). **If the equation**

$$(m^2-1)x=m^3-1$$

has no solutions then m **is equal to**

A. -1, **B.** 0, **C.** 1, **D.** 2, **E.** none of the above,

Problem (21). **If**

$$|a| + |b| = a + b$$

then
A. $ab > 0$, **B.** $ab \leq 0$, **C.** $ab = 0$, **D.** $ab \geq 0$, **E.** $ab < 0$,

Problem (22). **If**

$$x^2 + \sqrt{15},\ x^2 - \sqrt{15}$$

are reciprocal then x **is equal to**
A. -2, **B.** -1, **C.** 1, **D.** -2 **or** $+2$, **E.** none of the above,

Problem (23). **The vertical and horizontal asymptotes of**

$$f(x) = \frac{x+1}{x^2-1}$$

are
A. $x = 1,\ y = 0$, **B.** $x = 1,\ x = -1,\ y = 0$, **C.** $x = -1,\ y = 0$, **D.** $x = 1,\ y = 1$,
E. none of the above,

Problem (24). **The equation of the circle with center at the point** $K(1,2)$ **and tangential to the straight line** $\epsilon:\ 3x - 4y + 10 = 0$ **is**
A. $x^2 + y^2 - 3x - 4y + 5 = 0$, **B.** $x^2 + y^2 - 2x - 4y + 1 = 0$,
C. $(x-1)^2 + (y-2)^2 = 4$, **D.** $(x-1)^2 + (y-3)^2 = 2$, **E.** $(x-1)^2 + (y-2)^2 = 1$,

Problem (25). **The points of intersection of the ellipses**

$$4x^2 + 3y^2 = 14,\ 3x^2 + 4y^2 = 14$$

lie on a circle with radius R **equal to**
A. 1, **B.** $\sqrt{2}$, **C.** 2, **D.** 3, **E.** 4,

Problem (26). **If**

$$f(x) = \ln(-x^2 + 5x - 6)$$

the domain of f **is**
A. $x < -2$, **B.** $x > 3$, **C.** $x < 3$, **D.** $-2 < x < 3$, **E.** $2 < x < 3$,

Problem (27). **If the graph of the function**

$$f(x) = x^2 + mx + 3$$

has its vertex at the axis Ox **the value of** m **is**
A. 1, **B.** 2, **C.** 3, **D.** 4, **E.** 6,

Problem (28). **The values of** k **for which the equation**

$$x^2 + 2(k+2)x + 9k = 0$$

has no real roots are

A. $k < 1$, **B.** $k > 4$, **C.** $k < 1$ **or** $k > 4$, **D.** $0 < k < 4$, **E.** $1 < k < 4$,

Problem (29). The length of the radius of the sphere

$$x^2 + y^2 + z^2 + 4x + 6y - 8z = 0$$

is
A. 4, **B.** 5, **C.** 5.4, **D.** 6, **E.** 6.5,
and the correct answer is [C].

Problem (30). If the sphere

$$x^2 + y^2 + z^2 + 4x + 6y - 8z = 0$$

is inscribed in a cube, the volume V_{ext} **of the cube outside the sphere is**
A. 560, **B.** 561, **C.** 562, **D.** 562.2, **E.** 566,

Problem (31). A packet of 10 light bulbs contains 2 defective bulbs. Two bulbs are selected at random from the packet. The probability neither of these bulbs is defective is
A. $\frac{14}{45}$, **B.** $\frac{18}{45}$, **C.** $\frac{20}{45}$, **D.** $\frac{28}{45}$, **E. none of the above,**

Problem (32). If $p(A \cup B) = \frac{9}{10}$, $p(A) = \frac{3}{10}$, $p(B) = \frac{8}{10}$ **then** $p(A \cap B)$ **is equal to**
A. $\frac{1}{10}$, **B.** $\frac{2}{10}$, **C.** $\frac{3}{10}$, **D.** $\frac{4}{10}$, **E.** $\frac{5}{10}$,

Problem (33). The slope of the median AM **of a triangle** $\triangle ABC$ **with** $A(a, 5)$, $B(0, 4)$ **and** $C(6, 2)$, **is equal to 2. Then, the value of** a **is**
A. -4, **B.** -2, **C.** 0, **D.** 2, **E.** 4,

Problem (34). The limit

$$\lim_{x \to 3} \frac{x^2 - 9}{(x-3)^2}$$

is equal to
A. $-\infty$, **B.** 0, **C.** 3, **D.** $+\infty$, **E. none of the above,**

Problem (35). A root of

$$x^3 - 3x^2 + 7x - 5 = 0$$

is an integer, the other two roots are
A. $-1 \pm 2i$, **B.** $-2 \pm 4i$, **C.** $-2 \pm 2i$, **D.** $1 \pm 2i$, **E. none of the above,**

Problem (36). In a class there are 12 boys and 10 girls. The mean of the height of the boys is 172 and the corresponding mean for the girls is 168. The mean of the heights of all the students in this class is
A. 169.5, **B.** 170, **C.** 170.1, **D.** 170.2, **E.** 170.4,

Problem (37). **The domain of the function**

$$f(x) = \log \sqrt{-2 + \frac{x}{x-3}}$$

is
A. $3 < x < 4$, **B.** $3 \leq x \leq 6$, **C.** $3 < x \leq 6$, **D.** $3 < x < 6$, **E. none of the above,**

Problem (38). **If**

$$\log_4 x \log_7 4 = 2$$

then x **is equal to**
A. 7, **B.** 14, **C.** 21, **D.** 49, **E.** 56,

Problem (39). **If**

$$2kx^2 - 3x + k - 1 = 8x^2 - 6mx + 6m$$

for all $x \in \mathbb{R}$, **the values of** k, m **are**
A. $-2, 1$, **B.** $4, 1$, **C.** $4, -1$, **D.** $4, \frac{1}{2}$, **E. none of the above,**

Problem (40). **If**

$$y = \frac{x^2 + 2x - 11}{2(x-3)}$$

and x **is real, the values of** y **are**
A. $y < 2$, **B.** $y > 2$, **C.** $2 < y < 6$, **D.** $y \leq 2$ **or** $y \geq 6$, **E. none of the above,**

Problem (41). **If** $1 + i$ **is a root of**

$$x^2 - ax + b = 0$$

and a, b **are real numbers then the sum** $a + b$ **is**
A. -4, **B.** -2, **C.** 0, **D.** 2, **E.** 4,

Problem (42). **A circle is tangent to the straight line**

$$\epsilon:\ 4x + 3y + 20 = 0$$

and its center is at the origin. The radius of the circle is
A. 3, **B.** 4, **C.** 4.5, **D.** 5, **E.** 5.4,

Problem (43). **If**

$$a^2 + b^2 + c^2 = 2 \text{ and } a + b + c = 0$$

then $ab + bc + ca$ **is equal to**
A. -2, **B.** -1, **C.** 1, **D.** 2, **E. none of the above,**

Problem (44). **If**

$$f(x) = a \log_3 x,\ f^{-1}(1) = 9$$

then the value of a **is**
A. -2, **B.** -1, **C.** 0, **D.** $\frac{1}{2}$, **E.** 1,

Problem (45). **If**

$$\sin A = \frac{3}{5},\ 0 < A < 90,\ \cos B = \frac{-3}{5},\ 90 < B < 180$$

then $\cos(A+B)$ **is equal to**
A. $-\frac{24}{25}$, **B.** $-\frac{20}{25}$, **C.** $-\frac{12}{25}$, **D.** $\frac{12}{25}$, **E.** $\frac{24}{25}$,

Problem (46). **If** $\vec{u} = p\vec{i} + 6\vec{j}$, $\vec{v} = 4p\vec{i} - 6\vec{j}$ **and** $\vec{u} \perp \vec{v}$ **the value of** p **is**
A. -3, **B.** -3 **or** 3, **C.** 3, **D.** 9, **E. none of the above,**

Problem (47). **If**

$$A = \begin{bmatrix} 2 & 1 & x \\ -1 & 3 & 1 \end{bmatrix},\ B = \begin{bmatrix} -4 \\ x \\ 2 \end{bmatrix}$$

and

$$AB = \begin{bmatrix} -14 \\ 0 \end{bmatrix}$$

then, the value of x **is**
A. -3, **B.** -2, **C.** -1, **D.** 0, **E.** 1,

Problem (48). **How many triangles are there in the figure below?**

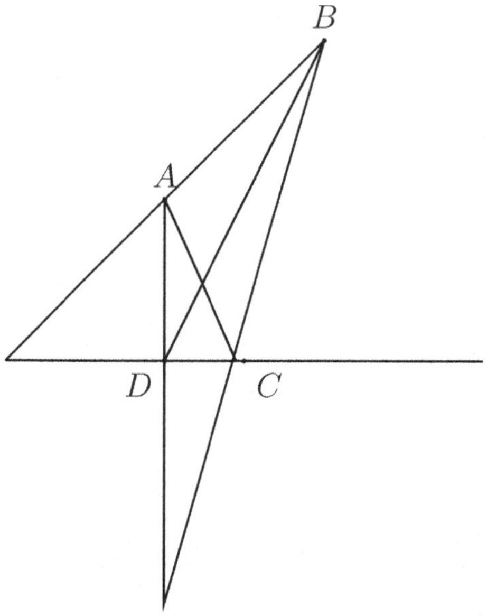

A. 8, **B.** 10, **C.** 12, **D.** 14, **E.** 16 ,

Problem (49). **Let** O, A, B, C **be points in the plane such that**

$$|\vec{OA}| = |\vec{OB}| = |\vec{OC}| = 1$$

and

$$\vec{OA} + \vec{OB} + \vec{OC} = \vec{0}.$$

Then, the triangle $\triangle ABC$ **is**
A. isosceles, B. orthogonal , C. equilateral, D. isosceles and orthogonal,
E. none of these.

Problem (50). Let $\triangle ABC$ be a triangle such that the medians BE and CZ are of equal length. Then, the triangle $\triangle ABC$ is
A. isosceles, B. orthogonal , C. equilateral, D. isosceles and orthogonal,
E. none of these.

9.2.1 Solutions of Practice Test B

Problem (1). *If*

$$x^2 + y^2 - 2x + 4y + 1 = 0$$

is the equation of a circle, then the coordinates of the center and the length of the radius are
A. $(1, 2)$ $r = 2$, *B.* $(-1, 2)$ $r = 1$, *C.* $(1, 2)$ $r = 4$, *D.* $(1, -2)$ $r = 2$,
E. none of these,
It is

$$x^2 + y^2 - 2x + 4y + 1 = 0 \Leftrightarrow (x - 1)^2 + (y + 2)^2 = 4$$

hence, the correct answer is [D].

Problem (2). *If*

$$x^2 + y^2 + 2ax + 2by + 2c = 0$$

is the equation of a circle, then
A. $a^2 + b^2 = 4c^2$, *B.* $a^2 + b^2 = 2c$, *C.* $a^2 + b^2 + 2c > 0$, *D.* $a^2 + b^2 > 2c$,
E. none of these,
The equation is

$$(x + a)^2 + (y + b)^2 = a^2 + b^2 - 2c > 0,$$

hence

$$a^2 + b^2 - 2c > 0 \Rightarrow a^2 + b^2 > 2c$$

hence [D] is the correct answer.

Problem (3). *If*

$$y^2 - x + 2y = 4$$

is the equation of a parabola, then
A. *vertex is* $(-5, -1)$ *and focus* $(\frac{1}{4}, 1)$, *B.* *vertex is* $(-5, -1)$ *and focus* $(\frac{-19}{4}, 1)$,
C. *vertex is* $(5, -1)$ *and focus* $(\frac{-19}{4}, 1)$, *D.* *vertex is* $(-5, -1)$ *and focus* $(\frac{-19}{4}, -1)$,
E. *none of these,*
The correct answer is [D] because the equation is

$$(y + 1)^2 = x + 5 \Rightarrow 2p = 1 \Rightarrow \frac{p}{2} = \frac{1}{4}$$

and

$$\textit{focus} \left(-5 + \frac{1}{4}, -1\right) = \left(-\frac{19}{4}, -1\right)$$

Problem (4). *One focus of the ellipse*

$$\frac{x^2}{2} + \frac{y^2}{3} = 1.$$

is

A. $(2, 3)$, *B.* $(0, \sqrt{2})$, *C.* $(\sqrt{3}, 0)$, *D.* $(-1, 0)$, *E.* $(0, -1)$.
The correct answer is [*E*] *because*

$$c^2 = 3 - 2 = 1 \Rightarrow c = \pm 1$$

Problem (5). *Three dice are tossed and it is given that at least one 4 appears. The probability that 1 will appear twice is*

A. $\frac{1}{72}$, *B.* $\frac{1}{36}$, *C.* $\frac{1}{24}$, *D.* $\frac{1}{12}$, *E.* $\frac{1}{6}$,
There are three possible outcomes: $4, 1, 1$ *or* $1, 4, 1$ *or* $1, 1, 4$ *so the probability is*

$$\frac{1}{6^2} + \frac{1}{6^2} + \frac{1}{6^2} = \frac{3}{36} = \frac{1}{12}$$

and the correct answer is [*D*]

Problem (6).

- *Six coins are flipped. The probability that 5 or 6 tails appear is*

  A. $\frac{7}{64}$, *B.* $\frac{1}{8}$, *C.* $\frac{5}{32}$, *D.* $\frac{7}{32}$, *E.* *none of these,*

- *A pair of dice is tossed. The probability that the sum of the sides facing up is an odd number is*

  A. 0.15, *B.* 0.25, *C.* 0.30, *D.* 0.40, *E.* 0.50,

- *For 5 tails the probability is*

  $$\frac{\binom{6}{5}}{2^6} = \frac{6}{64}$$

  and for 6 tails the probability is $\frac{1}{64}$, *hence, in total* $\frac{1+6}{64} = \frac{7}{64}$, *and the correct answer is* [*A*].

- *It is easy to count them* $(1, 2)$, $(1, 4)$, $(1, 6)$ *or* $(2, 1)$, $(4, 1)$, $(6, 1)$, $\ldots$ *etc. Finally we have that the probability is*

  $$\frac{18}{36} = 0.5$$

  and the correct answer is [*E*].

Problem (7).

- *A coin is tossed three times, the probability that exactly two heads will appear is*

  $A.\ 0.300, \quad B.\ 0.325, \quad C.\ 0.345, \quad D.\ 0.365, \quad E.\ 0.375,$

- *How many 5 - digit numbers can be formed from the numbers 1, 2, 3, 5, 0 if no digit is repeated?*

  $A.\ 72, \quad B.\ 76, \quad C.\ 80, \quad D.\ 90, \quad E.\ 96,$

- *The possible outcomes are:*

  $$HHH, TTT, HHT, HTT, TTH, THH, HTH, THT,$$

  so the probability is

  $$\frac{3}{8} = 0.375$$

  and the correct answer is $[E]$.

- *Since* 0 *cannot be the first digit, there are* $5! - 4! = 96$ *ways and the correct answer is* $[B]$.

Problem (8).

- *We flip 8 coins, the first three all come up tails. The probability that at least 1 more tail will be flipped is*

  $A.\ \frac{25}{32}, \quad B.\ \frac{27}{32}, \quad C.\ \frac{29}{32}, \quad D.\ \frac{30}{32}, \quad E.\ \frac{31}{32},$

- *Given that a multiple of 4 is less that 100, what is the probability that it is between 30 and 60?*

  $A.\ \frac{4}{25}, \quad B.\ \frac{5}{25}, \quad C.\ \frac{6}{25}, \quad D.\ \frac{7}{25}, \quad E.\ \frac{8}{25},$

- *At least* 1 *fail is equivalent to the event*

  $$1 - no\ fail = 1 - \left(\frac{1}{2}\right)^5 = 1 - \frac{1}{32} = \frac{31}{32}$$

  hence the correct answer is $[E]$.

- *There are* $\frac{100}{4} = 25$ *multiples of* 4 *and* 7 *of them are between* 30 *and* 60, *so the probability is*

  $$p = \frac{7}{25},$$

  hence $[D]$ *is the correct answer.*

Problem (9). *If*

$$f(x) = \frac{2}{x-2}, \quad g(x) = \sqrt{4-x^2}$$

the domain of the function $f \circ g$ *is*
A. $x > 2$, *B.* $x < 2$, *C.* $-2 \le x < 0$, *D.* $0 < x \le 2$, *E.* $-2 \le x < 0$ *or* $0 < x \le 2$,
It is

$$(f \circ g)(x) = \frac{2}{\sqrt{4-x^2}-2}$$

and

$$\left.\begin{array}{l} 4 - x^2 = (2-x)(2+x) \ge 0 \\ \sqrt{4-x^2} - 2 \ne 0 \end{array}\right\} \Rightarrow -2 \le x < 0 \;\; or \;\; 0 < x \le 2$$

and the correct answer is [*E*].

Problem (10). *If the remainder of the division of*

$$p(x) = x^4 + 3x^2 + x + a^4 - 1$$

by $x+1$, *is* 83, *the value of* a *is*
A. 3, *B.* 5, *C.* 6, *D.* 8, *E.* 30,
Since the remainder of the division of the polynomial $p(x)$ *by* $x+1$ *is* 83, *it follows:*

$$p(-1) = 83 \Rightarrow (-1)^4 + 3(-1)^2 + (-1) + a^4 - 1 = 83 \Rightarrow a = 3$$

hence, the correct answer is [*A*].

Problem (11). *If the function*

$$f(x) = \begin{cases} \dfrac{x^3-8}{x^2-4}, & if\ x \ne 2 \\ m, & if\ x = 2 \end{cases}$$

is continuous, the value of m *is*
A. 0, *B.* 1, *C.* 2, *D.* 3, *E.* 4,
Use the fact that f *is continuous for any* $x \ne 2$ *as a well defined fraction of polynomial functions and at the point* $x = 2$ *we have continuity if and only if*

$$\lim_{x \to 2} f(x) = f(2) = m \Leftrightarrow \lim_{x \to 2} \frac{x^3-8}{x^2-4} = m$$

so

$$\lim_{x \to 2} \frac{(x-2)(x^2+2x+4)}{(x-2)(x+2)} = m \Rightarrow \lim_{x \to 2} \frac{x^2+2x+4}{x+2} = m \Rightarrow x = 3$$

and the correct answer is [*D*].

Problem (12). *The value of*

$$\lim_{x \to 3} \frac{x^2-9}{(x-3)^3}$$

is

A. $-\infty$, *B.* -3, *C.* 3, *D.* $+\infty$, *E.* *none of these,*
We compute

$$\begin{aligned}\lim_{x\to 3}\frac{x^2-9}{(x-3)^3} &= \lim_{x\to 3}\frac{(x-3)(x+3)}{(x-3)^3}\\ &= \lim_{x\to 3}\frac{x+3}{(x-3)^2} = \frac{6}{0^+} = +\infty\end{aligned}$$

hence the correct answer is [D].

Problem (13). *If M is a point on the axis xx', $\widehat{AMB}=90^o$, $A(-1,7)$, $B(7,1)$ the coordinates of M are*

A. $(-6,0)$, $(0,0)$, *B.* $(-2,0)$, $(0,0)$, *C.* $(0,0)$, $(2,0)$, *D.* $(0,0)$, $(6,0)$,
E. none of the above,
Let $M(x,y)$, using the condition $\vec{AM}\perp\vec{MB}$ we get

$$\begin{aligned}&\vec{AM}\cdot\vec{BM}=0\Leftrightarrow (x+1)(x-7)+(y-7)(y-1)=0\Rightarrow\\ &x(x-6)=0\Rightarrow x=0 \text{ or } x=6\end{aligned}$$

and the correct answer is [D].

Problem (14). *The points $A(3,4)$ and $B(4,3)$ are symmetric with respect to*
A. the point $(0,0)$, *B. straight line* $x'x$, *C. straight line* $y'y$,
D. bisector of the angle $x\hat{O}y$, *E. none of the above,*
The points A, B are symmetric with respect to the bisector of xOy because the coordinates x,y interchange their roles, so the correct answer is [D].

Problem (15). *If the straight lines $y=(m^3-8)x+3$, $y=-2(m^2-2m)x+5$ are parallel then m is equal to*

A. -2 *or* -1, *B.* -2 *or* 2, *C.* 0, *D.* 2 *or* 0, *E.* 2 *or* 1,
By the parallelism condition

$$m^3-8=-2(m^2-2m)\Rightarrow (m-2)(m^2+4m+4)=0\Rightarrow m\in\{2,-2\}$$

and the correct answer is [B].

Problem (16). *If the distance OM, $O(0,0)$ and $M(x+1,x-1)$ is minimum then the value of x is*

A. -2, *B.* -1, *C.* 0, *D.* 1, *E.* 2,

$$d(M,O)=\sqrt{(x+1)^2+(x-1)^2},$$

where

$$(x+1)^2+(x-1)^2=2x^2+2\geq 2, \text{ for all } x\in\mathbb{R}$$

and by the minimality of $d(M,O)$ it follows that $x=0$, hence the correct answer is [C].

Problem (17). *If $A(1,9)$, $B(2,14)$ are two points of*

$$f(x)=x^2+ax+b$$

then $b-a$ is

A. -4, *B.* -2, *C.* 0, *D.* 2, *E.* 4,

$$\left.\begin{array}{r} A(1,9)\in \mathcal{G}_f \\ B(2,14)\in \mathcal{G}_f \end{array}\right\} \Rightarrow \left.\begin{array}{r} f(1)=9 \\ f(2)=14 \end{array}\right\} \Rightarrow \left.\begin{array}{r} 1+a+b=9 \\ 4+2a+b=14 \end{array}\right\} \Rightarrow a=2,\ b=6$$

and the correct answer is [*E*].

Problem (18). *If* r_1, r_2 *are roots of* $x^2-x-1=0$*, then the value*

$$\frac{r_1}{r_2}+\frac{r_2}{r_1}$$

is

A. -3, *B.* -2, *C.* -1, *D.* 0, *E.* 3,
It is

$$\frac{r_1}{r_2}+\frac{r_2}{r_1}=\frac{r_1^2+r_2^2}{r_1r_2}=\frac{(r_1+r_2)^2-2r_1r_2}{r_1r_2}=-3$$

and the correct answer is [*A*].

Problem (19). *The third term of a geometric sequence is 12 and the sixth is 96. Then, the sum of the first 10 terms is*

A. $2^{10}-1$, *B.* $2^{11}-2$, *C.* $2^{11}+1$, *D.* $3(2^{10}-1)$, *E.* $3(2^{10}+1)$,
Since $a_2=12=a_1r^2$, $a_6=96=a_1r^5$ *it follows that* $a_1=3$, $r=2$ *and consequently:*

$$S=\frac{3(2^{10}-1)}{2-1}=3(2^{10}-1)$$

hence, the correct answer is [*D*].

Problem (20). *If the equation*

$$(m^2-1)x=m^3-1$$

has no solutions then m *is equal to*

A. -1, *B.* 0, *C.* 1, *D.* 2, *E. none of the above,*
If $m=-1$ *then* $0=-1$ *i.e: impossible , the correct answer is* [*A*].

Problem (21). *If*

$$|a|+|b|=a+b$$

then

A. $ab>0$, *B.* $ab\le 0$, *C.* $ab=0$, *D.* $ab\ge 0$, *E.* $ab<0$,

$$|a|+|b|=a+b\Rightarrow |ab|=ab\Rightarrow ab\ge 0$$

and thus the correct answer is [*D*].

Problem (22). *If*

$$x^2+\sqrt{15},\ x^2-\sqrt{15}$$

are reciprocal then x *is equal to*

A. -2, *B.* -1, *C.* 1, *D.* -2 *or* $+2$, *E. none of the above,*

$$(x^2+\sqrt{15})(x^2-\sqrt{15}) = 1 \Rightarrow x^4 - 15 = 1 \Rightarrow x^4 = 16 \Rightarrow x = \pm 2$$

and the correct answer is [*D*].

Problem (23). *The vertical and horizontal asymptotes of*

$$f(x) = \frac{x+1}{x^2-1}$$

are

A. $x = 1$, $y = 0$, *B.* $x = 1$, $x = -1$, $y = 0$, *C.* $x = -1$, $y = 0$, *D.* $x = 1$, $y = 1$,
E. none of the above,

$$\lim_{x\to 1}\frac{x+1}{x-1} = \lim_{x\to 1}\frac{x+1}{(x-1)(x+1)} = \lim_{x\to 1}\frac{1}{x-1} = \pm\infty$$

i.e: $y = 1$ *is a vertical asymptote and*

$$\lim_{x\to\pm\infty}\frac{x+1}{x^2-1} = \lim_{x\to\pm\infty}\frac{1}{x} = 0$$

i.e: $x = 0$ *is a horizontal asymptote. The correct answer is* [*A*].

Problem (24). *The equation of the circle with center at the point* $K(1,2)$ *and tangential to the straight line* $\epsilon : 3x - 4y + 10 = 0$ *is*
A. $x^2 + y^2 - 3x - 4y + 5 = 0$, *B.* $x^2 + y^2 - 2x - 4y + 1 = 0$, *C.* $(x-1)^2 + (y-2)^2 = 4$,
D. $(x-1)^2 + (y-3)^2 = 2$, *E.* $(x-1)^2 + (y-2)^2 = 1$,

The distance of the center $K(1,2)$ *from the straight line* ϵ *is equal to the radius of the circle*

$$R = \frac{|3 - 8 + 10|}{\sqrt{3^2+4^2}} = 1$$

hence the equation of the circle is

$$(x-1)^2 + (y-2)^2 = 1$$

and the correct answer is [*E*]

Problem (25). *The points of intersection of the ellipses*

$$4x^2 + 3y^2 = 14, \; 3x^2 + 4y^2 = 14$$

lie on a circle with radius R *equal to*
A. 1, *B.* $\sqrt{2}$, *C.* 2, *D.* 3, *E.* 4,

Solving the system

$$\begin{aligned} 4x^2 + 3y^2 &= 14 \\ 3x^2 + 4y^2 &= 14 \end{aligned}$$

we find

$$x^2 = 2,\ y^2 = 2$$

hence

$$x^2 + y^2 = 4$$

which means that the points of intersection lie on a circle of radius 2 with center at the origin. So, the correct answer is $[C]$.

Problem (26). *If*

$$f(x) = \ln(-x^2 + 5x - 6)$$

the domain of f is
A. $x < -2$, *B.* $x > 3$, *C.* $x < 3$, *D.* $-2 < x < 3$, *E.* $2 < x < 3$,

$$-x^2 + 5x - 6 > 0 \Leftrightarrow -(x-2)(x-3) > 0 \Leftrightarrow (x-2)(x-3) < 0 \Leftrightarrow 2 < x < 3$$

so, the correct answer is $[E]$.

Problem (27). *If the graph of the function*

$$f(x) = x^2 + mx + 3$$

has its vertex at the axis Ox the value of m is
A. 1, *B.* 2, *C.* 3, *D.* 4, *E.* 6,

The vertex of f has coordinates $(\frac{-m}{2}, f(\frac{-m}{2}))$ *with* $f(\frac{-m}{2}) = 0$ *since the vertex lies on the xx' axis. Hence*

$$\frac{m^2}{4} - \frac{m^2}{2} + 8 = 0 \Rightarrow m^2 = 16$$

and the correct answer is $[D]$.

Problem (28). *The values of k for which the equation*

$$x^2 + 2(k+2)x + 9k = 0$$

has no real roots are
A. $k < 1$, *B.* $k > 4$, *C.* $k < 1$ *or* $k > 4$, *D.* $0 < k < 4$, *E.* $1 < k < 4$,

No real roots means $\Delta < 0$, *so:*

$$k^2 - 5k + 4 < 0 \Leftrightarrow (k-1)(k-4) < 0 \Leftrightarrow 1 < k < 4$$

The correct answer is $[E]$.

Problem (29). *The length of the radius of the sphere*

$$x^2 + y^2 + z^2 + 4x + 6y - 8z = 0$$

is

A. 4, *B.* 5, *C.* 5.4, *D.* 6, *E.* 6.5,

Let r be the radius:

$$x^2 + y^2 + z^2 + 4x + 6y - 8z = 0 \Leftrightarrow$$
$$(x+2)^2 + (y+3)^2 + (z-4) = 29 \Rightarrow r = \sqrt{29} \sim 5.4$$

and the correct answer is [C].

Problem (30). *If the sphere*

$$x^2 + y^2 + z^2 + 4x + 6y - 8z = 0$$

is inscribed in a cube, the volume V_{ext} of the cube outside the sphere is
A. 560, *B.* 561, *C.* 562, *D.* 562.2, *E.* 566,

The equation of the sphere is written

$$(x+2)^2 + (y+3)^2 + (z-4) = 29$$

so $r = \sqrt{29}$ and thus

$$V_{ext} = \left(2\sqrt{29}\right)^3 - \frac{4}{3}\pi\left(\sqrt{29}\right)^3 \sim 562.212$$

so, the correct answer is [D].

Problem (31). *A packet of 10 light bulbs contains 2 defective bulbs. Two bulbs are selected at random from the packet. The probability neither of these bulbs is defective is*
A. $\frac{14}{45}$, *B.* $\frac{18}{45}$, *C.* $\frac{20}{45}$, *D.* $\frac{28}{45}$, *E. none of the above,*

$$p = \frac{\binom{8}{2}}{\binom{10}{2}} = \frac{56}{90} = \frac{28}{45}$$

and the correct answer is [D].

Problem (32). *If $p(A \cup B) = \frac{9}{10}$, $p(A) = \frac{3}{10}$, $p(B) = \frac{8}{10}$ then $p(A \cap B)$ is equal to*
A. $\frac{1}{10}$, *B.* $\frac{2}{10}$, *C.* $\frac{3}{10}$, *D.* $\frac{4}{10}$, *E.* $\frac{5}{10}$,

It is

$$p(A \cup B) = p(A) + p(B) - p(A \cap B) \Rightarrow$$
$$p(A \cap B) = p(A) + p(B) - p(A \cup B)$$
$$= \frac{3}{10} + \frac{8}{10} - \frac{9}{10} = \frac{2}{10}$$

and the correct answer is [E].

Problem (33). *The slope of the median AM of a triangle $\triangle ABC$ with $A(a,5)$, $B(0,4)$ and $C(6,2)$, is equal to 2. Then, the value of a is*
A. −4, *B.* −2, *C.* 0, *D.* 2, *E.* 4,

The coordinates of the mid point M of BC, are $(\frac{0+6}{2}, \frac{4+2}{2}) = (3,3)$ *and using the formula for the slope of AM:*

$$\frac{3-5}{3-a} = 2 \Rightarrow 2a = 8 \Rightarrow a = 4$$

and the correct answer is $[E]$.

Problem (34). *The limit*

$$\lim_{x\to 3} \frac{x^2-9}{(x-3)^2}$$

is equal to

A. $-\infty$, *B.* 0, *C.* 3, *D.* $+\infty$, *E. none of the above,*

$$\lim_{x\to 3} \frac{x^2-9}{(x-3)^3} = \lim_{x\to 3} \frac{(x-3)(x+3)}{(x-3)^2(x-3)} = \lim_{x\to 3} \frac{x+3}{(x-3)^2} = +\infty$$

so, the correct answer is $[D]$.

Problem (35). *A root of*

$$x^3 - 3x^2 + 7x - 5 = 0$$

is an integer, the other two roots are

A. $-1 \pm 2i$, *B.* $-2 \pm 4i$, *C.* $-2 \pm 2i$, *D.* $1 \pm 2i$, *E. none of the above,*

The real - rational root is a divisor of the constant term 5 *and by trials we find that this root is* 1. *Factorizing we get*

$$x^3 - 3x^2 + 7x - 5 = 0 \Leftrightarrow (x-1)(x^2 - 2x + 5) = 0$$

hence, the other two complex conjugated roots are $1 \pm 2i$ *and the correct answer is* $[D]$.

Problem (36). *In a class there are 12 boys and 10 girls. The mean of the height of the boys is 172 and the corresponding mean for the girls is 168. The mean of the heights of all the students in this class is*

A. 169.5, *B.* 170, *C.* 170.1, *D.* 170.2, *E.* 170.4,

$$\text{Total Mean} = \frac{172 \times 12 + 168 \times 10}{12 + 10} = \frac{3944}{22} \sim 170.2$$

and the correct answer is $[D]$.

Problem (37). *The domain of the function*

$$f(x) = \log \sqrt{-2 + \frac{x}{x-3}}$$

is

A. $3 < x < 4$, *B.* $3 \le x \le 6$, *C.* $3 < x \le 6$, *D.* $3 < x < 6$, *E. none of the above,*

$$-2+\frac{x}{x-3}>0 \Leftrightarrow \frac{-x+6}{x-3}>0 \Leftrightarrow 3<x<6$$

and the correct answer is $[D]$.

Problem (38). *If*

$$\log_4 x \log_7 4 = 2$$

then x *is equal to*

A. 7, *B.* 14, *C.* 21, *D.* 49, *E.* 56,

Where is the subtle point? Let us try to have a formula for the changing of the base:

$$\log_b a = \frac{\log_c a}{\log_c b}$$

and thus

$$\frac{\log_7 x}{\log_7 4} \times \log_7 4 = 2 \Rightarrow \log_7 x = 2 \Rightarrow x = 7^2 = 49$$

and the correct answer is $[D]$.

Problem (39). *If*

$$2kx^2 - 3x + k - 1 = 8x^2 - 6mx + 6m$$

for all $x \in \mathbb{R}$, *the values of* k, m *are*

A. $-2, 1$, *B.* $4, 1$, *C.* $4, -1$, *D.* $4, \frac{1}{2}$, *E. none of the above,*

Where is the subtle point? Since it is true for all x *the coefficients* ***for terms of the same degree****, must be equal*

$$\left.\begin{aligned} 2k &= 8 \\ -3 &= -6m \\ k-1 &= 6m \end{aligned}\right\} \Rightarrow k=4,\ m=\frac{1}{2}$$

Problem (40). *If*

$$y = \frac{x^2+2x-11}{2(x-3)}$$

and x *is real, the values of* y *are*

A. $y<2$, *B.* $y>2$, *C.* $2<y<6$, *D.* $y \le 2$ *or* $y \ge 6$, *E. none of the above,*

For such problems there is a standard and simple technique:

$$y = \frac{x^2+2x-11}{2(x-3)} \Rightarrow 2yx - 6y = x^2 + 2x - 11$$

and since it holds true for all x, *we shall get:*

$$\Delta \ge 0 \Rightarrow y^2 - 8y + 12 \ge 0 \Leftrightarrow$$
$$(y-2)(y-6) \ge 0 \Rightarrow y \le 2 \text{ or } y \ge 6$$

and the correct answer is $[D]$.

Problem (41). *If $1+i$ is a root of*

$$x^2 - ax + b = 0$$

and a, b are real numbers then the sum $a+b$ is
A. -4, *B.* -2, *C.* 0, *D.* 2, *E.* 4,

$$(1+i)^2 - a(1+i) + b = 0 \Rightarrow (2-a)i - (a-b) = 0 \Rightarrow$$
$$\left.\begin{array}{l} 2-a=0 \\ b-a=0 \end{array}\right\} \Rightarrow a=2,\ b=2 \Rightarrow a+b=4$$

and the correct answer is [*E*].

Problem (42). *A circle is tangent to the straight line*

$$\epsilon : 4x + 3y + 20 = 0$$

and its center is at the origin. The radius of the circle is
A. 3, *B.* 4, *C.* 4.5, *D.* 5, *E.* 5.4,
The radius of the circle is equal to the distance r of the center from straight line

$$r = d(O,\epsilon) = \frac{|0+0+20|}{\sqrt{4^2+3^2}} = \frac{20}{5} = 4$$

and the correct answer is [*B*].

Problem (43). *If*

$$a^2 + b^2 + c^2 = 2 \text{ and } a+b+c=0$$

then $ab+bc+ca$ is equal to
A. -2, *B.* -1, *C.* 1, *D.* 2, *E. none of the above,*
It is

$$a+b+c=0 \Rightarrow a^2+b^2+c^2+2(ab+bc+ca)=0 \Rightarrow ab+bc+ca=-1$$

and the correct answer is [*B*].

Problem (44). *If*

$$f(x) = a\log_3 x,\ f^{-1}(1) = 9$$

then the value of a is
A. -2, *B.* -1, *C.* 0, *D.* $\frac{1}{2}$, *E.* 1,

$$f^{-1} = 9 \Rightarrow f(9) = 1 \Rightarrow$$
$$a\log_3 9 = 1 \Rightarrow 2a = 1 \Rightarrow a = \frac{1}{2}$$

and the correct answer is [*D*].

Problem (45). *If*

$$\sin A = \frac{3}{5},\ 0 < A < 90,\ \cos B = \frac{-3}{5},\ 90 < B < 180$$

then $\cos(A+B)$ is equal to

A. $-\frac{24}{25}$, *B.* $-\frac{20}{25}$, *C.* $-\frac{12}{25}$, *D.* $\frac{12}{25}$, *E.* $\frac{24}{25}$,

Use the formulas

$$\cos(A+B) = \cos A \cos B - \sin A \sin B$$

to obtain

$$\cos A = \sqrt{1-\sin^2 A} = \frac{4}{5}$$
$$\sin B = \sqrt{1-\cos^2 B} = \frac{4}{5}$$

so

$$\cos(A+B) = \frac{4}{5}\frac{-3}{5} - \frac{3}{5}\frac{4}{5} = \frac{-24}{25}$$

Problem (46). *If* $\vec{u} = p\vec{i} + 6\vec{j}$, $\vec{v} = 4p\vec{i} - 6\vec{j}$ *and* $\vec{u} \perp \vec{v}$ *the value of* p *is*
A. -3, *B.* -3 *or* 3, *C.* 3, *D.* 9, *E. none of the above,*

$$\vec{u} \perp \vec{v} \Leftrightarrow \vec{u}\cdot\vec{v} = 0 \Leftrightarrow 4p^2 = 36 \Rightarrow p^2 = 9$$

and the correct answer is [*B*].

Problem (47). *If*

$$A = \begin{bmatrix} 2 & 1 & x \\ -1 & 3 & 1 \end{bmatrix}, \quad B = \begin{bmatrix} -4 \\ x \\ 2 \end{bmatrix}$$

and

$$AB = \begin{bmatrix} -24 \\ 0 \end{bmatrix}$$

then, the value of x *is*
A. -3, *B.* -2, *C.* -1, *D.* 0, *E.* 1,

$$AB = \begin{bmatrix} 2 & 1 & x \\ -1 & 3 & 1 \end{bmatrix} \cdot \begin{bmatrix} -4 \\ x \\ 2 \end{bmatrix} = \begin{bmatrix} -14 \\ 0 \end{bmatrix} \Rightarrow \left.\begin{matrix} -8+3x = -14 \\ 6+3x = 0 \end{matrix}\right\} \Rightarrow x = -2$$

and the correct answer is [*B*].

Problem (48). *How many triangles are there in the figure below?*

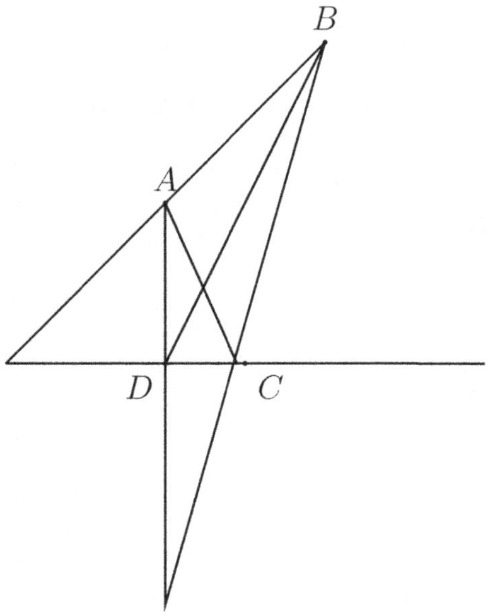

A. 8, *B*. 10, *C*. 12, *D*. 14, *E*. 16 ,

In the above figure they are given six lines, so the triangles we can form are

$$\binom{6}{3} = \frac{6!}{3! \cdot 3!} = 20.$$

But the four **tree - lines** passing through the points A or B, or C, or D, do not form any triangle. Hence, there are $20 - 4 = 16$ triangles and the correct answer is $[E]$.

Problem (49). *Let O, A, B, C be points in the plane such that*

$$|\vec{OA}| = |\vec{OB}| = |\vec{OC}| = 1$$

and

$$\vec{OA} + \vec{OB} + \vec{OC} = \vec{0}.$$

Then, the triangle $\triangle ABC$ is

A. isosceles, B. orthogonal , C. equilateral, D. isosceles and orthogonal,
E. none of these.

First, note that $\vec{OA} = -\vec{OB} - \vec{OC}$ and since $|\vec{OA}| = |\vec{OB}| = |\vec{OC}| = 1$, it follows that:

$$\left(\vec{OB} + \vec{OC}\right) \cdot \left(\vec{OB} + \vec{OC}\right) = 1 \Rightarrow |\vec{OB}|^2 + |\vec{OC}|^2 + 2\vec{OB} \cdot \vec{OC} = 1$$

So, if we denote by ϕ the angle of the vectors $\vec{OA}$, $\vec{OB}$, we get:

$$2\vec{OB} \cdot \vec{OC} = -1 \Rightarrow \vec{OB} \cdot \vec{OC} = \frac{1}{2} \Rightarrow |\vec{OB}||\vec{OC}| \cos\phi = -\frac{1}{2}$$

and thus

$$\cos\phi = -\frac{1}{2},$$

hence the angle

$$\widehat{BOC} = 120^o = \widehat{AOB} = \widehat{AOC}$$

hence the correct answer is $[C]$.

Problem (50). *Let $\triangle ABC$ be a triangle such that the medians BE and CZ are of equal length. Then, the triangle $\triangle ABC$ is*

A. isosceles, B. orthogonal , C. equilateral, D. isosceles and orthogonal, E. none of these.

With no loss of the generality let $A(0,a)$, $B(b,0)$ and $C(c,0)$ be the coordinates of the vertices A, B, C of the given triangle.

Then, the midpoint E of AC has coordinates $E\left(\frac{c}{2},\frac{a}{2}\right)$, the midpoint Z of AB has coordinates $\left(\frac{b}{2},\frac{a}{2}\right)$ therefore

$$\vec{BE}=\left(\frac{c}{2}-b,\frac{a}{2}\right) \quad \vec{CZ}=\left(\frac{b}{2}-c,\frac{a}{2}\right)$$

with corresponding lengths

$$\left|\vec{BE}\right|=\sqrt{\left(\frac{c}{2}-b\right)^2+\frac{a^2}{4}},\quad \left|\vec{CZ}\right|=\sqrt{\left(\frac{b}{2}-c\right)^2+\frac{a^2}{4}}$$

Use the hypothesis on the equality $|\vec{BE}|=|\vec{CZ}|$ to get:

$$\left(\frac{c}{2}-b\right)^2+\frac{a^2}{4}=\left(\frac{b}{2}-c\right)^2+\frac{a^2}{4}\Rightarrow\begin{cases}\frac{c}{2}-b=\frac{b}{2}-c\\ \text{or}\\ \frac{b}{2}-c=c-\frac{b}{2}\end{cases}\Rightarrow\begin{cases}3c=3b\\ \text{or}\\ c=b\end{cases}$$

hence, the correct answer is $[A]$.

Appendices

Appendix A

To the Editors

A.1 Comments regarding Kalomitsines' ideas

- **a.** W. Wickelgren - he has worked on problem solving at the MIT and he has written a book on problem solving. After reading the technique wrote to Mr. Kalomitsines:
  ...Let me emphasize that I believe your ideas are ingenious and of great value in improving peoples problem solving. The students of the world will be the losers if you are too modest to publish your ideas in prominent outlets ...

  Note:
  After the suggestion of professor J. Greeno in 1980 I made some computer programs in LRDC of Pittsburgh University, using methods and concepts of Cognitive Psychology which I was taught at the Carnegie - Mellon University by H. Simon (Nobel 1978). The computer solved a number of difficult problems from Euclidean Geometry and Group Theory. I was there as a Fullbright Scholar. Then J. Greeno wrote:

- **a.** J. Greeno - University of Pittsburgh and Stanford University:
  (i) *...The progress made in developing new theoretical analysis was much greater than any I have ever observed, due to Mr. Kalomitsines very intensive efforts and the thoroughness of the thinking he has previously done, on the topic of problem solving ...*
  ii *Many of us hope that the concepts and methods of cognitive psychology can be used profitably, by teachers and other developers of educational materials, to help provide clear and definite expressions of their ideas and methods. Kalomitsine's report gives an example that provides justification for this hope.*

- **c.** A. Posamentier - Journal: The Mathematics Teacher
  ...the instruction of Mathematics would be greatly strengthened if every Mathematics teacher would incorporate the many valuable ideas presented in this little book ...

  These comments refer to a book by the author, entitled *Attack your problem.* **Read also a review from Mathematical Association of America (MAA) (by David A. Hackaby, on 03/31/2009) for Kalomitsines' book: *How to solve problems: New methods and Ideas.***
  An extract from this review is: *Anyone who applies these methods to solve the many examples and problems in this book cannot help but become a better problem solver.*

Some more comments:
Professor Melas A., a talented mathematician, gold medalist in two Mathematical Olympiads, who is currently a celebrated mathematician at the University of Athens, when first read the technique, said:
This is the way I think when I solve problems.
An Olympiad medalist in Mathematics - once said to Spyros Kalomitsines:
I had gotten stuck with a problem in Functional Analysis. Then I decided to try your technique and I found the solution easily. This technique means to me: ***Pat the problem and it will respond!***
Applications of the methods to other areas:
1) G. Leondiadis, a researcher in bio-medicine in Canada, sent an e-mail to Mr. Kalomitsines
... *Your methods hooked me. They are ingenious, especially the* ***How to get out of loops****. I am sure they will be helpful in my research.*
2) The late H. Simon (Nobel 1978) researcher in Cognitive Psychology said to Mr. Kalomitsines *the technique of getting out of loops is very interesting.*

- **Searching the internet:**
  This book is unique in that it describes the actual method of problem solving. This is the classic book on problem solving.

- **Some comments on Spyros Kalomitsines research at the LRDC of Pittsburgh University:**
  ...we greatly admire the results of your work here ..., J. Greeno, 5 December 1980 (professor at Pittsburgh University and Stanford later)

- **Professor Steve Krantz (May 18, 2016):**
  ...quite interesting and different from anything else I have seen. **(e-mail to Sp. Kalomitsines)**

- **Professor D. Bergeles, first winner in a difficult mathematical competition, (November 10, 2014):**
  ...when we try to solve a difficult problem, Mr. Kalomitsines managed to describe, explicitely and succesfully, the steps that very often and unconciously we follow, trying to surpass dead ends or trying to get out of any loops ...

Bibliography

[1] Anderson, J. R., *Cognitive Psychology and its implications*, Freeman, 1980.

[2] Bambrough, R., *The philosophy of Aristotle*, The New American Library of World Literature, Inc., 1963.

[3] Kalomitsines, Sp., *Two methods for solving problems*, Educational Studies in Mathematics, Vol.**14, 3**, 1983.

[4] Kalomitsines, Sp., *The spiral method for solving problems*, International Journal of Mathematical Education in Science and Technology, Vol.**16, 4**, 1985.

[5] Kalomitsines, Sp., *How to solve problems: New methods and Ideas*,NOVA publishers, New York, 2008.

[6] Krantz, S., *Techniques for problem solving*, American Mathematical Society, 1997.

[7] Perkins, D., *THE EUREKA EFFECT*, Amazon, 2005.

[8] Polya, G., *How to solve problems*, Princeton University Press, 1945.

[9] Wickelgren, W. A., *How to solve it*, Freeman, 1974.

Made in United States
North Haven, CT
10 November 2026

22100373R00196